Dynamical Properties in Nanostructured and Low-Dimensional Materials (Second Edition)

Dynamical Properties in Nanostructured and Low-Dimensional Materials (Second Edition)

Michael G Cottam
Department of Physics and Astronomy, University of Western Ontario,
London, Ontario N6A 3K7, Canada

IOP Publishing, Bristol, UK

ISBN 978-0-7503-3903-2 (ebook)
ISBN 978-0-7503-3901-8 (print)
ISBN 978-0-7503-3904-9 (myPrint)
ISBN 978-0-7503-3902-5 (mobi)

DOI 10.1088/978-0-7503-3903-2

Version: 20220701

IOP ebooks

British Library Cataloguing-in-Publication Data: A catalogue record for this book is available from the British Library.

Published by IOP Publishing, wholly owned by The Institute of Physics, London

IOP Publishing, Temple Circus, Temple Way, Bristol, BS1 6HG, UK

US Office: IOP Publishing, Inc., 190 North Independence Mall West, Suite 601, Philadelphia, PA 19106, USA

To my wife Sandra for her love and also for her patience during the writing of this book.

Contents

Preface to the first edition

The last few years have seen dramatic advances in the growth, fabrication and characterization of low-dimensional materials, such as graphene on the one hand, and a wide variety of nanostructures on the other, such as those formed from ultrathin films, wires, disks and other 'dots'. The elements of the nanostructures may be formed either singly or in spatial arrays that can be periodic or otherwise (in one dimension or more) and may involve more than one type of material. Most studies of these artificially engineered materials have been driven by their potential for device applications that involve smaller and smaller physical dimensions and novel utilization of the nanostructuring. In particular, the dynamical properties of these materials are of fundamental interest for the devices that involve high-frequency operation and/or switching.

Consequently the different waves or excitations (vibrational, magnetic, optical, electronic and so on) need to be understood from the perspective of how their properties are modified in finite structures (especially on the nanometre to micrometre length scales) due to the presence of surfaces and interfaces. Recently the patterning of nanoelements, into periodic and other arrays, has become a focus of intense activity, leading for example to photonic crystals for optical signals (and their analogues such as phononic and magnonic crystals in other contexts). Here the control of the band gaps in the excitation spectrum is a basis for a variety of applications. Furthermore, the nonlinear properties of the excitations are increasingly a topic of interest, as well as their linear dynamics, and it is often the case that nonlinear excitations may arise that have no counterpart in the linear regime. Novel materials such as single-sheet graphene and related structures have provided the accurate realization of two-dimensional structures.

The target readership of this book includes physicists, chemists, materials scientists and engineers. They may be based in universities, industry and research laboratories. The level of presentation of this book is appropriate for final-year undergraduates, graduate students and researchers. The treatment given includes both the experimental and theoretical aspects of this expanding research field, but with an emphasis on the fundamental physical principles. Readers are assumed to have a basic background knowledge of quantum mechanics, statistical physics, electromagnetism and solid-state physics. More advanced topics are introduced and explained where needed.

The author is a Professor of Physics in the Department of Physics and Astronomy at the University of Western Ontario, Canada. He is a former Chair of the department and has also served as Associate Dean of Science (Research) and as the Director of the Western Institute for Nanomaterials Science. His research expertise is in the quantum theory of condensed-matter systems and, in particular, in the dynamical properties of the excitations, such as vibrational waves (or phonons) and magnetic excitations (spin waves or magnons), in these materials. Areas of focus in his recent work have included low-dimensional structures and nanostructured materials. He is author or co-author of around three hundred research papers and several books, including Cottam and Tilley 2004 *Introduction to Surface and Superlattice Excitations* 2nd edn (Bristol: IOP). The decade or more since that book was written has seen dramatic advances and new directions in this field, which are covered in the present publication. Notable examples are novel materials like graphene, the development of lateral arrays of interacting elements (giving rise to emerging topics such as phononic and magnonic crystals), interest in multiferroic materials, and further advances in nonlinear excitations.

I am indebted to the many friends, colleagues and collaborators who directly or indirectly have influenced this book and provided ideas.

Michael G Cottam
London, Ontario, Canada
November 2015

Preface to the second edition

The seven years since the writing of the first edition have been witness to many exciting new developments and advances in the field of the dynamical properties of low-dimensional materials. As an example, the discovery around 2016 of 'van der Waals' magnetic materials, which are two-dimensional (monolayer and a few layers) anisotropic materials exhibiting ferromagnetic and antiferromagnetic ordering, is a particular highlight. Also, the topic of graphene plasmonics has flourished recently, in which two-dimensional doped graphene layers can be inserted at the interfaces between dielectric layers in multi-component nanostructures. In the field of cavity magnonics, the dynamical effects of strong coupling of magnetic excitations to the electromagnetic modes (photons) in a cavity have attracted much attention, since the hybrid modes can exhibit nonlinear and nonclassical behaviours. These, and other, recent advances have provided the motivation and new material for inclusion in this second edition.

The general arrangement of material in the second edition is similar to previously, but there are now two additional chapters (nine instead of seven as previously, plus an appendix) to accommodate the new developments that were just mentioned, along with others. Specifically, there are now two chapters on magnetic nanostructures, one chapter with a focus on exchange-dominated dynamics (where advances concerning van der Waals material and Dzyaloshinskii–Moriya antisymmetric exchange effects have made profound changes) and the other chapter concerning the dynamical properties of dipolar and dipole-exchange systems. Also, the study of nonlinear processes in nanostructured materials has gained more attention in recent years, so this work is now presented in two chapters (covering the nonmagnetic and magnetic applications, respectively). The general level of presentation is similar to that in the first edition, and the target readership remains the same. As before, the emphasis is on a relatively concise but wide ranging treatment of the fundamental physical principles. Both theoretical and experimental topics are covered.

Again, I am indebted to many of my colleagues and collaborators for providing the stimulus and ideas leading to this book.

Michael G Cottam
London, Ontario, Canada
April 2022

Acknowledgements

It is a pleasure to acknowledge many colleagues and research collaborators who have provided the inspiration and background for the ideas presented in this book. These include, among others, Rodney Loudon, David Tilley, Mohamed Babiker, Robert Camley, Eudenilsen Albuquerque, Peter Grünberg, Alexei Maradudin, David Lockwood, Meng Hau Kuok, Andrei Slavin, Gianluca Gubbiotti, Dory Hélio Anselmo and Manoel Vasconcelos. Among my former graduate students and postdoctoral fellows, I would particularly like to mention and to thank Niu-Niu Chen, Raimundo Costa Filho, J Milton Pereira, Igor Rojdestvenski, Oleg Vassiliev, Eric Meloche, Trinh Nguyen, Hoa Nguyen, Tushar Das, Arash Akbari-Sharbaf, Zahra Haghshenasfard, Bushra Hussain and Peter Komorowski. I am grateful for the guidance provided by my editors at IOP Publishing (John Navas and Charlotte Ashbrooke for the first edition, and Emily Tapp for the second edition), as well as other editorial and production staff.

Author biography

Michael G Cottam

Michael Cottam has received his education in the UK, obtaining his Bachelor's and Master's degrees in mathematics and physics from Cambridge University and his doctorate (DPhil) in theoretical physics from Oxford University. He worked as a research scientist in semiconductor physics for the Plessey Company (UK) before becoming a faculty member at the University of Essex. In 1987 he moved to the University of Western Ontario, Canada, where he is a Professor (now emeritus) in the Department of Physics and Astronomy. He is a former Chair of the Department and also held appointments as Associate Dean of Science (Research) and Director of the Western Institute for Nanomaterials Science. He was honoured by the International Astronomical Union with the renaming of a main-belt asteroid (formerly numbered 273262) as 'Asteroid Cottam', in recognition of his scientific work on condensed-matter systems. His research interests over the years have spanned many aspects of the linear and nonlinear dynamics of materials, particularly on the nanoscale. He is married to Sandra Fox and they have four children/stepchildren.

Abbreviations

3D, 2D, 1D, 0D	three-dimensional, two-dimensional, one-dimensional, zero-dimensional
2DEG	two-dimensional electron gas
2DHG	two-dimensional hole gas
ABC	additional boundary condition
AFMR	antiferromagnetic resonance
ARPES	angle-resolved photoemission spectroscopy
ATR	attenuated total reflection
bcc	body-centred cubic
bct	body-centred tetragonal
BEC	Bose–Einstein condensation
BG	Bleustein–Gulyaev (equation)
BLS	Brillouin light scattering
BQ	biquadratic (exchange)
DE	Damon–Eshbach (mode)
DFPT	density functional perturbation theory
DFT	density functional theory
DFTS	differential Fourier-transform spectroscopy
DM	Dzyaloshinskii–Moriya (interaction)
ESR	electron spin resonance
EELS	electron energy loss spectroscopy
fcc	face-centred cubic
FMR	ferromagnetic resonance
GF	Green's function
GMR	giant magnetoresistance
GNR	graphene nanoribbon
H.c.	Hermitian conjugate
HREELS	high-resolution electron energy loss spectroscopy
HP	Holstein–Primakoff (transformation)
INS	inelastic neutron scattering
KDP	potassium dihydrogen phosphate
KdV	Korteweg–de-Vries (equation)
KP	Kronig–Penney (model)
L	longitudinal
LA	longitudinal acoustic
LLG	Landau–Lifshitz–Gilbert (equation)
LO	longitudinal optic
LRSP	long-range surface plasmon
LST	Lyddane–Sachs–Teller
MC	magnonic crystal
ML	monolayer
MQW	multi-quantum well
n–i–p–i	(n-type)–(insulator)–(p-type)–(insulator)
NL	nonlinear
NLSE	nonlinear Schrödinger equation
OOMMF	Object Oriented MicroMagnetic Framework (software)
PBG	photonic band gap
PC	photonic crystal

PL	photoluminescence
PR	photoreflectance
Py	permalloy ($Ni_{80}Fe_{20}$)
PZT	lead zirconate titanate
RKKY	Ruderman–Kittel–Kasuya–Yosida (interaction)
RPA	random-phase approximation
RW	Rayleigh wave
RWA	rotating wave approximation
RS	Raman scattering
sc	simple cubic
SF	spin–flop (transition)
sG	sine-Gordon
SHG	second-harmonic generation
SHO	simple harmonic oscillator
SNOM	scanning near-field optical microscopy
SPEELS	spin-polarized electron energy loss spectroscopy
STT	spin-transfer torque
SVA	slowly varying amplitude (approximation)
SW	spin wave
SWR	spin-wave resonance
T	transverse
TA	transverse acoustic
TDM	tridiagonal matrix
TE	transverse electric
THG	third-harmonic generation
TM	transverse magnetic
TO	transverse optic
TOF	time-of-flight
UV	ultraviolet
vdW	van der Waals (2D magnets)
YIG	yttrium iron garnet

IOP Publishing

Dynamical Properties in Nanostructured and Low-Dimensional Materials (Second Edition)

Michael G Cottam

Chapter 1

Introduction

In this chapter we will introduce many of the concepts that are central to the subject matter of this book. The topics will include a preliminary survey of the different waves or excitations in solids, along with a brief discussion of their linear and nonlinear (NL) dynamical properties. This chapter gives some examples of the low-dimensional structures and arrays that will be of importance for later applications, and also outlines the experimental and theoretical techniques that will be needed as tools in this study.

We will be less concerned here with the *static*, or time-independent, behaviour of these solids. Instead, we will focus on the temporal (and usually spatial) evolution of some characteristic property that can propagate through a solid or be localized in certain regions, such as near a surface or interface. Corresponding to this latter situation, the *excitations* mentioned above can be considered as just the wave-like Fourier components of a general propagating signal in the solid. Physically the signal might, for example, consist of optical waves, vibrations of the atoms in the solid, the fluctuating magnetization in a magnetic material, or the collective motion of charged particles in a plasma. In a finite-sized solid the excitations will be required to satisfy boundary conditions at the surfaces or interfaces, and so the excitation properties may become modified. The novel properties of the excitations in certain finite structures, frequently taken in conjunction with their NL dynamics as explained later, are of fundamental interest scientifically and may lead to a wide variety of practical device applications.

1.1 Types of excitations or waves

1.1.1 Fundamentals of the excitations

We start by briefly describing the 'bulk' (or 'volume') excitations that are characteristic of effectively unbounded solids, where there are no surface effects to be taken

doi:10.1088/978-0-7503-3903-2ch1

into account. Details may be found in the many standard textbooks on solid-state physics. The classic texts in this field include those by Kittel [1] and Ashcroft and Mermin [2]. Some examples of more recent books are references [3–5]. Next, the modifying effects of surfaces and interfaces will be introduced by following, for example, an approach based on symmetry arguments adopted for geometries of thin films and multilayers, as detailed in the book by Cottam and Tilley [6]. From a consideration of the symmetry and crystal lattices we shall arrive, via reciprocal lattices and Brillouin zones, at a form of Bloch's theorem for the elementary excitations.

A bulk crystalline solid may be regarded as a three-dimensional (3D) repetition in space of a *basis*, or building block, consisting of one or more atoms. The basis can be associated with the points on an infinitely extended lattice, which may be defined from symmetry in terms of three fundamental non-coplanar translational vectors that we denote by $\mathbf{a}_1$, $\mathbf{a}_2$ and $\mathbf{a}_3$. It then follows that the vector $\mathbf{R}$ connecting any two lattice points can be expressed as

$$\mathbf{R} = n_1\mathbf{a}_1 + n_2\mathbf{a}_2 + n_3\mathbf{a}_3, \tag{1.1}$$

where n_1, n_2 and n_3 take any integer values. The complete set of such vectors (forming a so-called *Bravais lattice*) may also possess other symmetry operations, such as rotations and reflections, that also leave the space lattice unchanged. We will see that the translations are particularly important here and they imply that any physical quantity $f(\mathbf{r})$ at a general position $\mathbf{r}$ in the system satisfies

$$f(\mathbf{r}) = f(\mathbf{r} + \mathbf{R}) \tag{1.2}$$

for any $\mathbf{R}$ given by equation (1.1). Depending on the context, the quantity $f(\mathbf{r})$ might refer to a component of a magnetization vector or an atomic displacement or a charge density, for example.

The *unit cell* of a lattice is just the smallest volume having the property that all the space can be filled by periodic repetitions of the unit cell for each lattice. Its specification is not unique, but one simple choice is the parallelepiped formed by the basic vectors $\mathbf{a}_1$, $\mathbf{a}_2$ and $\mathbf{a}_3$. Another construction is the *Wigner–Seitz* unit cell formed by drawing the perpendicular bisecting planes of the lattice vectors from any chosen lattice point to its neighbouring sites. Details are given in, e.g. [1, 2, 5]. Some choices of unit cell are shown in figure 1.1 for the special case in two dimensions (2D) of an oblique lattice with the basic vectors $\mathbf{a}_1$ and $\mathbf{a}_2$.

The next step is to introduce the *reciprocal lattice*. The basic vectors, denoted by $\mathbf{b}_1$, $\mathbf{b}_2$ and $\mathbf{b}_3$, of this lattice in reciprocal space have the properties in 3D that

$$\mathbf{a}_i \cdot \mathbf{b}_j = 2\pi\delta_{i,j}, \tag{1.3}$$

where i and j can take the values 1, 2 or 3, and $\delta_{i,j}$ is the Kronecker delta (equal to unity if the two subscripts have the same value and zero otherwise). For example, equation (1.3) can be satisfied if we write $\mathbf{b}_1 = (2\pi/V_0)(\mathbf{a}_2 \times \mathbf{a}_3)$ and so forth, with $V_0 = |\mathbf{a}_1 \cdot (\mathbf{a}_2 \times \mathbf{a}_3)|$ denoting the volume of the unit cell of the 3D space lattice.

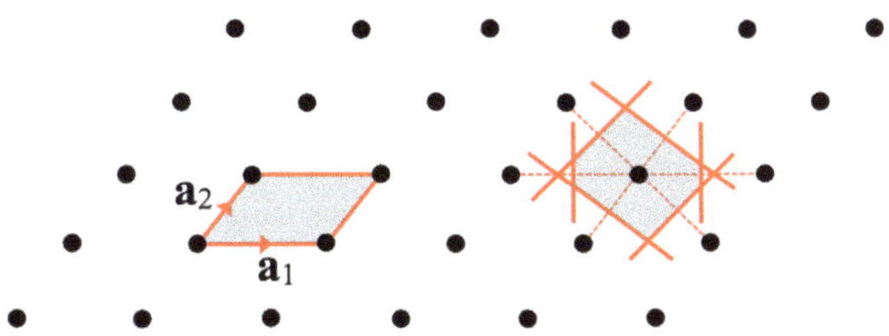

Figure 1.1. An oblique 2D space lattice with the basic vectors $\mathbf{a}_1$ and $\mathbf{a}_2$, showing two choices for the unit cell (shaded areas). The one on the left is a parallelepiped formed by the basic vectors, and the Wigner–Seitz cell (see the text) is illustrated on the right.

Since $f(\mathbf{r})$ introduced in equation (1.2) is a periodic function (generally in 3D) it can be expressed in terms of a Fourier series using complex exponentials as

$$f(\mathbf{r}) = \sum_{\mathbf{Q}} F(\mathbf{Q})\exp(i\mathbf{Q} \cdot \mathbf{r}), \tag{1.4}$$

where the $\mathbf{Q}$ vectors are required to satisfy $\exp(i\mathbf{Q} \cdot \mathbf{R}) = 1$. It is easily shown (see [1]) that these vectors are expressible in terms of the basic vectors of the reciprocal lattice as

$$\mathbf{Q} = \nu_1\mathbf{b}_1 + \nu_2\mathbf{b}_2 + \nu_3\mathbf{b}_3, \tag{1.5}$$

with ν_1, ν_2 and ν_3 denoting any integers. Finally we note that the unit cell of the reciprocal lattice, centred on one lattice point, is known as the *first Brillouin zone*, or simply the Brillouin zone. The results are particularly straightforward in the case of a simple-cubic (sc) space lattice with lattice parameter a, because the reciprocal lattice is also sc. We may choose for the space lattice

$$\mathbf{a}_1 = a(1, 0, 0), \qquad \mathbf{a}_2 = a(0, 1, 0), \qquad \mathbf{a}_3 = a(0, 0, 1), \tag{1.6}$$

whereupon the basic vectors of the reciprocal lattice are simply

$$\mathbf{b}_1 = \frac{2\pi}{a}(1, 0, 0), \qquad \mathbf{b}_2 = \frac{2\pi}{a}(0, 1, 0), \qquad \mathbf{b}_3 = \frac{2\pi}{a}(0, 0, 1). \tag{1.7}$$

Now we discuss how the translational symmetries of the space and reciprocal lattices influence the elementary excitations in an unbounded solid. The most fundamental result is embodied in *Bloch's theorem* [7], which provides information about the spatial behaviour of the variable that describes the amplitude of the excitations. This variable, which we will denote generically as a function of position as $\psi(\mathbf{r})$, depends on the type of excitation. For example, it could be the quantum-mechanical wave function in the case of electronic states, an atomic displacement in the case of lattice vibrations (or phonons), or the fluctuating magnetization in the case of spin waves (or magnons). These and other specific examples will be discussed in more depth in later chapters after we have considered some general properties here. Bloch's theorem can be expressed generally as

$$\psi(\mathbf{r}) = \exp(i\mathbf{q} \cdot \mathbf{r})\, U_{\mathbf{q}}(\mathbf{r}). \tag{1.8}$$

Here $U_{\mathbf{q}}(\mathbf{r})$ is a periodic amplitude function, just as in equation (1.2), and the phase-related factor $\exp(i\mathbf{q} \cdot \mathbf{r})$ describes a plane-wave variation (with wave vector $\mathbf{q}$) of the amplitude $\psi(\mathbf{r})$. We shall not provide a proof of this result, since this can be found in the solid-state physics textbooks cited earlier. The wave vector $\mathbf{q}$ is evidently not unique, since it can easily be checked that values of $\mathbf{q}$ differing by any reciprocal lattice vector $\mathbf{Q}$ as defined in equation (1.5) will equally satisfy the requirements of Bloch's theorem. Typically, as we will see later, an energy eigenvalue E for the excitation will be calculated for the dynamic variable $\psi(\mathbf{r})$ using its equation of motion, e.g. Schrödinger's equation in the case of an electronic excitation where $\psi(\mathbf{r})$ would be the stationary spatial wave function. On writing $E = \hbar\omega(\mathbf{q})$ with $\omega(\mathbf{q})$ representing the angular frequency of the excitation, it follows that we must have the property

$$\omega(\mathbf{q} + \mathbf{Q}) = \omega(\mathbf{q}). \tag{1.9}$$

In the idealized case of an unbounded crystal lattice the components of the vector $\mathbf{q}$ can take any real values, but equation (1.9) tells us that it is sufficient to take only $\mathbf{q}$ values that lie within the first Brillouin zone (the Wigner–Seitz cell centred at $\mathbf{q} = 0$). For the example of a sc space lattice mentioned earlier this gives $\mathbf{q} = (q_x, q_y, q_z)$ in Cartesian components with

$$-\pi/a < q_x \leqslant \pi/a, \quad -\pi/a < q_y \leqslant \pi/a, \quad -\pi/a < q_z \leqslant \pi/a. \tag{1.10}$$

There are analogous results to equations (1.6), (1.7) and (1.10) for other 3D space lattices [1, 3, 5], such as body-centred cubic (bcc) and face-centred cubic (fcc) lattices.

While there are many 3D space lattices, there are relatively few possibilities in 2D. In fact, there are only five in the 2D case, instead of fourteen [1, 3, 5]. These may be listed in terms of the lengths $|\mathbf{a}_1|$ and $|\mathbf{a}_2|$ of the basic vectors in 2D, along with the angle α between the vectors, as: *square* ($|\mathbf{a}_1| = |\mathbf{a}_2|$, $\alpha = \pi/2$), *rectangular* ($|\mathbf{a}_1| \neq |\mathbf{a}_2|$, $\alpha = \pi/2$), *hexagonal* ($|\mathbf{a}_1| = |\mathbf{a}_2|$, $\alpha = 2\pi/3$), *centred rectangular* ($|\mathbf{a}_1| \neq |\mathbf{a}_2|$, $\alpha = \cos^{-1}(|\mathbf{a}_1|/2|\mathbf{a}_2|)$) and *oblique* ($|\mathbf{a}_1| \neq |\mathbf{a}_2|$, $\alpha \neq \pi/2$). As we will see later, these 2D lattices will be important for thin films where there is translational symmetry in the 2D planes of atoms parallel to the surfaces, but not in the other dimension perpendicular to the surfaces. Consequently Bloch's theorem will be modified, as we discuss later.

As a specific 2D example we mention *graphene*, which is currently of immense interest for its several unique properties to be discussed in subsequent chapters. Graphene can be produced in the form of single sheets of atoms arranged in a 2D honeycomb space lattice (see e.g. [8, 9]). The lattice structure is illustrated in figure 1.2(a) and it can be described in terms of two basic vectors, $\mathbf{a}_1$ and $\mathbf{a}_2$, that define a unit cell with *two* carbon atoms per cell. Hence, there are two interpenetrating sublattices of carbon atoms, labelled for convenience as A and B, as shown. Relative to the chosen x- and y-coordinate axes we may write

$$\mathbf{a}_1 = \frac{\sqrt{3}a_0}{2}(\sqrt{3}, -1), \qquad \mathbf{a}_2 = \frac{\sqrt{3}a_0}{2}(\sqrt{3}, 1), \tag{1.11}$$

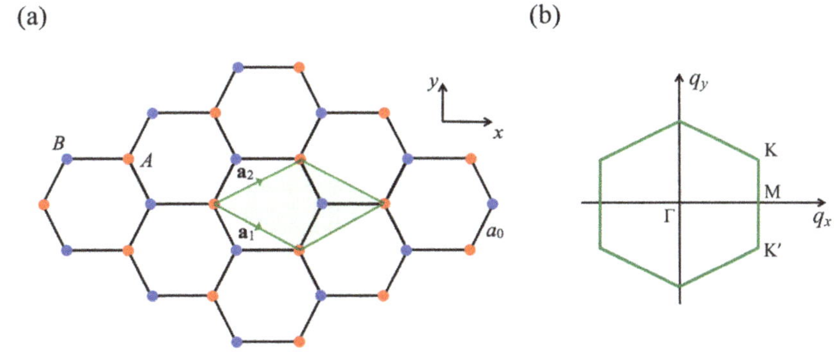

Figure 1.2. (a) The spatial structure of a single sheet of graphene. Each carbon atom lies on one of two interpenetrating sublattices labelled A and B (shown here as red and blue circles, respectively). A possible choice for the two basic lattice vectors $\mathbf{a}_1$ and $\mathbf{a}_2$ is indicated, together with the corresponding unit cell (shown shaded). (b) The usual choice of hexagonal Brillouin zone for graphene, showing the conventional labelling of some special symmetry points.

where a_0 denotes the nearest-neighbour carbon–carbon distance. Also the basic vectors of the reciprocal lattice are expressed simply [9] as

$$\mathbf{b}_1 = \frac{2\pi}{3a_0}(1, -\sqrt{3}), \qquad \mathbf{b}_2 = \frac{2\pi}{3a_0}(1, \sqrt{3}). \tag{1.12}$$

It can easily be verified that the above relationships satisfy the requirements of equation (1.3) in the 2D case. The reciprocal lattice of graphene is also hexagonal in form; the usual Brillouin zone is depicted in figure 1.2(b) along with the conventional labelling (which we will employ later) for some high-symmetry points. We remark that the honeycomb spatial lattice discussed here will also be of relevance for many of the so-called 2D van der Waals magnets (see chapter 3).

In some other nanosystems to be considered later we might have just one direction in which there is a translational symmetry. For example, long nanowires and nanotubes that have a uniform, finite cross-sectional shape will exhibit translational symmetry with respect to their length coordinate only. Also, on the atomic level, some 1D linear-chain materials (such as the ferromagnetic chain pyroxene $NaCrGe_2O_6$ [10]) will have this property. By contrast, nanostructures that are finite in all three spatial dimensions (such as cuboids and disks) will display no translational symmetry and are sometimes referred to as being zero-dimensional (or 0D).

1.1.2 Linear and nonlinear dynamics for excitations

Before looking further into the influence of the geometric structure on the excitations, it is useful to consider briefly how the excitations relate to the linear and NL aspects of the system dynamics. Although real physical systems might exhibit linear dynamics to a good approximation in many circumstances, it is nevertheless the case that NL behaviour will occur to some extent and may be enhanced under particular circumstances. It is to be expected that the excitations known to be associated with the linear regime will consequently have their properties modified when nonlinearities are taken into account. It is important to enquire into what ways and to what degree this takes place and if there are practical effects to be utilized (or avoided). Apart from this, it is particularly intriguing to find that there are some excitations associated with the NL regime of behaviour that have no counterpart in the linear approximation.

At this preliminary stage we will highlight some distinctions between the linear and NL dynamics of excitations, mainly by taking a few specific examples. These and other cases will then be explored in much more detail later, in particular in chapters 8 and 9. Some general references covering NL dynamical phenomena, mostly for bulk properties, are the books by Mills [11], Powers and Haus [12], and Boyd [13] for NL optical and dielectric materials. For the magnetization dynamics of magnetic materials some general books that cover both linear and NL aspects include those by White [14], Stancil and Prabakhar [15] and Demokritov [16].

To introduce some of the concepts, we start by considering an example of optical (electromagnetic) waves in a dielectric material, ignoring for the present any specifically magnetic effects. For now the material is assumed to be homogeneous and sufficiently large that we need not explicitly take account of boundary effects. The response of the material to an external position- and time-dependent electric field $\mathbf{E}(\mathbf{r}, \mathbf{t})$ is conventionally described in terms of a dielectric function ϵ by writing

$$\mathbf{D}(\mathbf{r}, t) = \epsilon_0 \int\int \epsilon(\mathbf{r} - \mathbf{r}', t - t')\, \mathbf{E}(\mathbf{r}', t')\, \mathrm{d}^3\mathbf{r}'\mathrm{d}t'. \tag{1.13}$$

Here ϵ_0 is the permittivity of free space and $\mathbf{D}(\mathbf{r}, \mathbf{t})$ is the electric displacement (see e.g. [17, 18]). The $\mathbf{D}$ and $\mathbf{E}$ vectors are also related to the polarization $\mathbf{P}$ induced in the material by

$$\mathbf{D}(\mathbf{r}, t) = \epsilon_0\mathbf{E}(\mathbf{r}, t) + \mathbf{P}(\mathbf{r}, t), \tag{1.14}$$

Rather than considering these quantities in terms of position and time, it is more useful for the dynamics to make Fourier transforms to a wave vector $\mathbf{q}$ and an angular frequency ω according to, for example,

$$\mathbf{E}(\mathbf{r}, t) = \int\int \mathbf{E}(\mathbf{q}, \omega)\, \exp(\mathrm{i}\mathbf{q} \cdot \mathbf{r} - \mathrm{i}\omega t)\mathrm{d}^3\mathbf{q}\mathrm{d}\omega. \tag{1.15}$$

Then equation (1.13) can be re-expressed more compactly and conveniently as

$$\mathbf{D}(\mathbf{q}, \omega) = \epsilon_0\epsilon(\mathbf{q}, \omega)\mathbf{E}(\mathbf{q}, \omega), \tag{1.16}$$

while the corresponding result for the polarization is

$$\mathbf{P}(\mathbf{q}, \omega) = \epsilon_0\, [\epsilon(\mathbf{q}, \omega) - 1]\, \mathbf{E}(\mathbf{q}, \omega). \tag{1.17}$$

The $\mathbf{q}$-dependence of the dielectric function in the above two equations is known as *spatial dispersion* and arises from the fact that the bulk dielectric term in equation (1.13) is assumed to have a dependence on the spatial separation $\mathbf{r} - \mathbf{r}'$. However, in many cases this spatial dispersion may vanish (by symmetry) or otherwise be negligible, and then we have simply a frequency-dependent dielectric function, denoted as $\epsilon(\omega)$, that is characteristic of the dynamical processes in the material. Examples will be given later, in particular in chapters 5 and 6.

Before going any further we need to elaborate on two points regarding the above discussion. The first is that, in general, the vectors **D** and **P** may not necessarily be collinear with **E**. The above relationships are still applicable provided the scalar dielectric function, $\epsilon(\mathbf{q}, \omega)$ or $\epsilon(\omega)$ as appropriate, is replaced by a second-rank tensor or equivalently a matrix in such a case. The second and more significant point is that for small values of the electric field there is a *linear* relationship, as expressed by equation (1.16), between the quantities **D** and **E**. In other words, $\epsilon(\mathbf{q}, \omega)$ is a constant function that does not depend on **E** and it is called the *linear dielectric function*. At larger electric-field values there is usually a *nonlinear regime* in which the dielectric function becomes dependent on **E**. A conventional approach in NL optics is to expand the dielectric function as a power series in terms of components of the electric field, as we will describe in chapter 8. Various different NL effects can then be identified depending on which terms in the expansion are important. For example, the *Kerr-type nonlinearity* involves an effective NL dielectric function [11, 13] that takes the form

$$\epsilon^{\mathrm{NL}} = \epsilon + \rho|\mathbf{E}|^2. \tag{1.18}$$

Here both the linear part ϵ and the coefficient ρ before the field-dependent term may, in general, be functions of both the wave vector $\mathbf{q}$ and the frequency ω.

To probe further some distinctions between the linear and NL dynamics that may arise due to the form of the dielectric function, we now turn to the basic equation satisfied by the electric field. This is just the electromagnetic wave equation coming from Maxwell's equations [17, 18]:

$$\epsilon(\omega)\frac{\partial^2\mathbf{E}}{\partial t^2} - c^2\nabla^2\mathbf{E} = 0. \tag{1.19}$$

This result is the form appropriate for a transverse wave (i.e., when the wave vector $\mathbf{q}$ that specifies the propagation direction is perpendicular to **E**) and c is the velocity of light in a vacuum. We note that the solution of this equation in the linear regime is relatively straightforward, because the linear dielectric function does not depend on **E** and so equation (1.19) is a linear differential equation. We are ignoring spatial dispersion here. The solution in this case where we assume an unbounded medium can be found in a plane-wave $\exp[\mathrm{i}(\mathbf{q}\cdot\mathbf{r} - \omega t)]$ form, yielding

$$q^2 = \epsilon(\omega)\frac{\omega^2}{c^2}. \tag{1.20}$$

This is just the implicit dispersion relation (the ω versus q relationship) arising when the electromagnetic wave is coupled to the polarization field of the medium. The excitation is generally referred to as a *polariton*. If the dielectric function is just a constant and written as n^2, where n is the refractive index of the medium, we recover the usual dispersion relation $\omega = (c/n)q$ for a photon (quantum of light) with speed c/n in the medium. In contrast, a simple example of a frequency-dependent dielectric function occurs for an electron plasma, realized as the 'gas' of electrons in a metal or semiconductor in the presence of positive ions than ensure the overall charge neutrality. The plasma oscillations [1–3] of the electron gas are known to define a characteristic angular frequency, the plasma frequency ω_p, which is given by $\omega_p{}^2 = n_0 e^2/\epsilon_0 m$. Here n_0 is the 3D concentration of electrons, where each have a mass m and charge $-e$. It can be shown (see also chapter 5) that

$$\epsilon(\omega) = \epsilon_\infty \left(1 - \frac{\omega_p{}^2}{\omega^2}\right), \tag{1.21}$$

where ϵ_∞ is the limiting dielectric constant for frequencies that are well above ω_p and damping has been neglected here. Substituting this dielectric function into the expression in equation (1.20) and rearranging gives us the plasmon–polariton dispersion relation explicitly as

$$\omega = \sqrt{\omega_p{}^2 + c^2 q^2}. \tag{1.22}$$

This simple dispersion relation is drawn as the solid line in figure 1.3. We have also included for comparison the dashed line $\omega = cq$ which corresponds to the light line for propagation in a vacuum. There are two limiting regions of interest for the polariton. One is for $cq/\omega_p \ll 1$ when $\omega \simeq \omega_p$, and the other is for $cq/\omega_p \gg 1$ when $\omega \simeq cq$. There is also a band gap, a region of forbidden frequencies, where no plasmon–polaritons exist, and this occurs here in the frequency range from 0 to ω_p.

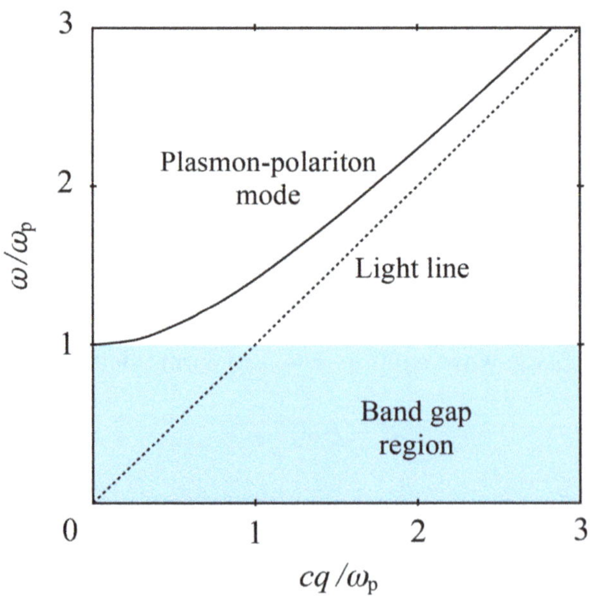

Figure 1.3. Dispersion relation for a linear plasmon–polariton (solid line) in a 3D material, as given by equation (1.22). The angular frequency, in terms of dimensionless ω/ω_p is plotted versus the dimensionless wave-number, in terms of cq/ω_p.

The above discussion constitutes an example of a *bulk* (or *volume*) excitation, since no account has been taken of any surface or boundary and the linear excitation propagates (or behaves in a wave-like fashion) in all three spatial dimensions. It turns out that other linear types of polaritons may occur that are associated with surfaces and interfaces, and this behaviour will be one of the main topics in chapter 6.

We discussed earlier in the section how NL effects could arise, for example, resulting in a Kerr-type form of the dielectric function, as in equation (1.18). Before concluding this section we will outline how an NL polariton can arise due to this term; more details will be given in chapter 8 (but see also [6]). The following example is related to the phenomenon known as *self-guiding* and can occur for polaritons at an interface. It is helpful to start with a qualitative description to provide some background. From linear optics we recall that the guiding condition for a light wave to remain localized within a film (through multiple internal reflections back and forth at the surfaces) is that the refractive index of the film should exceed the refractive index in the bounding medium on each side [17, 18]. It might be expected in the NL case that the situation will be broadly analogous to a guiding film with large refractive index, *provided* ϵ and ρ in equation (1.18) are both positive *and* the electric-field term $|\mathbf{E}|^2$ can be localized with a large value near a surface. The above conditions can, in fact, be realized as we now demonstrate by a simple calculation for an interface between a linear medium and a Kerr NL medium.

Specifically, we will assume that the NL medium fills the lower half-space $z < 0$, while a linear medium occupies the upper half-space $z > 0$. The interface between them is represented by the plane $z = 0$ and the system is assumed to be infinite in the x- and y-coordinate directions. We look for solutions for which the electric field is an s-polarized wave, which means that it can be written as $\mathbf{E} = (0, E, 0)$ and that its propagation direction is along the x-axis. From Bloch's theorem and assuming an angular frequency ω it follows that all electromagnetic field quantities are proportional to $\exp(\mathrm{i}q_x x - \mathrm{i}\omega t)$. Regarding the behaviour in the z-direction, the substitution of equation (1.18) into the wave equation (1.19) yields

$$\frac{\mathrm{d}^2 E}{\mathrm{d}z^2} - \left(\kappa^2 - \rho\frac{\omega^2}{c^2}E^2\right)E = 0, \tag{1.23}$$

where we define κ, which is real for large enough q_x, by

$$\kappa^2 = {q_x}^2 - \frac{\epsilon\omega^2}{c^2}. \tag{1.24}$$

If we assume the existence of solutions for which E is real, the NL differential equation can be solved by multiplying each side of equation (1.23) by $\mathrm{d}E/\mathrm{d}z$ and integrating to obtain

$$\left(\frac{\mathrm{d}E}{\mathrm{d}z}\right)^2 - \kappa^2 E^2 + \rho\frac{\omega^2}{2c^2}E^4 = 0. \tag{1.25}$$

Here a constant of integration has been set to zero on the assumption that both E and $\mathrm{d}E/\mathrm{d}z$ tend to zero as $z \to -\infty$ inside the NL medium. Next, equation (1.25) can be rearranged as an expression for $\mathrm{d}E/\mathrm{d}z$ as a function of z and directly integrated to give

$$E = \frac{2^{1/2} c\kappa}{\rho^{1/2} \omega \cosh[\kappa(z + z_0)]}. \tag{1.26}$$

This is the solution appropriate for $z < 0$, where z_0 is another constant of integration. To complete the calculation we also need to find the solution for E in the linear medium (labelled with subscript 1) for which $z > 0$. Here, in contrast to equation (1.23), we now have a much simpler second-order linear differential equation, for which the solution satisfying the requirement that $E \to 0$ as $z \to \infty$ is

$$E = E_1 \exp(-\kappa_1 z). \tag{1.27}$$

We have denoted $\kappa_1 = (q_x^2 - \epsilon_1 \omega^2/c^2)^{1/2}$ by analogy with equation (1.24), with ϵ_1 denoting the dielectric constant of medium 1, and as before we take κ_1 to be real. The constant E_1 is easily found from the electromagnetic boundary condition at the interface $z = 0$ that the tangential electric-field components must be continuous. We are still left, however, with the problem of finding the other constant z_0.

This presents us with the kind of situation that is typical of many problems involving NL excitations, and in the present case it can be handled by considering the power flow in the propagating wave, which leads to another equation involving z_0. In electromagnetism the power flow comes from the time-averaged Poynting vector [17, 18], and further discussion of this detail will be deferred to chapter 8. The main message to keep in mind is that we have demonstrated through equation (1.26) that a spatially localized electromagnetic wave can occur as a consequence of the nonlinearity of the medium.

1.2 Survey of different nanostructures

In the examples given so far, we have considered excitations occurring in unbounded (bulk) media, in a 2D sheet (for graphene), or else localized at the planar interface between two media. More generally, there may be multiple boundaries (surfaces or interfaces) that drastically affect the dimensionality and symmetries of the system, thus modifying the form and applicability of Bloch's theorem. Also the boundaries introduce different length scales, and here we will be mainly interested in effects that are in the range of a few nanometres (nm) up to micrometres (μm). We will now describe briefly some of the different nanostructures that will be considered in the following chapters, starting with the examples represented schematically in figure 1.4. The formation of such structures has been driven by recent advances in materials fabrication techniques and the potential of these structures for a variety of device applications (see e.g. [19, 20]).

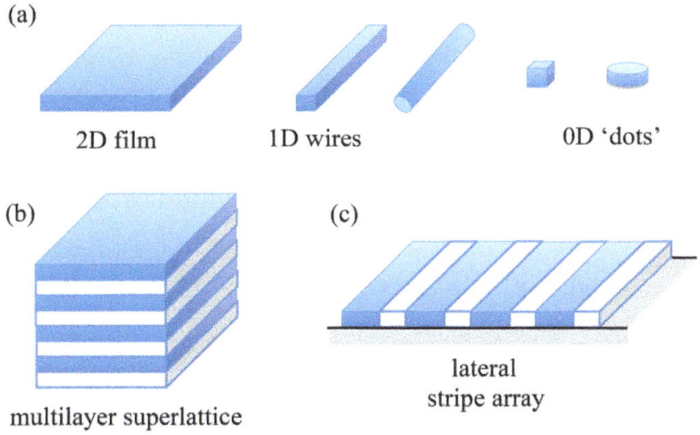

Figure 1.4. Schematic illustration of some different nanostructures: (a) low-dimensional structures with one material, (b) periodic multilayers or superlattices of two materials and (c) laterally patterned periodic structures of two materials.

1.2.1 Low-dimensional structures

In this category we have materials with reduced dimensionality as expressed in terms of the number of spatial dimensions in which there is overall translational symmetry. They include the 2D cases of thin films with finite thickness L and parallel planar surfaces, as well as sheets of atoms such as the graphene example mentioned earlier. They are characterized by a 2D space lattice associated with the planes of atoms parallel to the surfaces, as mentioned in section 1.1.1. The previous form of Bloch's theorem will be modified slightly compared with equation (1.8) to become in this case

$$\psi(\mathbf{r}_{\|}, z) = \exp(\mathrm{i}\mathbf{q}_{\|} \cdot \mathbf{r}_{\|})\, U_{\mathbf{q}_{\|}}(\mathbf{r}_{\|}, z). \tag{1.28}$$

We have assumed that the z-axis is chosen to be perpendicular to the surfaces, so the in-plane vectors are $\mathbf{r}_{\|} = (x, y)$ and $\mathbf{q}_{\|} = (q_x, q_y)$. The challenge in solving for the excitations in films with a nonzero thickness is to find the z-dependence of the amplitude term $U_{\mathbf{q}_{\|}}(\mathbf{r}_{\|}, z)$.

In the case of linear excitations in films we will often find later that this z-dependence is expressible as a linear combination of terms having the form $\exp\{\mathrm{i}q_z^{(j)}z\}$, where $q_z^{(j)}$ may be *complex* with a series of allowed values labelled by j (see e.g. [6]). Broadly, there are two possible situations that may arise. One is that $q_z^{(j)}$ is real, and thus it represents the third wave-vector component of a bulk excitation in the film with total wave vector $\mathbf{q} = (\mathbf{q}_{\|}, q_z^{(j)})$. This bulk excitation is affected in various ways, as we will see later, by the film surfaces because it must satisfy boundary conditions there. The other possibility is that $q_z^{(j)}$ may have an imaginary part, which we denote as $\mathrm{i}\kappa$. Then the $\exp\{\mathrm{i}q_z^{(j)}z\}$ term will involve a real factor such as $\exp(-\kappa z)$, which is characteristic of a localized surface excitation that decays with distance from one of the film surfaces. In such cases, a surface excitation can, therefore, be described in terms of a 2D wave vector $\mathbf{q}_{\|}$ along with a decay length that is equal to κ^{-1}.

The category of low-dimensional structures also includes nanowires (of different cross sections) and nanotubes, which all have just one direction of translational symmetry (the axis along the length). Therefore, the excitations are characterized by a 1D wave vector in this case. The straightforward modification of Bloch's theorem yields

$$\psi(x, y, z) = \exp(\mathrm{i}q_z z)U_{q_z}(x, y, z), \tag{1.29}$$

where it is assumed here that the z-axis is along the translational symmetry axis and the wave vector is simply q_z. Finally, we may have structures that are finite in all three spatial dimensions, such as a small cuboid or a finite disk. These have no translational symmetry, and so Bloch's theorem does not apply. The structures may be labelled as being 0D, like the so-called quantum dots that have been of particular interest for semiconductors (see chapter 5) or flat platelets, disks and rings for magnetic materials (see chapters 3 and 4).

Simple examples of these low-dimensional nanostructures, considered to be formed from a single homogeneous material as described here, are illustrated schematically in figure 1.4(a).

1.2.2 Multilayers and superlattices

If several materials are utilized, then more elaborate cases of nanostructures can arise than those in the previous subsection. One simple possibility is to build up multilayers with alternating thin films of two materials, denoted as A and B, to form a $ABABAB$... pattern of growth on a substrate material. Such an example is provided in figure 1.4(b), where the materials are shown in different colours. The simplest case would be a bilayer AB, but multilayers with a large number of layers (e.g. hundreds or more) can now be fabricated. In all cases these systems would have the property of translational symmetry for lattice directions parallel to the planar surfaces and interfaces. Thus the 2D Bloch's theorem in equation (1.28) still applies, with the only essential difference being that the term $U_{\mathbf{q}_{\|}}(\mathbf{r}_{\|}, z)$ may have a much more complicated dependence on z. There may, for example, be excitations that are localized only near the interfaces between A and B layers.

Another aspect of multilayers that makes them of special interest occurs when all the A layers have identical properties (composition, thickness, etc) to one another and likewise for all the B layers. The structure then consists of repeats of the same basic AB building block. If there is a large number of these repeats, then a new symmetry operation emerges corresponding to translations in the z-direction through a periodicity length $D = d_A + d_B$, where d_A and d_B denote the thicknesses of the individual A and B layers. We then have an example of a *periodic superlattice*, for which the inbuilt artificial periodicity in the z-direction may be associated with a new 1D Brillouin zone that extends in reciprocal space from $-\pi/D$ to π/D. Since D is typically designed to be much larger than the intrinsic lattice parameter of either constituent material, this new Brillouin zone is smaller than the crystal Brillouin zone. This can have striking consequences for the excitations that propagate as waves in the perpendicular (or z) direction of the superlattice. A smaller Brillouin zone can sometimes be advantageous as regards the experimental techniques for studying the excitations, and also it leads to a 'folding' of the dispersion curves, as we will discuss through examples in section 1.5 and in later chapters.

The concept of a superlattice was originally introduced by Esaki and Tsu [21] for ultrathin semiconductor multilayers where the layer thicknesses were less than the electron mean free path, but nowadays the term is applied more generally for multilayers (whether of semiconductors or other materials) that are either periodic or alternatively are generated according to certain mathematical rules. In the former category we may also have periodic arrangements with three (or more) components, such as *ABCABCABC*..., as well as the simpler two-component superlattice described above. In the latter category we may include *quasiperiodic superlattices*, where the sequencing of layers along the growth direction is quasiperiodic in the sense that it involves two different fundamental periods whose ratio is an irrational number. One widely used means of achieving this property is through the Fibonacci sequence of numbers, $\{F_n\}$ with integer $n = 0, 1, 2, \ldots$. These numbers can be defined by the recursion relation

$$F_{n+2} = F_{n+1} + F_n, \tag{1.30}$$

with the initial conditions taken as $F_1 = F_0 = 1$. Hence the first few numbers in the sequence are 1, 1, 2, 3, 5, 8, 13, 21, ... It is straightforward to prove that as n becomes very large ($n \to \infty$) the ratio F_{n+1}/F_n tends to the value $(1 + \sqrt{5})/2 \simeq 1.618$, which is known as the *golden mean*. For the application to Fibonacci superlattices some fundamental physical property of the layered growth is developed by analogy with equation (1.30), as explained in later chapters. The excitations in quasiperiodic superlattices have various novel properties (e.g. regarding their spatial localization, scaling laws, multifractal behaviour, etc) that make them distinctive and of special interest [22].

1.2.3 Laterally patterned structures

In contrast with the geometry in the previous subsection, we will now discuss some artificial nanostructures produced by a lateral arrangement (usually a periodic sequence) of elements such as that represented in figure 1.4(c). This structure consists of an alternating sequence of stripes of two different materials deposited side-by-side on a planar surface, rather than being stacked vertically. Interest in such structures was given impetus in 1987 through seminal work by Yablonovitch [23] and John [24], who studied artificial electromagnetic materials, typically in low dimensions, that are designed to have a periodic modulation of the refractive index. This was originally done with a view to control the propagation, localization and scattering of light, and it led to periodic structures known as *photonic crystals* (or alternatively *photonic band-gap materials*) for applications in optoelectronics. Some consequences for the photonic band structure in a periodic case are developed later in this chapter (see section 1.5).

In recent years, aided by the rapid advances in materials growth and fabrication on the sub-micron scale, other types of laterally patterned materials have been utilized to study phonons (hence *phononic crystals*), spin waves or magnons (*magnonic crystals*), and plasmons (*plasmonic crystals*), among others. One of the characteristic features of these artificial materials is the tailoring of the band structure of the excitations and the forbidden-frequency gaps (or band gaps) occurring in the spectrum. We shall discuss several such cases in more detail in later chapters. Just as with the superlattices described in the previous subsection, effects will arise due to a modified Bloch's theorem and the occurrence of mini-Brillouin zones due to the artificial periodicity. The example depicted in figure 1.4(c) with a striped array is a 1D crystal in the lateral plane, but 2D patterning (analogous to a chess or checker board array) provides another case of interest. Other examples will be described later.

1.3 Experimental techniques for dynamic properties

Although a wide range of different physical processes (e.g. electromagnetic, electronic, vibrational, magnetic, etc) may contribute to the overall dynamic properties of materials, many of the same experimental techniques will be applicable to probe the excitations. A similar remark may be made as regards the theoretical tools needed to study the excitations, and so it is useful to give a broad introduction in this chapter and to provide some general references. Other more specialized methods will be introduced later, as needed. The experimental aspects are covered first in this section, and the theoretical topics will be described in section 1.4.

We will emphasize here the techniques that are most useful as regards the dynamical properties and characterization of the excitations. These include inelastic light scattering, inelastic particle scattering, electromagnetic techniques and magnetic (or spin) resonance methods, etc. We do not describe the sample growth and preparation techniques here, or the analytical tools for the *structural* (or static) characterization of surface and interface, since these topics are not the focus of this book; they are covered extensively elsewhere (see e.g. [19, 20, 25, 26]).

1.3.1 Inelastic light scattering

Both Raman and Brillouin scattering of light by dense media involve an inelastic process whereby the incident light with angular frequency ω_I (or a photon of energy $\hbar\omega_I$) and wave vector $\mathbf{k}_I$ interact with the medium by creating or absorbing an excitation and producing a scattered light beam. Although these effects were discovered in the early part of the twentieth century, it was not until much later with the advent of the laser and other technical developments that they were widely used for studying bulk and surface excitations. The essential difference between the two techniques lies in the method used to analyse the frequency of the scattered light. This method involves a grating spectrometer for Raman scattering (RS), where the wave-number shifts of the light are typically in the range 5–4000 cm^{-1}, and a Fabry–Pérot interferometer for Brillouin light scattering (BLS), where the wave-number shifts are below about 5 cm^{-1}. Note that a wave-number shift of 1 cm^{-1} corresponds

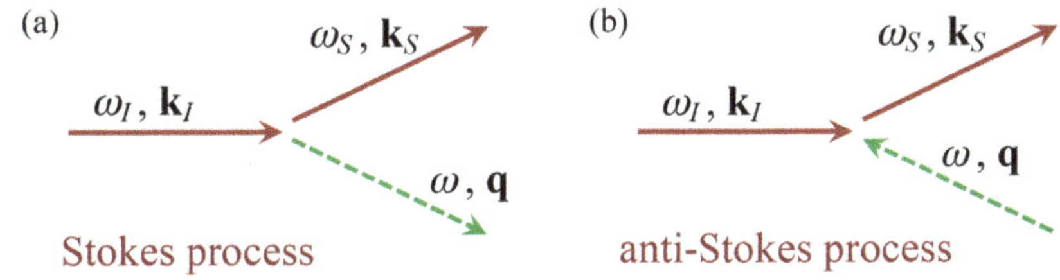

Figure 1.5. Representation of the (a) Stokes and (b) anti-Stokes processes of inelastic light scattering. The incident and scattered photons are depicted by solid lines and the created or absorbed excitation by a dashed line.

to a 29.98 GHz frequency shift. We label the angular frequency and wave vector of the excitation by ω and $\mathbf{q}$, respectively, and the corresponding quantities for the scattered light by ω_S and $\mathbf{k}_S$. Then the two simplest processes, known as Stokes scattering (where an excitation is emitted) and anti-Stokes scattering (where an excitation is absorbed) can be represented schematically as in figure 1.5. When the light scattering takes place in a bulk (effectively infinite) 3D transparent material the energy and momentum are conserved, and so

$$\hbar\omega_I = \hbar\omega_S \pm \hbar\omega, \qquad \hbar\mathbf{k}_I = \hbar\mathbf{k}_S \pm \hbar\mathbf{q}. \tag{1.31}$$

Here the upper and lower signs correspond to the Stokes and anti-Stokes case, respectively. The photon dispersion relations of the incident and scattered light yield $|\mathbf{k}_I| = \eta_I\omega_I/c$ and $|\mathbf{k}_S| = \eta_I\omega_S/c$, where η_I and η_S denote the corresponding values of the refractive index. The above relations, together with the conservation properties in equation (1.31), imply that only excitations with wave vectors near the centre of the Brillouin zone in a bulk material are excited.

There are several factors to take into account for the application of light scattering to low-dimensional and/or to optically absorptive materials. First, for thin films and multilayers, the conservation of momentum will apply only in the directions where the system has translational symmetry, i.e. in the plane of the surfaces. In the other dimension, the direction away from the surface(s), there will be a spread of wave-vector components for the excitation, implying in general a spread of frequencies in the measured spectrum. The extent to which this occurs depends on the optical absorption (or the imaginary part of the refractive index) in the material and the geometry. If the absorption is large, which is typical for metals and some semiconductors, the light will penetrate only close to the surface of the material, so the light scattering process can be sensitive to the surface properties and hence to the excitations localized there. In the case of 0D materials (or dots) none of the wave-vector components will be conserved precisely.

As well as providing information about the frequencies and dispersion properties of excitations, inelastic light scattering also yields results for the intensities of the excitations. The light scattering cross section is, in fact, related to the correlation functions between components of the electric-field vector for the scattered light, and this gives the important link between theory and experiment, as we describe later. Several general references provide a good background to inelastic light scattering (see e.g. [27–30]), but continuing developments have led to improvements in sensitivity and applicability. There are two other aspects of inelastic light scattering that we will discuss in later chapters. First, *micro-focused* light scattering spectroscopy, in which the area of laser beam spot on the sample surface is made sufficiently small (e.g. of order 250 nm across), enables the spatial variations in intensity across a nanostructure to be sampled. Second, *stimulated* Raman and Brillouin scattering allow the study of NL processes involving the light beams and the excitation in the sample.

1.3.2 Inelastic particle scattering

Next we turn to the inelastic scattering of particles, rather than light, by the crystal excitations. In principle, depending on the type of particle, this scattering process may involve larger momentum changes than encountered with light scattering, so by contrast particle scattering may be useful for probing excitations throughout the entire Brillouin zone.

In electron energy loss spectroscopy (EELS), and in its high-resolution variant (known as HREELS), a beam of electrons is scattered from the surface of a material (see e.g. [26, 31]). A typical geometry for the scattering of particles from a planar surface is shown in figure 1.6, where a surface excitation such as a phonon or plasmon is emitted or absorbed in the process. The condition for conservation of energy yields

$$E_I = E_S \pm \hbar\omega, \tag{1.32}$$

where we follow the notation in figure 1.6 and the upper (lower) sign refers to the creation (annihilation) case for the excitation. The other scattering condition corresponds to the conservation of the momentum component in the direction parallel to the surface plane, giving

$$|\hbar k_I \sin\theta_I - \hbar k_S \sin\theta_S| = \hbar q_{\|}. \tag{1.33}$$

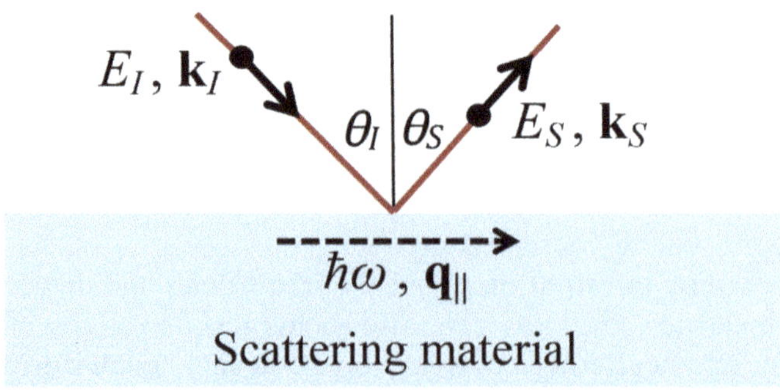

Figure 1.6. Geometry for inelastic particle scattering from a planar surface of a material, where the incident and scattered beam are in the same vertical plane. An excitation (dashed line) is created or annihilated in the material during the process.

We also have the classical energy–momentum relations $E_I = (\hbar k_I)^2/2M$ and $E_S = (\hbar k_S)^2/2M$ with M denoting the particle mass. Generally for EELS or HREELS we can approximate by assuming $E_I \gg \hbar\omega$, which is usually well satisfied (e.g. if $E_I \sim 100$ eV and $\hbar\omega \sim 100$ meV), implying $|\mathbf{k}_S| \simeq |\mathbf{k}_I| \equiv k_I$. For the same reason we can also ignore Umklapp processes in equation (1.33). It follows that, by varying the angles θ_I and θ_S, the in-plane wave vector can be varied across its 2D Brillouin zone and the dispersion relation giving ω versus $\mathbf{q}_{\|}$ for the surface excitation can be investigated. Examples of this type of analysis applied to phonons localized at a surface will be given later. A variant of the technique is spin-polarized EELS (or SPEELS), which has been used for magnetic excitations at surfaces. It involves employing a spin-polarized incident electron beam together with a polarization analyser in the scattered beam (e.g. [32]).

The inelastic scattering of particles that are much heavier than electrons allows for larger momentum changes and hence offers the prospect of making it easier (in principle) to study excitations with larger wave vectors. Inelastic neutron scattering (INS) is well-established for studying bulk excitations. However, neutrons interact only very weakly with materials and have macroscopically large penetration depths. Consequently, the sample sizes have to be rather large (often with dimensions of the order of 1 cm or more) and the method is relatively insensitive to surface effects. This drawback may be mitigated if the surface or interface area is effectively increased (e.g. by using a short-period superlattice) or if a neutron beam is used at grazing incidence to a surface (see e.g. [33]).

In contrast, in the context of *surface* excitations, the inelastic scattering of small neutral atoms, such as He atoms, has proved to be more useful. Their penetration into a solid is typically very small. For example, He atoms at incident energies of order 20 meV will be scattered back from a surface by essentially just the first monolayer of the material. Also the atoms in the incident beam will have a de Broglie wavelength ($= h/\sqrt{2ME_I}$) that is typically comparable with the lattice constant, so excitations in the entire 2D Brillouin zone can be studied. The technique was pioneered by Brusdeylins *et al* [34] using time-of-flight (TOF) spectroscopy to study surface phonons. The basic equations for the scattering are essentially the same as for EELS, equations (1.32) and (1.33), except that it is no longer true that $k_I \simeq k_S$ because of the much larger He particle mass here. In the case of a 90° scattering geometry (when $\theta_I + \theta_S = \pi/2$ in figure 1.6) it is straightforward to show that ω and $q_{\|}$ are related by

$$\omega = \frac{\hbar^2 k_I{}^2}{2M}\left[\left(\frac{q_{\|} + k_I \sin\theta_I}{k_I \cos\theta_I}\right)^2 - 1\right]. \tag{1.34}$$

This allows the dispersion relation of a surface excitation to be deduced. An up-to-date implementation of the He scattering technique, as applied to surface phonons in Bi, is described for example in [35, 36] (see also section 2.3).

1.3.3 Other techniques

To conclude this section we briefly mention some experimental techniques that are specific to certain types of excitations. Others will be introduced in later chapters. Magnetic resonance techniques have become well-established for studying excitations in magnetic materials, both in the bulk and for low-dimensional geometries (see e.g. [29] for a general review). At low power levels a microwave magnetic field (in a direction transverse to the magnetization) will couple linearly to the zero wave-vector spin waves (magnons) in a ferromagnet, leading to the ferromagnetic resonance (FMR) condition being achieved when the frequencies of the field and excitation match one another. The analogous effect in antiferromagnets is called antiferromagnetic resonance (AFMR) and typically occurs at higher frequencies. In spin-wave resonance (SWR) in ferromagnetic thin films the technique is used to excite selectively the standing spin waves across the film thickness that have small but nonzero wave vectors. At higher power levels the oscillating (or 'pumping') magnetic field can be used to promote various NL spin-wave processes in ferromagnets, as we will discuss in chapter 9. In the emerging field of cavity magnonics, a magnetic sample would be placed inside a resonant microwave cavity to study the enhanced coupling (hybrid modes) formed between the magnons and cavity photons (see e.g. [37, 38]).

Another area that involves specialized techniques occurs with excitations that generate electromagnetic waves. These can be studied using optical methods that employ, for example, evanescent wave coupling in the region adjacent to closely positioned interfaces between media of different refractive index. This is the basis of the method of attenuated total reflection (ATR), which has been successfully applied to plasmons and polaritons (see chapters 5 and 6). Evanescent wave properties are also exploited in the scanning near-field optical microscopy (SNOM) method, which is a form of scanning probe microscopy.

1.4 Theoretical methods for dynamic properties

Some of the theoretical considerations were mentioned already in earlier parts of this chapter, particularly in section 1.1.2, where we presented simple examples of linear and NL excitations. There we emphasized the role of an equation of motion for the excitation amplitude, e.g., involving Maxwell's wave equation for the electric field as in equation (1.19), along with boundary conditions. We start with such an approach, as described below, and then we generalize to more powerful techniques (involving linear response theory and Green's functions). In all cases the use of Bloch's theorem to deal with spatial periodicity, where applicable, is important.

1.4.1 Equations of motion for excitations

From such an approach one would normally be able to deduce the excitation frequency of the system and hence a dispersion relation, as well as information about the spatial variation of the amplitude in the case of a localized excitation. As discussed earlier in section 1.2 a bulk excitation will depend on a 3D wave vector $\mathbf{q}$, whereas a surface or interface excitation at a planar interface will be characterized by a 2D wave vector $\mathbf{q}_{\parallel}$ and one or more attenuation factors (or reciprocal decay lengths). Other considerations apply in 1D and 0D systems, as mentioned before. Calculating the excitation frequencies in a low-dimensional system normally involves having a set of *homogeneous* equations of motion satisfied by the amplitude variable(s) in the different parts of the system.

These equations may either be differential equations if a macroscopic (or continuum) approach is being followed, as in the previous examples given for electromagnetic excitations, or they may be finite-difference equations if a microscopic approach is used, as in a discrete-lattice model. An example of the latter is the calculation in lattice dynamics for the vibrational modes (phonons) in a linear (1D) chain of atoms, which will be given in chapter 2. Another approach for phonons, however, is based on using the differential equations for a continuous elastic medium; this is another example of the former approach mentioned above and it is employed in the next paragraph, as well as later. Generally, both macroscopic and microscopic approaches will be used for topics throughout the following chapters, as appropriate. We shall be emphasizing cases where an analytic solution is feasible. However, it is worthwhile mentioning that, in some cases, specialized numerical simulation packages have been developed to carry out these calculations. For example, the OOMMF software is applicable to studying a range of properties of micromagnetic low-dimensional systems [39].

Next, we briefly quote some other examples that involve a homogeneous equation of motion, further to the electromagnetic example given earlier. First, we take the case of an infinite, isotropic elastic medium with a longitudinal vibrational wave (an acoustic phonon) propagating in one direction, which we label as the z-direction. If we use $u(z, t)$ to denote the local displacement in the z-direction of the medium, ρ for the uniform density of the medium, and v_L for the longitudinal sound wave velocity, the usual linear wave equation of motion has the form (see e.g. [1, 3])

$$\rho\frac{\partial^2 u}{\partial t^2} + \rho\gamma\frac{\partial u}{\partial t} - \rho {v_L}^2\frac{\partial^2 u}{\partial z^2} = 0. \tag{1.35}$$

where weak damping effects are included through the term proportional to a damping constant $\gamma > 0$. A solution is easily found by taking $u(z, t) \propto \exp[\mathrm{i}(qz - \omega t)]$ which yields the condition $\omega^2 + \mathrm{i}\omega\gamma = {v_L}^2 q^2$. In the limit of zero damping ($\gamma \to 0$) this gives the usual acoustic-phonon dispersion relation $\omega = v_L q$, as expected. This solution is applicable for an effectively infinite material and is consistent with the requirements of the 1D Bloch's theorem in equation (1.29). Boundary effects, e.g. leading to a reflected wave, can be introduced as discussed later in chapter 2.

Another example is for the electronic states in matter, where the appropriate equation of motion is Schrödinger's time-dependent equation and the electron is described by the quantum-mechanical wave function $\Psi(\mathbf{r}, t)$. We have [40, 41] the equation

$$\mathrm{i}\hbar\frac{\mathrm{d}}{\mathrm{d}t}\Psi = \mathcal{H}\Psi, \tag{1.36}$$

where $\mathcal{H}$ is the Hamiltonian, typically taking the simple form

$$\mathcal{H}(\mathbf{r}) = -\frac{\hbar^2}{2m}\nabla^2 + V(\mathbf{r}). \tag{1.37}$$

Here m denotes the electron mass and $V(\mathbf{r})$ is the potential energy, assumed to be independent of time. If we are able to write $\Psi(\mathbf{r}, t) = \psi(\mathbf{r})\exp\{-\mathrm{i}Et/\hbar\}$ for a stationary state with energy E, then Schrödinger's time-independent equation is simply

$$\mathcal{H}(\mathbf{r})\psi(\mathbf{r}) = E\psi(\mathbf{r}). \tag{1.38}$$

The time-dependent and time-independent forms of Schrödinger's equation will be extensively employed for applications coming in chapter 5.

When a discrete-lattice approach is followed, leading to finite-difference equations for the spatial variables instead of differential equations, some special techniques are useful. One of these techniques is the *tridiagonal matrix* (or TDM) method, which is useful for surface and interface problems when the interactions are of a short-range nature, often just between nearest neighbours. Some applications to introduce the method are discussed later for the lattice dynamics of phonons in chapter 2, for exchange-dominated spin waves in ferromagnets in chapter 3, and for the electronic excitations in graphene in chapter 5. Also a general outline of this method is given in the appendix.

1.4.1.1 Quantum-mechanical operators and second quantization

A somewhat different situation from the above examples will occur if the excitation is described in terms of a quantum-mechanical operator, rather than a scalar amplitude. This will be so, for example, in the case of a magnetic system consisting of a lattice of atoms, where each atom is associated with a spin angular-momentum operator (see chapters 3 and 4). If the quantum-mechanical system is represented by a Hamiltonian operator $\mathcal{H}$, then the equation of motion for any operator A of the system is obtained from (see [40, 41])

$$\mathrm{i}\hbar\frac{\mathrm{d}A}{\mathrm{d}t} = [A, \mathcal{H}], \tag{1.39}$$

where $[A, \mathcal{H}] \equiv (A\mathcal{H} - \mathcal{H}A)$ denotes the commutator between the two operators. Equation (1.39) may be either a linear or NL equation in A, depending on whether the outcome of evaluating the commutator of the operator A (which describes the excitation) with the Hamiltonian leads to an expression that is linear in A or if has some other more complicated dependence. Examples of both cases will arise in later in this book, but it is probably most clearly illustrated for magnons in ferromagnets (see chapters 3, 4 and 9), where the Hamiltonian can be expressed in terms of products of non-commuting spin operators.

At this stage it is appropriate to mention the concept of creation and annihilation operators that can often be used to describe the excitations within the formalism known as *second quantization* of the operators [40, 41]. The basic ideas are often introduced in terms of the quantum-mechanical simple harmonic oscillator (SHO) in 1D as an example. The Hamiltonian in this case can be written as

$$\mathcal{H} = \frac{\hat{p}^2}{2m} + \frac{1}{2}m\omega^2 z^2, \tag{1.40}$$

where $\hat{p} = -i\hbar\partial/\partial z$ is the momentum operator for motion of the particle (with mass m) for 1D motion along the z-direction. The last term represents the potential energy, which is proportional to z^2, and the factor ω denotes the angular frequency for a classical oscillator. In the second quantization technique a Hermitian conjugate pair of operators are defined by $a^\dagger = (2m\omega)^{-1/2}(m\omega z + i\hat{p})$ and $a = (2m\omega)^{-1/2}(m\omega z - i\hat{p})$. It is then straightforward to prove that

$$\mathcal{H} = \hbar\omega\left(a^\dagger a + \frac{1}{2}\right) \tag{1.41}$$

and that $[a, a^\dagger] = 1$ for the commutator of the operators. Several convenient properties can be established next (see e.g. [40, 41]). First, there is a hierarchy (or 'ladder') of eigenstates of $\mathcal{H}$ that we may denote as $|n\rangle$ where n is an integer taking the possible values 0, 1, 2, …. Moreover, the corresponding eigenvalues are $(n + \frac{1}{2})\hbar\omega$, so there is a series of energy levels with spacing $\hbar\omega$. Second, the role of $a^\dagger$ is as a 'creation' operator, in the sense that its action on an eigenstate $|n\rangle$ is to take it up to the next level; the formal expression is that $a^\dagger|n\rangle = \sqrt{n+1}\,|n+1\rangle$ for normalized eigenfunctions. The operator a has a similar role but it acts as an 'annihilation' operator, since it satisfies $a\,|n\rangle = \sqrt{n}\,|n-1\rangle$. These basic properties are represented schematically in figure 1.7.

It will be seen later that the concept of creation and annihilation operators, which we have illustrated mathematically for an SHO, generalizes to other systems. This

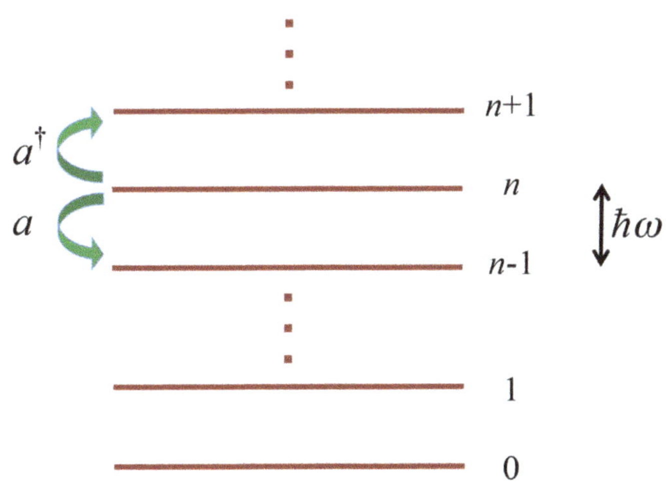

Figure 1.7. Schematic representation of the roles for the creation operator $a^\dagger$ and the annihilation operator a on the eigenstate $|n\rangle$ for the ladder of equally spaced energy levels of a 1D simple harmonic oscillator.

occurs most directly in the lattice dynamical treatment of phonons in linear atomic chains (as coupled harmonic oscillators) in chapter 2. In terms of the statistics, the above quantities are 'boson' operators satisfying the commutation property stated after equation (1.41); there are also 'fermion' operators satisfying anticommutation relations that we will introduce for electronic excitations in chapter 5.

1.4.2 Linear response and Green's functions for excitations

(Note: this subsection may optionally be omitted in a first reading.) In many cases we are interested in other properties of the excitations apart from their dispersion relations. For example, we may need to know the relative statistical weighting (or spectral intensity) of the excitations or to find a scattering cross section to compare with experiment. This information can be provided by linear response theory, and we will now outline this method and then illustrate it with a preliminary example. The basic idea is to calculate the response of the system to a small applied stimulus, taking care to choose the appropriate classical or quantum variables (which are related to the excitation being studied). The *response functions* provide not only the dispersion relations for the excitations but also the intensity-related information mentioned above. General accounts of linear response theory are to be found, for example, in [6, 42, 43] where the close connection to *Green's functions* is also explained. In the following paragraphs (along with additional material in the appendix) we present, for later reference, the important results needed in the context of low-dimensional systems.

We suppose that a time-dependent perturbation corresponding to the Hamiltonian $\mathcal{H}_1(t) = -B f(t)$ is applied to the system, where $f(t)$ is an external scalar field variable that couples linearly to a system variable denoted by the operator B. A simple physical example is when B is the displacement z from equilibrium of a particle moving in 1D, and $f(t)$ is a mechanical force acting on the particle. The total Hamiltonian can then be written as

$$\mathcal{H} = \mathcal{H}_0 - Bf(t), \tag{1.42}$$

where $\mathcal{H}_0$ denotes the Hamiltonian for the unperturbed system, taken to be in thermal equilibrium at time $t = -\infty$. At any later time t the response of the system due to the perturbation can be expressed in terms of the change, denoted by $\bar{A}(t)$, that is produced in any other chosen system variable specified by an operator A. The procedure for doing this is outlined in section A.2 of the appendix using properties of the density matrix operator in statistical physics. The main conclusion is that we can define a *response function* (also called a *generalized susceptibility*) by the ratio

$$\chi(\omega) = \frac{\bar{A}_\omega}{F(\omega)}, \tag{1.43}$$

where $\bar{A}_\omega$ and $F(\omega)$ are defined to be the frequency Fourier components of $\bar{A}(t)$ and $f(t')$, respectively. The above result holds within a linear approximation in terms of the perturbation term, and it can be shown that $\chi(\omega)$ may be identified (apart from a numerical factor) with the Fourier component of a quantum-mechanical Green's function (GF) written as $G(A; B \mid \omega)$. This quantity describes correlations between the operators A and B. In our notations, the formal relationship between $\chi(\omega)$ and $G(A; B \mid \omega)$ is given by equation (A.23) in the appendix. The theory establishes the role of GFs as response functions, in the sense that they provide the proportionality factor between an applied stimulus to the system, represented by $F(\omega)$, and the response that it produces, represented by $\bar{A}_\omega$. The formal definitions and some basic properties of GFs are given in the appendix.

In order to illustrate more clearly the formalism for response functions and GFs, we return to the physical example of excitations in an infinite elastic medium as considered earlier in this section. The homogeneous equation of motion for the displacement $u(z)$ for a longitudinal acoustic wave in the z-direction was given by equation (1.35), to which we now add a driving term (a perturbation term) which is chosen to represent a harmonic point force at any position z' in the system, i.e. we take

$$f(z, t) = (f_0/A_0)\exp(-\mathrm{i}\omega t)\,\delta(z - z'), \tag{1.44}$$

where f_0 is an amplitude and A_0 is the area presented by the system in the xy-plane. The total interaction energy with the system is

$$\begin{aligned} H_1(t) &= -\iiint u(z)f(z, t)\mathrm{d}x = -f_0\exp(-\mathrm{i}\omega t)\int u(z)\delta(z - z')\mathrm{d}z \\ &= -f_0\exp(-\mathrm{i}\omega t)\,u(z'), \end{aligned} \tag{1.45}$$

so it takes the general form for $\mathcal{H}_1$ that was assumed earlier. The equation of motion obtained by generalizing equation (1.35) is

$$\rho\frac{\partial^2 u}{\partial t^2} + \rho\gamma\frac{\partial u}{\partial t} - \rho v_L{}^2\frac{\partial^2 u}{\partial z^2} = \frac{f_0}{A_0}\exp(-\mathrm{i}\omega t)\delta(z - z'). \tag{1.46}$$

As we might have anticipated, the outcome is that we now have an *inhomogeneous* differential equation to solve. We may first take a common time dependence such as $\exp(-\mathrm{i}\omega t)$, leaving the equation representing the spatial dependence as

$$\frac{\mathrm{d}^2 u}{\mathrm{d}z^2} + q^2 u = -\frac{f_0}{\rho v_L^2 A_0}\delta(z - z'). \tag{1.47}$$

Here the complex q^2 is defined in the same way as was found following equation (1.35). We seek the solution that is bounded as $z - z' \to \pm\infty$ and that has the correct behaviour as $z - z' \to 0$ consistent with the delta function term. It may easily be verified that the result is

$$u(z) = \frac{\mathrm{i}f_0}{2\rho v_L^2 q A_0}\exp(\mathrm{i}q|z - z'|). \tag{1.48}$$

In the spirit of linear response theory, this result is the driven equation for $u(z)$ giving its response to a harmonic force applied at point $z = z'$. Using equations (1.43) and (A.23) we obtain the GF as

$$G(u(z);\, u(z') \mid \omega) = \frac{\mathrm{i}}{2\pi\rho v_L^2 A_0} \exp(\mathrm{i}q|z - z'|). \tag{1.49}$$

The above expression is our main result in this example, but we can also use it to illustrate some other properties. Since the result in equation (1.49) is a function of the spatial separation $z - z'$, as would be expected for a system with translational symmetry, we can re-express it by making a Fourier transform from the spatial variables to a wave-number component (denoted as k) in the z-direction:

$$G(u;\, u \mid k,\, \omega) \equiv \int_{-\infty}^{\infty} G(u(z);\, u(z') \mid\omega)\mathrm{e}^{-\mathrm{i}k(z-z')}\mathrm{d}(z - z'). \tag{1.50}$$

The last step is to carry out the integration in the above expression; this can be accomplished by splitting the range of integration into two parts with $z < z'$ and $z > z'$, leading eventually to

$$G(u;\, u \mid k,\, \omega) = \frac{1}{2\pi A_0(v_L^2 k^2 - \omega^2 - \mathrm{i}\omega\gamma)}. \tag{1.51}$$

We note that the condition for the energy denominator to be zero yields the dispersion relation $\omega_k = v_L k$ for the excitation (an acoustic phonon) if the damping is ignored. This illustrates a general property for the denominator of GFs in yielding the dispersion relations of the excitations in the system.

Finally we may use equation (1.51) together with the fluctuation-dissipation theorem described in the appendix to obtain the correlation function, denoted as $\langle |u|^2 \rangle_{k,\omega}$, which is a measure of the spectral intensity (or mean-square amplitude) of the excitation as a function of k and ω. A numerical example of the result appears in figure 1.8, where the peaks occur for ω close to $\pm\omega_k$ provided the damping is relatively small ($\gamma \ll \omega_k$). The peaks are unequal in height due to their statistical-

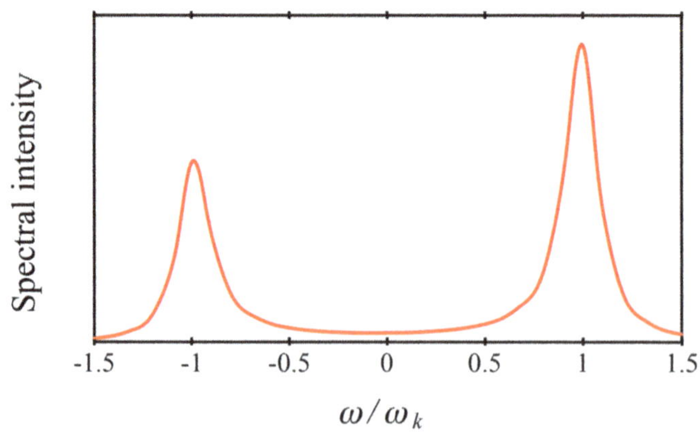

Figure 1.8. The spectral intensity $\langle |u|^2 \rangle_{k,\omega}$ versus reduced frequency ω/ω_k for the acoustic phonons in an infinite elastic medium (see the text). We denote $\omega_k \equiv v_L k$ for any fixed value of wave-vector component k. The assumed numerical values of the parameters correspond to taking the dimensionless ratios $k_B T/\hbar\omega_k = 2$ and $\gamma/\omega_k = 0.2$.

mechanical weighting. This difference has consequences, for example, regarding the difference in intensities observed between Stokes and anti-Stokes peaks in inelastic light scattering, as we will discuss in chapters 2, 3 and 4.

1.5 Photonic band gaps in periodic structures

We have already briefly mentioned periodic structures in sections 1.2.2 and 1.2.3 in terms of periodic structures, typically formed from two or more materials, either as superlattices or as lateral arrays. The bands of excitations in such structures (i.e. their frequency versus wave-vector relationships) can be very complicated in general, and particular features may sometimes arise, such as the appearance of band gaps at the mini-Brillouin zone boundaries mentioned earlier. The properties for the occurrence and manipulation of such bands have led to the tremendous interest in *band-gap materials*, initially in the photonics case in optics (see [44] for a thorough review) but subsequently for other excitations as well. In this introductory chapter we present a simple formulation for a 1D photonic band-gap material (a photonic crystal or PC) and discuss some extensions to 2D or 3D. This treatment will serve as a template for carrying out other analogous calculations for periodic (and quasi-periodic) structures in the later chapters.

1.5.1 Bulk slab model

Here we consider alternating layers of medium A and medium B to form the superlattice structure depicted in figure 1.9. The layer thicknesses are d_A and d_B, both assumed to be large compared with the atomic lattice parameters of the materials. The length associated with the artificial periodicity is $D = d_A + d_B$, and so the 1D wave vector Q (the so-called Bloch wave vector) in the z-direction associated with

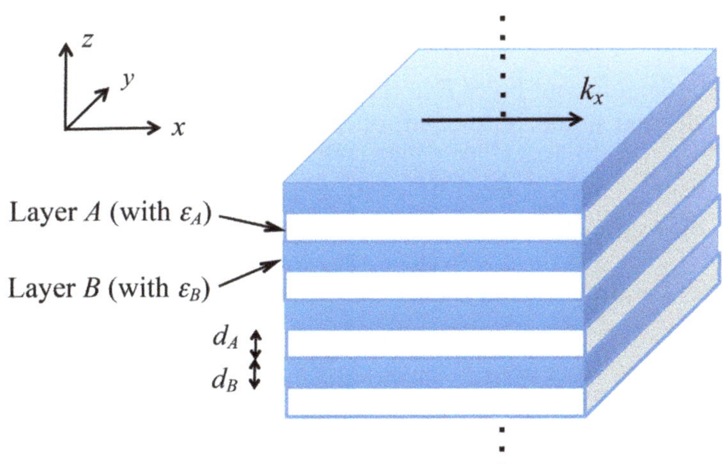

Figure 1.9. Geometry and notation for the calculation of the photonic (optical) band structure of a two-component alternating periodic superlattice. The component materials are labelled A and B, with thicknesses d_A and d_B and dielectric functions ϵ_A and ϵ_B, respectively. The periodic length D in the z-direction is equal to $d_A + d_B$.

this translational symmetry has a first Brillouin zone corresponding to $-\pi/D < Q \leqslant \pi/D$. Further it will be assumed that the dielectric functions are ϵ_A and ϵ_B, taken for simplicity to be the bulk dielectric constants (independent of the excitation angular frequency ω). This is sometimes referred to as the *bulk slab model* of a superlattice.

We now employ standard results from electromagnetism for the optical wave propagation in the layers. There are basically two cases, depending on whether the waves have s-polarization (with electric-field vector **E** in the y-direction) or p-polarization (with **E** in the xz-plane). The wave-vector component Q enters into the calculations through Bloch's theorem in the form

$$\mathbf{E}(z + D) = \exp(\mathrm{i}QD)\, \mathbf{E}(z). \tag{1.52}$$

Also, all field components will be taken to have dependences on x and t in accordance with the factor $\exp(\mathrm{i}q_x - \mathrm{i}\omega t)$, where for completeness we have included a wave vector q_x in the x-direction to allow for the case of oblique incident of the optical waves at each interface.

The calculation for s-polarization will be shown first. The solution of the wave equation (1.19) for E_y in any layer will be a superposition of a forward- and a backward-travelling wave in the z-direction. Alternative forms can be written down, depending on whether the phases are taken relative to the lower (L) or upper (U) interface of each layer, giving

$$\begin{aligned} E_y &= a_l^L \exp[\mathrm{i}q_{Az}(z - lD)] + b_l^L \exp[-\mathrm{i}q_{Az}(z - lD)] \\ &= a_l^U \exp[\mathrm{i}q_{Az}(z - lD - d_A)] + b_l^U \exp[-\mathrm{i}q_{Az}(z - lD - d_A)] \end{aligned} \tag{1.53}$$

in an A layer (with $lD \leqslant z \leqslant lD + d_A$) where q_{Az} must satisfy

$$q_{Az}^2 + q_x^2 = \epsilon_A \omega^2 / c^2. \tag{1.54}$$

For the field in the adjacent layer B in the same cell l (with $lD + d_A \leqslant z \leqslant (l + 1)D$) and with q_{Bz} defined similarly to equation (1.54), we have

$$\begin{aligned} E_y &= d_l^L \exp[\mathrm{i}q_{Bz}(z - lD - d_A)] + e_l^L \exp[-\mathrm{i}q_{Bz}(z - lD - d_A)] \\ &= d_l^U \exp[\mathrm{i}q_{Bz}(z - (l + 1)D)] + e_l^U \exp[-\mathrm{i}q_{Bz}(z - (l + 1)D)]. \end{aligned} \tag{1.55}$$

Using a matrix notation the amplitudes in equation (1.53) are related by

$$|u_l^U\rangle = F_A |u_l^L\rangle, \tag{1.56}$$

where we denote

$$|u_l^{L,U}\rangle = \begin{pmatrix} a_l^{L,U} \\ b_l^{L,U} \end{pmatrix} \tag{1.57}$$

and F_A is a 2×2 matrix given by

$$F_A = \begin{pmatrix} f_A & 0 \\ 0 & f_A^{-1} \end{pmatrix} \tag{1.58}$$

with phase term $f_A = \exp(\mathrm{i}q_{Az}d_A)$. Likewise for the other layer we write

$$|w_l^U\rangle = F_B|w_l^L\rangle \tag{1.59}$$

with analogous notations and definitions.

The electromagnetic boundary conditions at the $z = lD + d_A$ and $z = (l + 1)D$ interfaces require that E_y and H_x are continuous (implying that E_y and $\partial E_y/\partial z$ are continuous). These conditions can be used to give additional relationships between the column matrices of coefficients. The results can eventually be expressed in matrix form as

$$X_A|u_l^U\rangle = X_B|w_l^L\rangle \text{ and } X_B|w_l^U\rangle = X_A|u_{l+1}^L\rangle, \tag{1.60}$$

where

$$X_i = \begin{pmatrix} 1 & 1 \\ q_{iz} & -q_{iz} \end{pmatrix}, \qquad i = A, B. \tag{1.61}$$

The equations (1.56), (1.59) and (1.60) may now be combined to give

$$|u_{l+1}^L\rangle = T|u_l^L\rangle, \tag{1.62}$$

which relates the amplitudes in cell l to the equivalent part in cell $l + 1$, and it introduces the *transfer matrix* T given by

$$T = X_A^{-1}X_BF_BX_B^{-1}X_AF_A. \tag{1.63}$$

The transfer matrix has some useful properties, as we now explain. First, it is unimodular in the sense that det $T = 1$, which can be proved from the form of matrix products in equation (1.63). Second, the statement of Bloch's theorem in equation (1.52) is equivalent to

$$|u_{l+1}^L\rangle = \exp(\mathrm{i}QD)|u_l^L\rangle, \tag{1.64}$$

and therefore we have the property that

$$[T - \exp(\mathrm{i}QD)I]|u_l^L\rangle = 0, \tag{1.65}$$

where I is the unit 2×2 matrix. There is a similar equation obtained by relating $|u_{l-1}^L\rangle$ to $|u_l^L\rangle$, yielding

$$[T^{-1} - \exp(-\mathrm{i}QD)I]|u_l^L\rangle = 0. \tag{1.66}$$

Then, by adding equations (1.65) and (1.67) we find

$$[T + T^{-1} - 2\cos(QD)I]|u_l^L\rangle = 0, \tag{1.67}$$

which holds for any cell l. Therefore, from the property that $(T + T^{-1}) = (\mathrm{tr}T)I$, where tr denotes the trace, we must have

$$\cos(QD) = \frac{1}{2}\mathrm{tr}T. \tag{1.68}$$

This is a general result, providing an implicit dispersion relation for the excitation frequencies in terms of the Bloch wave vector Q and the transfer matrix T.

Explicit evaluation shows that the diagonal elements of T in the present example are

$$\begin{aligned} T_{11} &= f_A[f_B(q_{Az} + q_{Bz})^2 - f_B^{-1}(q_{Az} - q_{Bz})^2]/4q_{Az}q_{Bz}, \\ T_{22} &= f_A^{-1}[f_B^{-1}(q_{Az} + q_{Bz})^2 - f_B(q_{Az} - q_{Bz})^2]/4q_{Az}q_{Bz}. \end{aligned} \tag{1.69}$$

When these results are substituted into equation (1.68) we find after some algebraic manipulation that the dispersion relation in this case of s-polarization is

$$\cos(QD) = \cos(q_{Az}d_A)\cos(q_{Bz}d_B) - g_s \sin(q_{Az}d_A)\sin(q_{Bz}d_B), \tag{1.70}$$

where

$$g_s = \frac{1}{2}\left(\frac{q_{Bz}}{q_{Az}} + \frac{q_{Az}}{q_{Bz}}\right). \tag{1.71}$$

The calculation for the case of p-polarization follows very similar steps, including the use of a transfer matrix. It is found that equation (1.70) is still applicable provided g_s is replaced by g_p where

$$g_p = \frac{1}{2}\left(\frac{\epsilon_B q_{Az}}{\epsilon_A q_{Bz}} + \frac{\epsilon_A q_{Bz}}{\epsilon_B q_{Az}}\right). \tag{1.72}$$

The dispersion relation in the general form represented by equation (1.70) occurs commonly for excitations in periodic structures, as we will see in later chapters. It is sometimes called the *Rytov equation*, since its first derivation is attributed to Rytov in the context of acoustic waves [45]. A full discussion of the results obtained for the above optical superlattice, or photonic crystal (PC), can be found in [44, 46]. A more recent review of the optical properties of 1D photonic band-gap (PBG) materials has been given by Bellingeri *et al* [47].

In figure 1.10 we present a numerical example of a dispersion relation obtained from equation (1.72) in the special case of normal incidence where $q_x = 0$. This makes the distinction between s- and p-polarizations vanish. The plot is in dimensionless units for the frequency ω versus Q. In the reduced-zone scheme used here the dispersion curves are 'folded back' and gaps open up at $QD = 0$ and π. In this example the band gaps (or stop bands) are relatively large because we have assumed a significant difference, or mismatch, between the dielectric constants of the adjacent layers. If the dielectric constants are taken to be close in value, the gaps shrink to a small value.

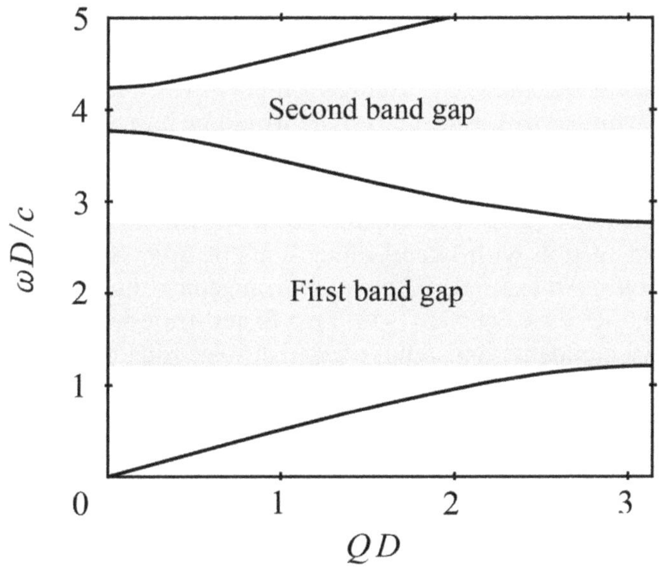

Figure 1.10. Dispersion curves for the frequency (in terms of $\omega D/c$) versus the Bloch wave vector (in terms of QD) for normal-incidence optical waves in a 1D photonics crystal consisting of a two-component alternating periodic superlattice. The assumed parameters are $d_A = d_B = 0.5D$, $\epsilon_A = 3$ and $\epsilon_B = 2$. The lowest two photonic band gaps are shown as the shaded areas.

We remark, finally, that the transfer-matrix method can be modified in a straightforward manner to apply to other types of excitations, as well as to study the localized surface and/or interface modes (when they exist) in periodic and quasiperiodic structures that are either semi-infinite or finite in extent (in the z-direction). Results to illustrate this method more generally will be covered in some of the later chapters.

1.5.2 Photonic crystals in 2D and 3D

In the previous subsection it was demonstrated that stop bands would arise as a result of the 1D periodicity in a superlattice. The interesting question immediately arises as to whether it is possible to achieve arrays with varying dielectric properties in 2D, or even 3D, that would have stop bands or band gaps for optical signals at some frequencies of interest and for both polarizations. The first investigations relating to this were made by Yablonovitch [23] in 1987, while related light localization studies were reported at around the same time by John [24]. A motivation for these early studies was the realization that a stop band with the properties mentioned would prohibit spontaneous emission of radiation in optical transitions and this could be utilized for semiconductor lasers and devices. Subsequently, however, a wide range of other quantum-optical device applications has emerged and an excellent reference is the book by Joannopoulos *et al* [44], which also covers the modelling techniques used to calculate the stop bands and the transmission characteristics.

The first realization of a full PBG material in 3D was due to Yablonovitch *et al* [48] who formed a structure that absorbed 15 GHz radiation for all spatial directions and polarizations. It was made by drilling air holes at various angles into a dielectric block. Such an arrangement, whilst effective, would be impractical for scaling to the higher frequencies (or shorter wavelengths) required for applications in the optical region. Versatile alternative structure have been based on the so-called 'woodpile' periodic arrangements of semiconductor rods, in which each rod may have a rectangular cross section with lateral dimensions of order 1 μm and a length that is very much larger. An example of such an arrangement using rods of polycrystalline silicon (with dielectric constant ~13) on a Si substrate was given by [49]. In this case the spatial alignments and separations between rods constituted an effective face-centred tetragonal lattice symmetry, and the stop band corresponded to the 10 to 14.4 μm range of wavelengths. Another PBG structure that has been investigated (see [44]) is the 'inverse opal' arrangement. The latter can be formed by utilizing the property of closely spaced microscopic spheres (e.g. of silica, SiO_2) to self-assemble into a fcc lattice from a colloidal solution. The structure can then be inverted by filling the voids between spheres with a high dielectric material and finally dissolving away the spheres.

Recent efforts at producing practical and more economical PBG materials in 3D with a greater number of repeat unit cells have employed a variety of techniques, including photolithography, self-assembly and holographic lithography, among others (see e.g. [50]). Notably, the fabrication of large photonic crystals by controlling the self-assembly processes based on liquid crystal blue phases was reported by Chen *et al* [51].

As well as the intense activity mentioned above with its focus on PBG materials in 3D, there have also been advances in developing 2D structures [44]. In particular, there have been studies of structures with artificial 2D periodicity, including cases where the dielectric functions are dispersive, giving rise to plasmons, phonons and polaritons, for example. We will discuss some of these 2D cases in the later chapters.

References

[1] Kittel C 2004 *Introduction to Solid State Physics* 8th edn (New York: Wiley)

[2] Ashcroft N W and Mermin N D 1976 *Solid State Physics* (New York: Saunders College)

[3] Cohen M L and Louie S G S 2016 *Fundamentals of Condensed Matter Physics* (Cambridge: Cambridge University Press)

[4] Ketterson J B 2016 *The Physics of Solids* (Oxford: Oxford University Press)

[5] Snoke D W 2020 *Solid State Physics: Essential Concepts* 2nd edn (Cambridge: Cambridge University Press)

[6] Cottam M G and Tilley D R 2005 *Introduction to Surface and Superlattice Excitations* 2nd edn (Bristol: IOP Publishing)

[7] Bloch F 1928 *Z. Phys.* **52** 555

[8] Novoselov K S, Geim A K, Morozov S V, Jiang D, Katsnelson M I, Grigorieva I V, Dubonos S V and Firsov A A 2005 *Nature* **438** 197

[9] Castro Neto A H, Guinea F, Peres N M R, Novosolov K S and Geim A K 2009 *Rev. Mod. Phys.* **81** 109

[10] Nénert G, Ritter C, Isobe M, Isnard O, Vasiliev A N and Ueda Y 2009 *Phys. Rev.* B **80** 024402
[11] Mills D L 1998 *Nonlinear Optics* 2nd edn (Berlin: Springer)
[12] Powers P E and Haus J W 2017 *Fundamentals of Nonlinear Optics* (Boca Raton: CRC Press)
[13] Boyd R W 2020 *Nonlinear Optics* 4th edn (Boston: Academic)
[14] White R M 2007 *Quantum Theory of Magnetism: Magnetic Properties of Materials* 3rd edn (Heidelberg: Springer)
[15] Stancil D D and Prabhakar A 2009 *Spin Waves: Theory and Application* (Heidelberg: Springer)
[16] Demokritov S O (ed) 2017 *Spin Wave Confinement: Propagating Waves* 2nd edn (Singapore: Pan Stanford)
[17] Jackson J D 1998 *Classical Electrodynamics* 3rd edn (New York: Wiley)
[18] Griffiths D J 2012 *Introduction to Electrodynamics* 4th edn (Boston: Addison-Wesley)
[19] De Teresa J M (ed) 2020 *Nanofabrication: Nanolithography Techniques and their Applications* (Bristol: IOP Publishing)
[20] Dahman Y 2017 *Nanotechnology and Functional Materials for Engineers* (Amsterdam: Elsevier)
[21] Esaki L and Tsu R 1970 *IBM J. Dev.* **14** 61
[22] Albuquerque E L and Cottam M G 2003 *Phys. Rep.* **376** 225
[23] Yablonovich E 1987 *Phys. Rev. Lett.* **58** 2059
[24] John S 1987 *Phys. Rev. Lett.* **58** 2486
[25] Vollath D 2013 *Nanomaterials: An Introduction to Synthesis, Properties and Applications* 2nd edn (New York: Wiley)
[26] Goncharova L 2018 *Basic Surfaces and their Analysis* (Bristol: IOP Publishing)
[27] Cottam M G and Lockwood D J 1986 *Light Scattering in Magnetic Solids* (New York: Wiley)
[28] Cardona M and Güntherodt G (ed) 1989 *Light Scattering in Solids V* (Berlin: Springer)
[29] Dutcher J 1995 *Linear and Nonlinear Spin Waves in Magnetic Films and Superlattices* ed M G Cottam (Singapore: World Scientific) p 287
[30] Hillebrands B and Ounadjela K (ed) 2003 *Spin Dynamics in Confined Magnetic Structures II* (Berlin: Springer)
[31] Egerton R F 2011 *Electron Energy-Loss Spectroscopy in the Electron Microscope* 3rd edn (Heidelberg: Springer)
[32] Vasilyev D and Kirschner J 2016 *Rev. Sci. Instrum.* **87** 083902
[33] Schreyer A, Schmitte T, Siebrecht R, Bödeker P, Zabel H, Lee S H, Erwin R W, Majkrzak C F, Kwo J and Hong M 2000 *J. Appl. Phys.* **87** 5443
[34] Brusdeylins G, Doak R B and Toennies J P 1981 *Phys. Rev. Lett.* **46** 437
[35] Mayrhofer-Reinhartshuber M, Tamtögl A, Kraus P, Rieder K H and Ernst W E 2012 *J. Phys.: Condens. Matter* **24** 104008
[36] Benedek G and Toennies P 2018 *Atomic Scale Dynamics at Surfaces: Theory and Experimental Studies with Helium Atom Scattering* (Berlin: Springer)
[37] Bai L, Harder M, Chen Y P, Fan X, Xiao J Q and Hu C-M 2015 *Phys. Rev. Lett.* **114** 227201
[38] Leo A, Rizzato S, Monteduro A G and Maruccio G 2019 *Three-Dimensional Magnonics* ed G Gubbiotti (Singapore: Jenny Stanford) p 347
[39] ITL/NIST The Object Oriented Micromagnetic Framework (OOMMF) project http://math.nist.gov/oommf

[40] Griffiths D J and Schroeter 2018 *Introduction to Quantum Mechanics* 3rd edn (Cambridge: Cambridge University Press)
[41] Sakurai J J and Napolitano J 2020 *Modern Quantum Mechanics* 3rd edn (Cambridge: Cambridge University Press)
[42] Nolting W 2009 *Fundamentals of Many-Body Physics: Principles and Methods* (Berlin: Springer)
[43] Cottam M G and Haghshenasfard Z 2020 *Many-Body Theory of Condensed Matter Systems: An Introductory Course* (Cambridge: Cambridge University Press)
[44] Joannopoulos J D, Johnson S G, Winn J N and Meade R D 2008 *Photonic Crystals: Molding the Flow of Light* 2nd edn (Princeton, NJ: Princeton University Press)
[45] Rytov S M 1956 *Sov. Phys. Acoust.* **2** 68
[46] Yariv A and Yeh P 1984 *Optical Waves in Crystals* (New York: Wiley)
[47] Bellingeri M, Chiasera A, Kriegel I and Scotognella F 2017 *Opt. Mater.* **72** 403
[48] Yablonovitch E, Gmitter T J and Leung K M 1991 *Phys. Rev. Lett.* **67** 2295
[49] Lin S Y, Fleming J G, Hethterington D L, Smith B K, Biswas R, Ho K M, Sigalas M M, Zubrzycki W, Kurtz S R and Bur J 1998 *Nature* **394** 251
[50] von Freymann G, Ledermann A, Thiel M, Staude I, Essig S, Busch K and Wegener M 2010 *Adv. Funct. Mater.* **20** 1038
[51] Chen C W *et al* 2017 *Nat. Commun.* **8** 727

IOP Publishing

Dynamical Properties in Nanostructured and Low-Dimensional Materials (Second Edition)

Michael G Cottam

Chapter 2

Phonons

In this and the next several chapters we will describe a range of different types of excitations in low-dimensional systems, starting here with phonons, or quantized vibrational excitations. We build on many of the concepts and methods introduced in chapter 1. The focus in the first instance will be on the *linear* properties of the excitations, but many of the results will be extended to the regime of nonlinear dynamics in chapter 8.

There are two main approaches followed in describing phonons in materials. One of these involves the lattice dynamics for the discrete array of atoms in the material. This technique may be difficult to apply and is sometimes idealized in the form of 1D models, but it has an advantage of yielding results that apply across the entire Brillouin zone. The recent interest in 2D materials such as graphene on various substrates has spurred on efforts to use lattice dynamics in higher dimensions. The other broad approach is to deal with the regime of small wave vectors (or, equivalently, large wavelengths compared with the lattice constants of the material). Then a *continuous-medium approximation* can be adopted in which the underlying lattice structure is ignored and the macroscopic variables of elasticity theory are used. This leads to a simplified description of the long-wavelength acoustic phonons, but it does not account for the higher-frequency optic phonons. In this chapter we will follow both approaches, and both will be employed in the interpretation of experimental data. General accounts of phonons are to be found in the recent text books on condensed-matter physics cited earlier (see e.g. [1, 3, 4]) and in more specialized books such as [4, 5]. Here we will emphasize applications to phonons in a variety of low-dimensional structures, including periodic and quasiperiodic arrays of nanoelements and the recent developments for phononic crystals.

2.1 One-dimensional models of lattice dynamics for films

We begin with examples for the vibrational modes of 1D chains of atoms, which we initially treat according to classical physics. Then we follow up with a quantum-mechanical

doi:10.1088/978-0-7503-3903-2ch2

treatment in terms of boson creation and annihilation operators. Later we generalize the method to more complex structures in 1D as well as in higher dimensions.

2.1.1 Monatomic chains

The simplest example is a 1D infinite lattice of identical atoms located at $z = na$ (with $-\infty < n < \infty$ for integer n) along the z-axis as in figure 2.1(a). It is assumed that each atom has mass m and is coupled to its two nearest neighbours by an elastic spring with force constant denoted by C. Then Newton's equation of motion for the longitudinal displacement u_n of the nth atom is

$$m\frac{\mathrm{d}^2u_n}{\mathrm{d}t^2} = C[(u_{n+1} - u_n) - (u_n - u_{n-1})]. \tag{2.1}$$

This equation can be solved by seeking normal mode solutions with an angular frequency ω and a spatial dependence chosen in accordance with Bloch's theorem as in equation (1.29), namely we take $u_n \propto \exp[\mathrm{i}(qna - \omega t)]$, yielding

$$\omega^2 = \frac{2C}{m}[1 - \cos(qa)] = \frac{4C}{m}\sin^2(qa/2). \tag{2.2}$$

A plot of this dispersion relation is shown in figure 2.2 for q ranging from 0 to π/a in the positive half of the first Brillouin zone. Some limiting cases are $\omega = v_L|q|$ when $|qa| \ll 1$, where we denote $v_L = a\sqrt{C/m}$, and also $\omega = 2\sqrt{C/m}$ when $|qa| = \pi$. The first case is for the regime of long wavelengths (compared to a) where we have essentially a sound wave propagating longitudinally with speed v_L and the discrete lattice structure has become unimportant. The second case represents the maximum phonon frequency, which corresponds to a standing wave at the Brillouin-zone boundary.

The above simple calculation can be generalized to include effects of surfaces and interfaces. First we consider what happens if the monatomic chain in figure 2.1(a) is semi-infinite, i.e., it has a termination point at one end (taken to be on the left) such

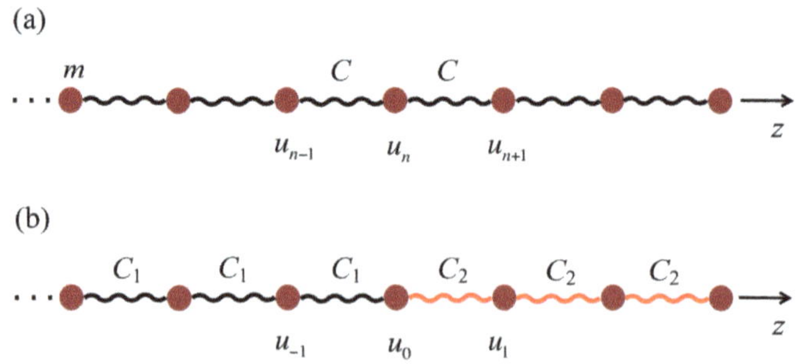

Figure 2.1. Models in 1D used for the lattice dynamics of monatomic chains, denoting the longitudinal displacement of atom n by u_n: (a) a simple chain of equally spaced masses m with the same spring constant C providing nearest-neighbour coupling; (b) an interface between two chains where the masses are all the same but the spring constants are different to the right and left of the interface mass.

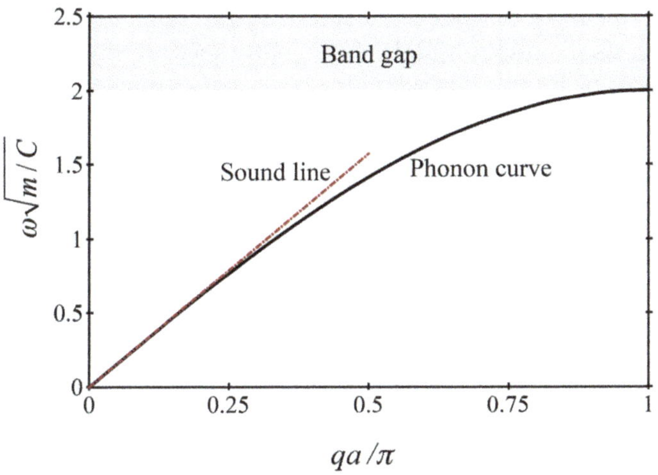

Figure 2.2. Phonon dispersion relation for a monatomic chain showing a plot of angular frequency ω versus the 1D wave vector for q ranging from 0 to the first Brillouin-zone boundary at π/a, according to equation (2.2). The tangential line at the origin represents the dispersion for a sound wave.

that now $n = 1, 2, \ldots, \infty$. Bloch's theorem will not apply, but we may still seek normal mode solutions with a time dependence like $\exp(-i\omega t)$ whereupon we have

$$\begin{aligned} -m\omega^2 u_1 &= C(u_2 - u_1),\ (n = 1), \\ -m\omega^2 u_n &= C(u_{n+1} - 2u_n + u_{n-1}),\ (n \geqslant 2). \end{aligned} \tag{2.3}$$

The first line above represent a modified equations (a boundary condition) for the 'surface' atom at the end of the chain. It can be shown, e.g. by direct substitution, that the solutions for the displacements are of the form

$$u_n = f e^{iqna} + g e^{-iqna}, \tag{2.4}$$

where f and g are constants independent of n. This is just a superposition of waves travelling to the right and left with amplitudes f and g respectively, and application of the boundary equation (the above expression for $n = 1$) yields a result for the reflection coefficient corresponding to f/g. In this example it is found that no additional solutions exist for the lattice dynamics corresponding to localized modes with decay-like characteristics from the free end (see further discussion in the appendix).

Next we briefly examine the situation depicted in figure 2.1(b) which models the interface in 1D between two media. We assume a fully infinite chain where the masses are all the same, but the spring constant changes abruptly from C_2 to C_1 at one of the masses labelled as $n = 0$. The equations of motion for the masses with $n < 0$ and $n > 0$ have the same form as equation (2.1) when the appropriate value of the spring constant has been inserted, but the equation for the interface ($n = 0$) mass has the modified form

$$m\frac{d^2 u_0}{dt^2} = C_2(u_1 - u_0) - C_1(u_0 - u_{-1}). \tag{2.5}$$

The solutions to the left and right of the interface are found to take the same form as discussed above, as in equation (2.4), and again there are no localized or spatially decaying modes to be found. It is useful for later applications to consider the form of the interface condition in a continuum approximation. In this regime the finite differences in equation (2.5) may be replaced by spatial derivatives giving

$$\rho a \frac{\partial^2 u}{\partial t^2} = K_2 \frac{\partial u_+}{\partial z} - K_1 \frac{\partial u_-}{\partial z}, \tag{2.6}$$

where $\rho = m/a$ is a density and $K = Ca$ is an elastic modulus. The quantities u_- and u_+ denote the continuum displacements to the left and right of the interface, respectively. This result simplifies further if we observe that the left-hand side is of second order in the small quantity qa, and so can be neglected compared with the right-hand side of order qa, giving

$$K_2 \frac{\partial u_+}{\partial z} = K_1 \frac{\partial u_-}{\partial z} \tag{2.7}$$

for the interface condition.

Finally, suppose we have a monatomic chain of finite length, i.e. the situation where the chain shown in figure 2.1(a) is terminated at both ends, so that $n = 1, 2, \ldots, N$, corresponding to a chain of length $L = (N - 1)a$. Most of the mathematics proceeds as before using equation (2.4) for the displacement u_n. However, we now get two expressions for the ratio f/g, one when $n = 1$ and another when $n = N$. Equating them provides us with a consistency condition to be satisfied by qa, the length L and the spring constant C. The actual form of this condition can be worked out, but is unimportant here; the main point is that for given values of L and C it will be satisfied only for certain discrete values of qa (in fact, there are N values as expected for the N normal modes of vibration for the system). This contrasts with the situation for infinite or semi-infinite chains, where we found that qa was unrestricted. This 'quantization' of the wave vector frequently occurs in finite systems, as we will see throughout later examples.

2.1.1.1 Quantum-mechanical treatment

Here we will show that the same results for the phonon dispersion in a monatomic chain may be obtained from a quantum-mechanical calculation. This will be based on an extension of the concept of boson creation and annihilation operators introduced for a simple harmonic oscillator (SHO) in section 1.4.1. This should not be surprising since the monatomic chain is essentially just a series of coupled SHOs in 1D.

We start with the Hamiltonian, which has the following form in terms of the classical variables:

$$\mathcal{H} = \sum_n \frac{p_n^2}{2m} + \frac{1}{2} C \sum_n (u_n - u_{n-1})^2 \tag{2.8}$$

for the infinitely extended monatomic chain in figure 2.1(a) with p_n denoting the momentum of mass n. After expanding the square in the last term and carrying out some relabelling, this result can be re-expressed as

$$\mathcal{H} = \sum_n \frac{p_n^2}{2m} + \frac{1}{2}C\sum_{n,n'} u_n u_{n'}(\delta_{n,n'} - \delta_{n,n'+1}), \tag{2.9}$$

where $\delta_{n,n'}$ is the Kronecker delta defined in section 1.1.

The next step is to transform to normal mode variables, which in this case is achieved by making a Fourier transform to wavenumber label q, defining

$$u_q = \frac{1}{\sqrt{N}}\sum_n u_n\, \mathrm{e}^{-\mathrm{i}qan}, \quad p_q = \frac{1}{\sqrt{N}}\sum_n p_n\, \mathrm{e}^{-\mathrm{i}qan}, \tag{2.10}$$

where N is the macroscopically large number of atoms in the chain. From these results, it can be shown after some straightforward but lengthy algebra that the Hamiltonian in the wavenumber representation becomes

$$\mathcal{H} = \sum_q \left(\frac{p_q p_{-q}}{2m} + C[1 - \cos(qa)]u_q u_{-q} \right) = \sum_q \left(\frac{|p_q|^2}{2m} + \frac{1}{2}m\omega_q^2 |u_q|^2 \right). \tag{2.11}$$

The angular frequency ω_q in the last expression denotes the dispersion relation previously found in equation (2.2). Also we have used the properties for the Fourier components that $p_{-q} = (p_q)^*$ and $u_{-q} = (u_q)^*$, along with

$$\delta_{n,\,n'} = \frac{1}{N}\sum_q \mathrm{e}^{\mathrm{i}q(n-n')}. \tag{2.12}$$

Equation (2.11) is significant in formally establishing that the Hamiltonian for the coupled chain is equivalent to a description in terms of a sum of independent SHOs. This follows on comparing equation (2.11) with equation (1.40) for an SHO. It is now a simple matter to quantize this result, following the procedure outlined in section 1.4.1. Briefly, this involves replacing p_q by the differential operator $\hat{p}_q = \mathrm{i}\partial/\partial u_q$, in units with $\hbar = 1$. Then the creation and annihilation operators for a phonon (with frequency ω_q and wavenumber q) are defined here by analogy with the SHO as $a_q^\dagger = (2m\omega_q)^{-1/2}(m\omega_q u_q + \mathrm{i}\hat{p}_q)$ and $a_q = (2m\omega_q)^{-1/2}(m\omega_q u_q - \mathrm{i}\hat{p}_q)$, respectively. It is then straightforward to prove that

$$\mathcal{H} = \sum_q \omega_q \left(a_q^\dagger a_q + \frac{1}{2} \right), \tag{2.13}$$

where we have $[a_q, a_{q'}^\dagger] = \delta_{q,q'}$ for the commutators.

To sum up, we have established consistency between the classical and quantum approaches for calculating the phonon dispersion relation of an infinite monatomic chain. The same can be demonstrated for the other chain structures that are considered here in the case of elastic interactions between nearest neighbours.

2.1.2 Diatomic chains

We now extend the calculations in the previous subsection to the more interesting case of a diatomic chain modelled as in figure 2.3 with alternating masses denoted by m_1 and m_2 and with the nearest neighbours coupled by identical springs with elastic constant C. If the spacing between the atoms is a, it follows that the periodic length for an infinitely extended chain is $2a$ so the 1D Brillouin zone in the z-direction has a wave-vector component q extending from $-\pi/2a$ to $\pi/2a$. We take the masses to be located at $z = (2n + 1)a$ and $z = 2na$ for the m_1 and m_2 sites, respectively, where integers $-\infty < n < \infty$. By analogy with the monatomic chain case, and seeking normal modes with a time dependence such as $\exp(-i\omega t)$, we now have two sets of coupled equations for the two masses:

$$\begin{aligned} -m_1\omega^2 u_{2n+1} &= C(u_{2n+2} - 2u_{2n+1} + u_{2n}), \\ -m_2\omega^2 u_{2n} &= C(u_{2n+1} - 2u_{2n} + u_{2n-1}). \end{aligned} \tag{2.14}$$

The solution for the infinitely extended system is found easily using Bloch's theorem, which implies here that $u_{2n+1} = U_1 \exp[iq(2n + 1)a]$ and $u_{2n} = U_2 \exp[iq2na]$, where the amplitude terms U_1 and U_2 are independent of the spatial variables. Substitution into equation (2.14) leads to two expressions for the ratio U_1/U_2, which may then be equated to obtain the phonon dispersion relation in the form of a quadratic equation in ω^2. This may be solved to obtain the usual textbook result (see [1, 3, 4]) that

$$\omega^2 = C\left(\frac{1}{m_1} + \frac{1}{m_2}\right) \pm C\left[\left(\frac{1}{m_1} + \frac{1}{m_2}\right)^2 - \frac{4\sin^2(qa)}{m_1 m_2}\right]^{1/2}. \tag{2.15}$$

This has given us two distinct phonon branches, as might be expected since there are two masses in the unit cell. Taking the upper sign, we have an *optic phonon branch* occurring in a frequency range $(2C/m_2)^{1/2} \leqslant \omega \leqslant [2C(m_1 + m_2)/m_1 m_2]^{1/2}$ assuming $m_1 > m_2$. By comparison, taking the lower sign, we have an *acoustic phonon branch* occurring in a frequency range $0 \leqslant \omega \leqslant (2C/m_1)^{1/2}$. There is an intervening region with frequencies from $(2C/m_1)^{1/2}$ up to $(2C/m_2)^{1/2}$ where there are no phonon excitations, and this is referred to as a *stop band* or *band gap*. The phonon dispersion relations are illustrated in figure 2.4.

The above-mentioned band gap and the higher-frequency band gap for which $\omega > [2C(m_1 + m_2)/m_1 m_2]^{1/2}$ are where, in principle, any localized modes might be expected to occur, e.g. if there are surfaces and interfaces introduced that remove the translational symmetry in the z-direction. To illustrate this point we consider the case where the diatomic chain terminates at the left-hand end with a mass m_2 at

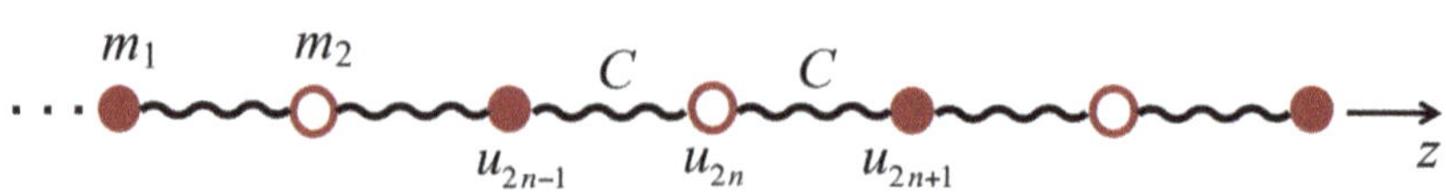

Figure 2.3. Model in 1D used for the lattice dynamics of a diatomic chain, where the notations are similar to figure 2.1. We depict a simple diatomic chain with alternating masses m_1 and m_2 and with the spring constant C taken to be the same between all nearest neighbours.

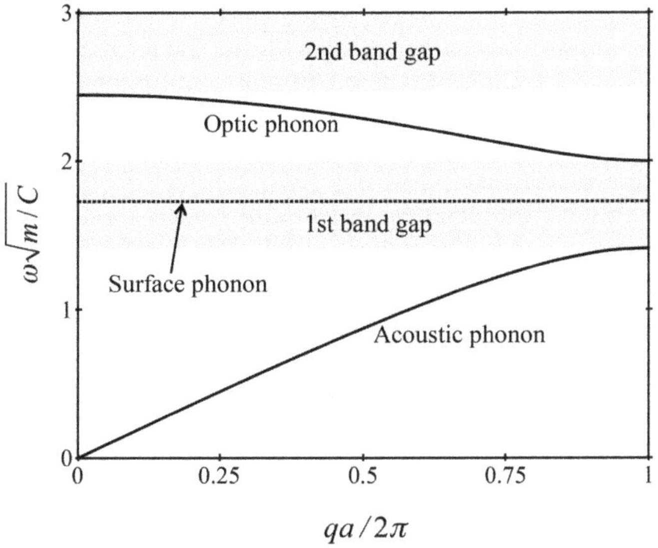

Figure 2.4. Phonon dispersion relation for a diatomic chain showing a plot of angular frequency ω versus 1D wave vector for q ranging from 0 out to the first Brillouin-zone boundary at $\pi/2a$, according to equation (2.15). The acoustic and optic phonon branches are shown, as well as the dispersionless surface phonon branch for a semi-infinite chain, which occurs within the lower band gap (see equation (2.17) and the text).

$n = 0$. The coupled equations for all $n > 0$ are formally the same as in (2.14), whereas the 'surface' $n = 0$ mass satisfies the boundary equation

$$-m_2\omega^2 u_0 = C(u_1 - u_0). \tag{2.16}$$

There are now two possible ways that we could proceed with the calculation. One way is to assume the same form of solutions as we employed following equation (2.14), except that solutions with q complex (as well as real) become allowable since Bloch's theorem is no longer applicable. For localized solutions that decay with distance away from the surface (as n becomes large), we require $\mathrm{Im}(q) > 0$. It is then straightforward to show that there is such a surface phonon solution only when $m_2 < m_1$, i.e. when the chain terminates with the lighter mass. Also it is found that the angular frequency of this mode is given by

$$\omega_S = \left[C\left(\frac{1}{m_1} + \frac{1}{m_2}\right)\right]^{1/2}, \tag{2.17}$$

so it occurs within the first band gap between the acoustic and optic phonons (see figure 2.4). The other way to arrive at the same result more rigorously, i.e. without making assumptions about the form of the solution, is to follow the tridiagonal-matrix method as outlined in section A.1 of the appendix.

Various extensions may be made to these calculations for diatomic chains. One possibility is to consider a finite-length chain with N atoms. If N is an even integer, then one termination of the chain will necessarily occur with the heavier mass and the other termination will be with the lighter mass. It is found in this case that a surface mode occurs at just one end (that with the lighter mass), as expected, and there are $N - 1$ other normal modes in the form of quantized bulk modes. If N is odd we may have either two surface modes or no surface modes, depending on the termination conditions. A careful analysis of this problem was given by Wallis [6]. Another extension of the method would be to consider an interface between two diatomic chains, but we will defer this topic until section 2.4 on superlattices where we examine the behaviour in the context of multiple interfaces.

2.2 Higher-dimensional lattice dynamics for films

In the previous section, by considering monatomic and diatomic 1D chains for modelling solids, we were examining wave propagation effects in the direction perpendicular to the surface. In particular, we focused on systems where there is no wave vector $\mathbf{q}_{\|}$ in a surface plane, as would be the case more generally in solids. The lattice dynamics approach can be extended to 2D and 3D (see e.g. [7, 8]) and to include situations where $\mathbf{q}_{\|} \neq 0$. The calculations, however, rapidly become laborious and often require intensive computational resources and/or the use of numerical simulation packages. This is, to some extent, due to the multiplicity of the different phonon branches and also due to complexities in the lattice structures.

In general, in a periodic structure composed of one material, we would expect 3 acoustic phonon branches and $3(N_p - 1)$ optic phonon branches in a crystal, where N_p denotes the number of atoms per primitive cell; the factor of 3 arises from there being a longitudinal wave and two transverse waves in terms of the atomic displacements. For a layered structure made up of several constituent materials, there would be more phonon branches. A recent review article by Wang *et al* [9] stresses the first-principles calculations and the available software packages (that include ABINIT, CASTEP, VASP, etc) for simulating the electronic structure and the phonon dispersion relations. A range of examples for phonons in 2D and 3D systems will be given later in section 2.4 in the context of some experimental studies, but we start here by considering the phonon modes arising in a 2D graphene sheet.

The honeycomb crystal lattice of graphene with sublattices A and B was described briefly in section 1.1.1. Graphene has many remarkable properties, in particular the electronic properties which we will examine in chapter 5, but the phonon modes in this 2D material have also attracted a great deal of attention. On the theory side, a detailed analysis was made by Grüneis *et al* [10] using a force constant approach, i.e. the direct generalization of our linear chain calculations in section 2.1. Since there are two carbon atoms per unit cell and the vibrations in all three dimensions need to be considered, it follows that there are six phonon modes. Their frequencies can be obtained by studying a 6×6 dynamical matrix for the system. The interactions up to fourth nearest neighbours were modelled in [10], leading to a satisfactory overall interpretation of the phonon branches. However, in order to improve the results close to the Γ and K symmetry points (see figure 1.2) in the Brillouin zone, Yan *et al* [11] carried out a perturbation analysis using density-functional theory (DFT). Nowadays, DFT is a standard computational tool, in which the electron density plays a key role,

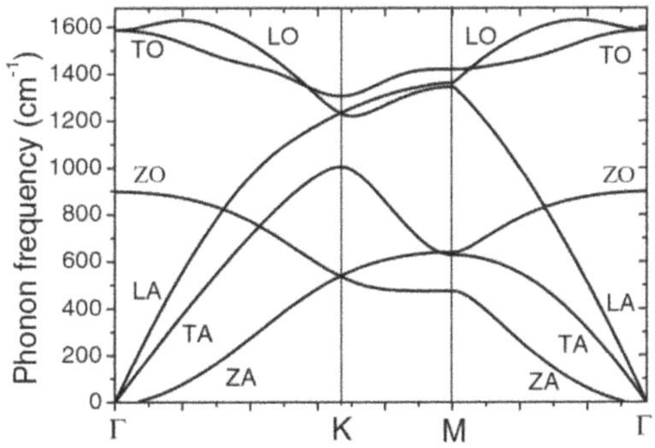

Figure 2.5. Phonon dispersion relations showing frequency versus wave vector (along lines joining the symmetry points) for monolayer graphene as obtained from DFT calculations. The notations for the six branches are explained in the text. (Reproduced with permission from [11]. Copyright 2008 American Physical Society.)

that can be used to obtain an approximate (usually ground-state) solution of Schrödinger's equation for a many-body system (see e.g. [12] for an introductory account). The improved calculations [11] for the phonon dispersion in a single graphene sheet are shown in figure 2.5, where LA and TA label the longitudinal and in-plane transverse acoustic branches, respectively, and LO and TO refer to the corresponding optic branches. The out-of-plane acoustic and optic modes are ZA and ZO, respectively. The agreement between theory and experiment is much improved compared with previous calculations, as we will discuss in section 2.4.

Yan *et al* [11] also employed their DFT method to study the phonon dispersion in bilayer and trilayer graphene sheets, assuming there to be a weak interlayer coupling. They deduced that the phonon modes of monolayer graphene are influenced to varying extents, depending sensitively on the wave vector. Typically, the splitting and shifting of the phonon modes in multilayer graphene were found to be relatively small (e.g. of order 20 cm^{-1} or less). Other graphene-related calculations of phonons, mostly using DFT and/or some simulation packages, include applications to graphene grown on metallic (and other) substrates. Some recent reviews of this topic are given in [13–16].

2.3 Elastic waves for single surfaces and films

An introductory example involving elastic waves was outlined in section 1.4, taking the case of an infinitely extended isotropic medium. We now develop this approach more formally and with the effects of surfaces and interfaces included. Some simple cases of elastic properties are covered in general books on solid-state physics (see e.g. [1, 3, 4]), while a more specialized account can be found in [17].

The starting point is to introduce the *strain* components $\bar{u}_{ij}$ defined in terms of the displacement $\mathbf{u}(\mathbf{r}, t)$ at position $\mathbf{r}$ and time t by

$$\bar{u}_{ij} = \frac{1}{2}\left(\frac{\partial u_i}{\partial r_j} + \frac{\partial u_j}{\partial r_i}\right), \tag{2.18}$$

where u_i and r_i are the components of the vectors $\mathbf{u}$ and $\mathbf{r}$, with the i and j labels denoting 1, 2, 3 (or equivalently x, y, z). Next the *stress* σ_{ij} is defined by the relationship

$$\delta F_i = \sum_k \sigma_{ik} \hat{n}_k \delta s, \tag{2.19}$$

where $\delta \mathbf{F}$ is the force on an elementary surface area δs measured normal to the unit vector $\hat{\mathbf{n}}$. In general, both the stress and strain are second-rank tensors (or matrices) and they are symmetric, so $\sigma_{ij} = \sigma_{ji}$, etc. They are taken to be linearly related through *Hooke's law*, which states that

$$\sigma_{ij} = \sum_{kl} \lambda_{ijkl} \bar{u}_{kl}. \tag{2.20}$$

Here λ_{ijkl} is the *elasticity tensor*. It is of fourth rank and is symmetric under interchange of labels i and j and of k and l, as well as satisfying $\lambda_{ijkl} = \lambda_{klij}$. From these symmetries it may be inferred that the elasticity tensor may have up to 21 independent components. For isotropic media, however, it can be shown that there are actually only two independent components and these are expressed conventionally in terms of Young's modulus E and Poisson's ratio σ. The stress–strain relationships then take the form

$$\begin{aligned} \sigma_{xx} &= \frac{E}{(1+\sigma)(1-2\sigma)}[(1-\sigma)\bar{u}_{xx} + \sigma(\bar{u}_{yy} + \bar{u}_{zz})], \\ \sigma_{xy} &= \frac{E}{(1+\sigma)}\bar{u}_{xy}, \end{aligned} \tag{2.21}$$

with analogous expressions for σ_{yy}, σ_{yz}, etc. We note that typically $0 \leqslant \sigma \leqslant \frac{1}{2}$.

Using the above definitions, the equation of motion of an element of mass can be expressed as (see [17])

$$\rho \frac{\partial^2 \mathbf{u}}{\partial t^2} = \frac{E}{2(1+\sigma)} \nabla^2 \mathbf{u} + \frac{E}{2(1+\sigma)(1-2\sigma)} \nabla(\nabla \cdot \mathbf{u}), \tag{2.22}$$

where ρ is the density. There is a simplification in the cases of waves that propagate either longitudinally (with a wave vector $\mathbf{q}$ parallel to the displacement denoted as $\mathbf{u}_L$) or transversely (with $\mathbf{q}$ perpendicular to the displacement $\mathbf{u}_T$). Then there is a separation into two wave equations, one with the form

$$\rho \frac{\partial^2 \mathbf{u}_L}{\partial t^2} = v_L^2 \nabla^2 \mathbf{u}_L, \tag{2.23}$$

where the longitudinal velocity v_L is given by

$$v_L^2 = \frac{E(1-\sigma)}{\rho(1+\sigma)(1-2\sigma)}. \tag{2.24}$$

There is an analogous wave equation satisfied by the transverse displacement $\mathbf{u}_T$, but the velocity v_T is now given by

$$v_T^2 = \frac{E}{2\rho(1 + \sigma)}. \tag{2.25}$$

It follows from the above expressions, together with the inequality mentioned for σ, that $v_L \geqslant \sqrt{2}\, v_T$. As in section 1.4.1 the dispersion relations for the two types of bulk acoustic waves (without damping) are $\omega = v_L q$ and $\omega = v_T q$.

The other basic properties that we will require relate to the conditions satisfied by the stress and strain at any boundary (either a terminating surface or an interface between two different materials). As the simplest example, we start by considering the reflection of bulk acoustic waves at a planar surface, taken as the plane $z = 0$ with the isotropic elastic medium filling the half space $z < 0$. If the surface plane is assumed to be stress-free, then $\sigma_{iz} = 0$ (for $i = x, y, z$), which implies using equation (2.21) that

$$\begin{aligned} &\bar{u}_{xz} = 0, \\ &\bar{u}_{yz} = 0, \\ &\sigma[(\bar{u}_{xx} + \bar{u}_{yy})] + (1 - \sigma)\bar{u}_{zz} = 0. \end{aligned} \tag{2.26}$$

To illustrate the use of these three boundary conditions we consider, as in [19], the problem of solving for the reflected signal when a *transverse* bulk acoustic wave is incident on the planar surface from the region $z < 0$. The case of a *longitudinal* bulk acoustic wave can be treated in an analogous fashion, but will not be presented. If the incident plane (as defined by $\mathbf{q}$ and the normal to the surface) is taken to be the xz-plane, then there are two distinct cases arising for a transverse wave depending on the polarization of the displacement vector $\mathbf{u}$. In one case (known as s-polarization) $\mathbf{u}$ is in the y-direction perpendicular to the incident plane, and in the other case (known as p-polarization) $\mathbf{u}$ is in the incident plane, as illustrated in figure 2.6.

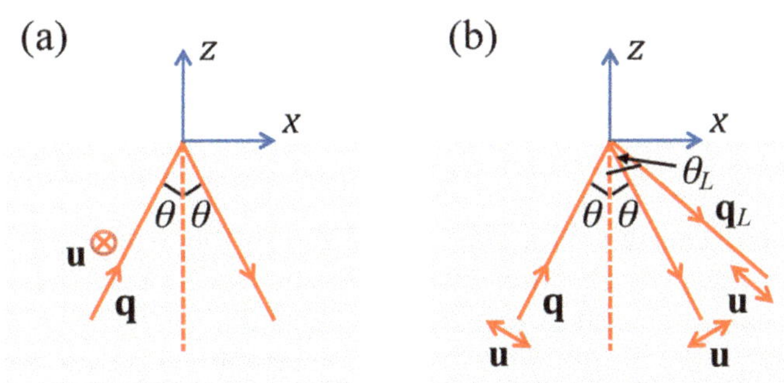

Figure 2.6. Reflection of a transverse bulk acoustic wave at a planar free surface ($z = 0$). The isotropic elastic medium fills the lower half plane ($z < 0$), while the region $z > 0$ is vacuum. The two cases are for (a) s-polarization and (b) p-polarization.

In the case of s-polarization the incident and reflected waves are straightforwardly described by

$$\begin{aligned}\mathbf{u} &= (0,\ a,\ 0)\exp[\mathrm{i}(q_x x + q_z z)]\exp(-\mathrm{i}\omega t)\ \text{(incident)},\\ \mathbf{u} &= (0,\ b,\ 0)\exp[\mathrm{i}(q_x x - q_z z)]\exp(-\mathrm{i}\omega t)\ \text{(reflected)},\end{aligned} \tag{2.27}$$

where a and b are the incident and reflected amplitudes, respectively. It follows from the relationships $\tan\theta = q_x/q_z$ and $\omega^2 = v_T^2(q_x^2 + q_z^2)$ that the q_z values are of the same magnitude and the angles of incidence and reflection are the same (equal to θ) as in figure 2.6(a). Also, by using the boundary conditions in equation (2.26), we conclude that $b = a$.

It quickly becomes evident that the case of p-polarization is more complicated. Suppose we were to start in the same manner by writing down the incident and reflected waves, which now would have the form

$$\begin{aligned}\mathbf{u} &= (a\cos\theta,\ 0,\ -a\sin\theta)\exp[\mathrm{i}(q_x x + q_z z)]\exp(-\mathrm{i}\omega t)\ \text{(incident)},\\ \mathbf{u} &= (b\cos\theta,\ 0,\ -b\sin\theta)\exp[\mathrm{i}(q_x x - q_z z)]\exp(-\mathrm{i}\omega t)\ \text{(reflected)},\end{aligned} \tag{2.28}$$

with a and b again representing amplitudes. It is obvious, however, that this is an insufficient description of the reflection process, because there are two boundary conditions to be satisfied from equation (2.26) but so far only one reflected amplitude. The matter is resolved by noting that another reflected wave, having a longitudinal character and a different propagation angle, is additionally allowed, as shown in figure 2.6(b). This longitudinal wave (with an amplitude denoted by c) can be expressed as

$$\mathbf{u} = (c\sin\theta_L,\ 0,\ -c\cos\theta_L)\exp[\mathrm{i}(q_x x - q_z z)]\exp(-\mathrm{i}\omega t). \tag{2.29}$$

It is straightforward to deduce for this longitudinal wave that

$$\frac{\sin\theta_L}{\sin\theta} = \frac{v_L}{v_T} > 1, \tag{2.30}$$

so we must have $\theta_L > \theta$. There is an interesting effect if the angle of incidence θ exceeds a critical value θ_c satisfying $\sin\theta_c = v_T/v_L$. Then equation (2.30) would imply a value for $\sin\theta_L$ greater than unity. The propagation equations can then be shown to require that $q_x^2 > \omega^2/v_L^2$, which can only be possible if q_L^2 is negative. To examine this situation, if we now put $q_L = \mathrm{i}\kappa_L$, where κ_L is real and positive, then equation (2.29) for the longitudinal wave is modified to take the form

$$\mathbf{u} = (c\sin\theta_L,\ 0,\ -c\cos\theta_L)\exp[\mathrm{i}(q_x x)\exp(\kappa_L z)]\exp(-\mathrm{i}\omega t). \tag{2.31}$$

We can now recognize this as a wave travelling in the x-direction parallel to the surface and decaying with distance into the elastic medium. In these circumstances the reflected signal is sometimes referred to as a *pseudo-surface wave*. The ratios b/a and c/a between the amplitudes, which represent reflection coefficients, can straightforwardly be deduced, if required, by further use of the boundary conditions.

Before moving on to calculate surface and interface modes in a more general context, we note in passing that the linear response calculation of a displacement-displacement

GF given in section 1.4.2 for an infinite elastic medium can be extended to apply to the acoustic phonon modes in the semi-infinite geometry depicted in figure 2.6 (see e.g. [18]).

2.3.1 Planar surfaces and films

For the first realistic example of an elastic surface wave we continue to consider the planar free surface as in figure 2.6 to derive the dispersion relation for the so-called *Rayleigh wave*. It is a localized surface wave that can be formed from a mixture of waves with p-transverse and longitudinal components. The total displacement of a wave localized in the lower half plane $z < 0$ can be expressed as the vector sum $\mathbf{u}_L + \mathbf{u}_T$, where the longitudinal and transverse parts must each satisfy the wave equation and have z-dependences such as $\exp(\kappa_{L,T}\, z)$ with $\kappa_L > 0$ and $\kappa_T > 0$. The in-plane component q_x of the wave vector must be the same for each term in order for the boundary conditions at $z = 0$ to be satisfied for all x. By using the wave equation we deduce that

$$\kappa_{L,T} = (q_x^2 - \omega^2/v_{L,\,T}^2)^{1/2}. \tag{2.32}$$

To proceed with the calculation for the properties of the surface wave, we first note that the solutions for $\mathbf{u}_L$ and $\mathbf{u}_T$ must have the form

$$\begin{aligned} \mathbf{u}_L &= (aq_x,\, 0,\, -\mathrm{i}a\kappa_L)\exp(\mathrm{i}q_x x)\exp(\kappa_L z)\exp(-\mathrm{i}\omega t),\\ \mathbf{u}_T &= (b\kappa_T,\, 0,\, -\mathrm{i}bq_x)\exp(\mathrm{i}q_x x)\exp(\kappa_T z)\exp(-\mathrm{i}\omega t), \end{aligned} \tag{2.33}$$

in order that the conditions $\nabla \times \mathbf{u}_L = 0$ and $\nabla \cdot \mathbf{u}_T = 0$ can be satisfied. A relationship between the constants a and b can be found through application of the first boundary condition in equation (2.26) as

$$2\kappa_L q_x a + (q_x^2 + \kappa_T^2)b = 0. \tag{2.34}$$

The second boundary condition is automatically satisfied here, while the third eventually leads to

$$(q_x^2 + \kappa_T^2)a + 2\kappa_T q_x b = 0 \tag{2.35}$$

after using the above definitions of κ_L and κ_T, as well as equations (2.24) and (2.25). Then, after eliminating a and b between equations (2.34) and (2.35), we arrive at an implicit dispersion relation in the form

$$\left(1 - \frac{\omega^2}{q_x^2 v_L^2}\right)\left(1 - \frac{\omega^2}{q_x^2 v_T^2}\right) = \left(1 - \frac{\omega^2}{2q_x^2 v_T^2}\right)^4. \tag{2.36}$$

Although this result looks complicated, it is just a quartic equation in $(\omega/q_x)^2$. Therefore, a solution may readily be found numerically. There is only one physically realistic solution, which can be written simply as

$$\frac{\omega}{q_x} = \xi\, v_T, \tag{2.37}$$

where ξ is a dimensionless constant with a value that depends on v_T/v_L (and hence on Poisson's ratio σ). In fact, a numerical solution of equation (2.36) shows that, for the physical range of the elastic parameters, ξ has a value less than unity and typically is close to 0.9. Further details can be found in [19]. The conclusion from the above analysis is that, for a given value of q_x, the frequency of the Rayleigh surface wave falls below that of the longitudinal (L) and transverse (T) bulk elastic waves for the same value of q_x.

Next, it is interesting to consider the elastic waves that can exist in a film of finite thickness L. This system can be formed when a second free surface is introduced at $z = -L$ in figure 2.6. The elastic medium now occupies the region of space for which $-L < z < 0$, and there are additional boundary conditions to implement at the second surface. There are two notable features regarding the elastic waves in this case. One is that a Rayleigh-type mode can be shown to exist at each surface. These two modes are essentially degenerate in frequency for large enough thickness (i.e., when both $\kappa_L L \gg 1$ and $\kappa_T L \gg 1$), but for a smaller film thickness the modes form symmetric and antisymmetric combinations (in terms of the mode displacements across the film) and the mode frequencies are shifted and split. The general results are quite complicated, although an approximate expression giving the Rayleigh-wave frequencies in a film of moderate thickness is quoted in [19].

The other modification found for a finite-thickness film is the occurrence of modes where the z component of a wave vector (i.e., the component perpendicular to the surfaces) is real but quantized with discrete values (depending on the boundary conditions). This occurs in a similar manner to the quantization for a finite monatomic chain described in section 2.1.1 except that we now have the wave polarization to take into account. The simplest case is for s-transverse polarization, where the solution for the y-component of the displacement has the form of a standing wave with respect to its z-dependence. It can be written as

$$u_y = \left[a_1 \exp(\mathrm{i}q_z z) + a_2 \exp(-\mathrm{i}q_z z)\right]\exp(\mathrm{i}q_x x)\exp(-\mathrm{i}\omega t), \tag{2.38}$$

where q_z is real and satisfies $q_x^2 + q_z^2 = \omega^2/v_T^2$. The application of the boundary conditions, as in equation (2.26), allows the coefficients a_1 and a_2 to be eliminated to yield a dispersion relation.

The situation with the coupled p-transverse and longitudinal modes is more complex and can give rise to what are often called *slab modes* in acoustics (or analogously *guided waves* in optics). Briefly, we could form solutions similar to the combinations in equation (2.33), except that some or all of the $\exp(\kappa_{L,T} z)$ terms for the z-dependence would be replaced by factors such as $\exp(\mathrm{i}q_{L_z} z)$ and $\exp(\mathrm{i}q_{T_z} z)$ with real wave-vector components. In this way it is possible to have mixtures of waves in which, for example, the T part is like a standing wave and the L part is like a localized wave at the surfaces. Collectively, all the elastic waves of a film are often referred to as *Lamb waves*.

So far, we have considered only stress-free surfaces, but other localized modes will arise if we extend some of our results to interfaces between two different elastic media. There are various different possibilities, depending on the polarizations, elastic constants and geometries. We mention some examples below. One case is the direct extension of the Rayleigh wave to a planar interface between two media, where it is found that a localized mode called a *Stoneley wave* can exist provided the elastic constants of the two media satisfy certain criteria. Its dispersion relation can be written in the same form as equation (2.37), but the expression for ξ is more complicated (see e.g. [20]). The Stoneley wave involves coupled p-transverse and longitudinal components, while the simplest example of an interface-related mode in s-polarization is the *Love wave*. This occurs for an elastic film of thickness L on a semi-infinite substrate of another elastic material, provided the transverse acoustic velocity v'_T of the substrate material exceeds the corresponding value v_T in the film. This condition allows solutions to be found for displacements that are oscillatory, or wave-like, functions of z inside the film and decaying functions of z with distance into the substrate [21]. Another type of surface acoustic wave that is associated with a film deposited on a substrate is the *Sezawa wave* mentioned later in section 2.4.

The analysis of surface elastic waves in situations where there are *non-parallel* planar surfaces (as in edges or wedges) is surprisingly complicated, mainly because of the nature of the stress–strain relationships in even the most straightforward cases, e.g. see equation (2.21). Notably the elastic modes localized at a 90° edge in a solid were studied by Maradudin *et al* [22]. The modes are characterized by a 1D wave vector for propagation parallel to the edge and a decaying amplitude with respect to the two directions away from the planar surfaces. The decay properties were expressed in terms of Laguerre polynomials, and it was found that the lowest-frequency edge mode has a slightly lower acoustic speed than that for the Rayleigh surface mode in equation (2.37). Subsequently, generalizations were made to edges (or wedges) with other angles between the planar surfaces (see [23] for a review).

2.3.2 Spheres and cylinders

Here we consider phonon (elastic wave) excitations in systems with smooth curved surfaces, taking for simplicity the cases of spheres and long cylinders with a circular cross section. These provide us with simple examples of phonons in 0D nanodots and 1D nanowires, respectively, as sketched in figure 1.4.

The calculations of acoustic waves in a solid sphere of an isotropic elastic medium have a long history, dating back to work by Lamb [24] in 1882 for the so-called spheroidal and torsional modes. Because there is no direction of translational symmetry for a sphere (a 0D structure), the modes have no well-defined wave vector. A clear description of the underlying theory is given in the paper by Kuok *et al* [25], which we will mention again in the upcoming experimental section 2.4. The starting point is the elastic wave equation, which is quoted in equation (2.22) in its general vector form. On substituting for the ∇^2 operator in terms of spherical polar coordinates (r, θ, ϕ), as conventionally defined (see e.g. [26]), with r as the radial distance from the centre of the sphere, θ as the polar angle and ϕ as the azimuthal angle, we obtain coupled partial differential equations satisfied by the components of the displacement vector. The method of separation of variables can then be employed to find solutions that are linear combinations of products such as

$$j_\ell(\omega r/v_{L,\,T})\; Y_\ell^m(\theta,\,\phi)\exp(-\mathrm{i}\omega t). \tag{2.39}$$

Here $j_\ell(\omega r/v_{L,T})$ is a spherical Bessel function of the first kind, where $j_\ell(x)$ for any real variable x satisfies

$$x^2\frac{\mathrm{d}^2 j_\ell}{\mathrm{d}x^2} + 2x\frac{\mathrm{d}j_\ell}{\mathrm{d}x} + \{x^2 - \ell(\ell+1)\}j_\ell = 0. \tag{2.40}$$

Also $Y_\ell^m(\theta,\,\phi)$ is a spherical harmonic describing the angular dependences of any mode, ω is an angular frequency, and ℓ and m are the usual quantum numbers associated with the angular momentum [26]:

$$\ell = 0,\,1,\,2,\,\dots\,,\qquad m = -\ell,\,-\ell+1,\,\dots\,,\,\ell-1,\,\ell. \tag{2.41}$$

The mode frequencies and eigenfunctions are deduced by constructing combinations of solutions of the form appearing in equation (2.39) that satisfy the stress-free boundary conditions at $r = R$ for all values of θ and ϕ, where R is the radius of the sphere. Two types of solutions arise, corresponding to either

$$\frac{\eta j_{\ell+1}(\eta)}{j_\ell(\eta)} - (\ell - 1) = 0 \qquad (\ell \geqslant 1) \tag{2.42}$$

or

$$\begin{aligned}&[\eta^2 - 2\ell(\ell-1)(\ell+2)]\frac{\eta j_{\ell+1}(\eta)}{j_\ell(\eta)} + \frac{1}{2}\eta^4 + (\ell-1)(2\ell+1)\eta^2\\ &+ 2\left[\eta^2 + (\ell-1)(\ell+2)\left(\frac{\eta j_{\ell+1}(\eta)}{j_\ell(\eta)} - (\ell+1)\right)\right]\frac{\xi j_{\ell+1}(\xi)}{j_\ell(\xi)} = 0 \quad (\ell \geqslant 0).\end{aligned} \tag{2.43}$$

Here we are using a shorthand in which $\eta = \omega R/v_T$ and $\xi = \omega R/v_L$ denote parameters that depend on the mode frequency ω. Equation (2.42) describes toroidal oscillations without dilation, where the displacement vectors are always parallel to the surface. Equation (2.43) describes spheroidal oscillations with dilation, i.e. the oscillations involve volume expansions and contractions of the sphere. The expressions for the mode frequencies have no wave-vector dependence, as mentioned. Also it can be noted that the equations for the frequencies are independent of the quantum number m, so each mode has a degeneracy factor of $2\ell + 1$. For each value of ℓ there are multiple solutions that may be labelled as $\omega_{n,\ell}$, where the integer n (taking the values 1, 2, …) labels the modes for a given ℓ starting with the lowest frequency. There will be further discussion of these results in the next section in the context of Brillouin light scattering experiments.

Next we turn briefly to the case of acoustic waves in a solid cylinder of an isotropic elastic medium, where the length is taken to be very large (effectively infinite compared with the radius R). In this limiting case we have translational symmetry in 1D along the length axis. In accordance with this geometry, it is appropriate to use cylindrical polar coordinates (r, θ, z), where z refers to the coordinate along the cylindrical axis while r and θ are the usual circular polar variables in the xy-plane perpendicular to the length (with $x = r\cos\theta$ and $y = r\sin\theta$). The solutions of the elastic wave equation can now be investigated by using the expression for the ∇^2 operator in the cylindrical polar coordinates [26]. It is found that the solutions are linear combinations of products such as

$$f(r)\exp(\pm im\theta)\exp(iq_z z)\exp(-i\omega t). \tag{2.44}$$

Here m takes integer values (= 0, 1, 2, …), which ensures the single-valuedness with respect to θ, and q_z is a propagation wave vector along the length of the cylinder as required from Bloch's theorem. The radial function $f(r)$ satisfies a form of Bessel's equation, and so solutions with different kinds of Bessel functions can be constructed. The solutions for the allowed values of the angular frequency ω are found after applying the stress-free boundary conditions at all points on the curved surface at $r = R$. In general, there are multiple solutions for any value of m, so the modes can be labelled with an addition integer n as $\omega_{n,m}$ with n = 1, 2, 3 … . The work by Hudson [27] is notable among the early calculations applied to surface and bulk elastic waves in solid cylinders, and later generalizations were made by others [28, 29] for Rayleigh-type modes propagating circumferentially rather than along z. An interesting property is that the Rayleigh-type surface waves exist only if $q_z R$ exceeds a cut-off value, otherwise they are no longer localized and they decay into bulk elastic waves.

2.4 Experimental studies

At this stage we present a few examples of experimental measurements that relate to the phonon modes discussed so far. As mentioned in the introductory survey of experimental techniques given in section 1.3, the inelastic scattering of light, by means of either Brillouin light scattering (BLS) or Raman scattering (RS), can provide a surface-sensitive technique with good resolution for detecting the surface- and interface-related modes. This is particularly so if optically opaque materials are being investigated, in which case the light penetrates only a limited distance into the sample where the amplitude of any surface wave will be largest. Extensive results for phonons are now also available from various forms of particle scattering, which allow for studies to be made across the entire Brillouin zone of excitation wave vectors.

A comprehensive review of some of the earlier RS studies for surface phonons was given by Esser and Richter [8], and recent developments and examples are reviewed in [30]. The surface RS technique has been used extensively to study materials (often III–V semiconductors such as GaAs or InP) with an ultrathin film of another material (such as Sb) deposited on their planar surface [8]. Typically the film might have a thickness of only one or a few monolayers (MLs), or sometimes might have a sub-monolayer coverage. A much-cited example comes from the RS spectra obtained by Hünermann *et al* [31] for a InP sample where different thicknesses of Sb, ranging from fractional coverage up to one ML, were deposited on the surface. The TO and LO phonon peaks of InP were clearly seen, as well as the surface phonons of the Sb layer. The observed evolution of the surface phonon peaks with increasing Sb, and their dependence on the polarizations of the incident and scattered light, served to motivate further studies. These have included Raman measurements for various 1D and 2D surface reconstructions, e.g. with a patterning of self-organized Au sub-monolayers on the surface of a Si substrate [30].

The higher spectral resolution available with the BLS technique led to pioneering studies of the acoustic phonons at planar surfaces by Sandercock [32]. He employed an oblique-incidence backscattering geometry to excite phonon modes with $q_{\|} \neq 0$ at the (001) surface of a thick Si sample. An example of a BLS spectrum is shown in figure 2.7, where the peaks corresponding to the longitudinal (L) and transverse (T) optic phonons are indicated, as well the surface Rayleigh (R) phonon. The occurrence of these peaks on both the anti-Stokes and Stokes side of the spectrum is in accordance with the theory discussed in section 2.3.1. Also, the peak heights on the two sides are essentially the same in this case because $k_B T \gg \hbar\omega$ for the room-temperature measurements, where ω is the angular frequency shift. In figure 2.8 we show a BLS spectrum for oblique-incidence scattering from a thin film of hexagonal boron nitride on a Si substrate [33]. Sharp peaks were observed for Stokes and anti-Stokes scattering involving the Rayleigh and Sezawa surface wave. The recent innovation of micro-focused BLS, with the capability of focusing the laser beam onto a smaller surface area than previously, has enabled new studies of coherent interference effects between the Rayleigh and Sezawa surface wave in a system of GaN on a Si substrate [34].

As another example, we consider BLS from the acoustic phonons at a curved surface. Figure 2.9 shows experimental Stokes and anti-Stokes scattering results

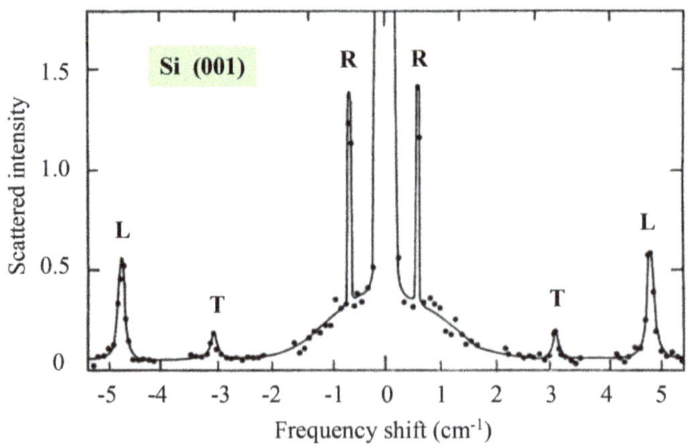

Figure 2.7. Brillouin light scattering (BLS) spectrum from a Si surface in an oblique-angle backscattering geometry peaks on both the Stokes and anti-Stokes side show scattering from the longitudinal optic (L) and transverse optic (T) phonons and from the surface Rayleigh (R) phonon. (Reproduced with permission from [32]. Copyright 1978 Elsevier.)

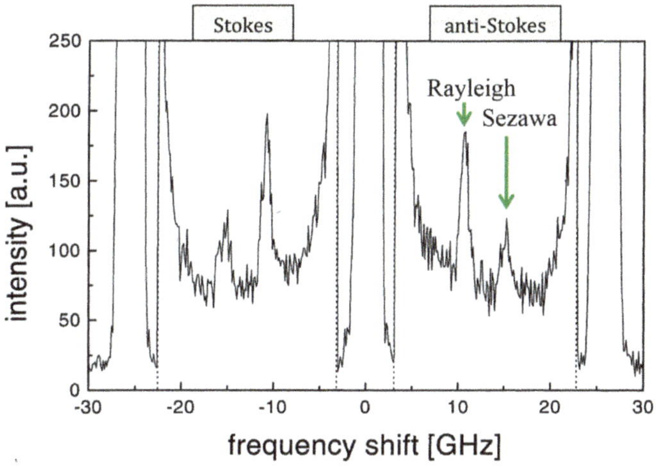

Figure 2.8. Brillouin spectrum for light scattering from phonons at the planar surface of a hexagonal boron nitride film grown on a Si substrate. The results were obtained in an oblique-incidence backscattering geometry. The sharp peaks are due to scattering from the Rayleigh mode and from the first Sezawa mode. The vertical dotted lines in the spectra indicate the frequency ranges of the suppressed elastically scattered light at small frequency shifts. (Reproduced with permission from [33]. Copyright 2000 Elsevier.)

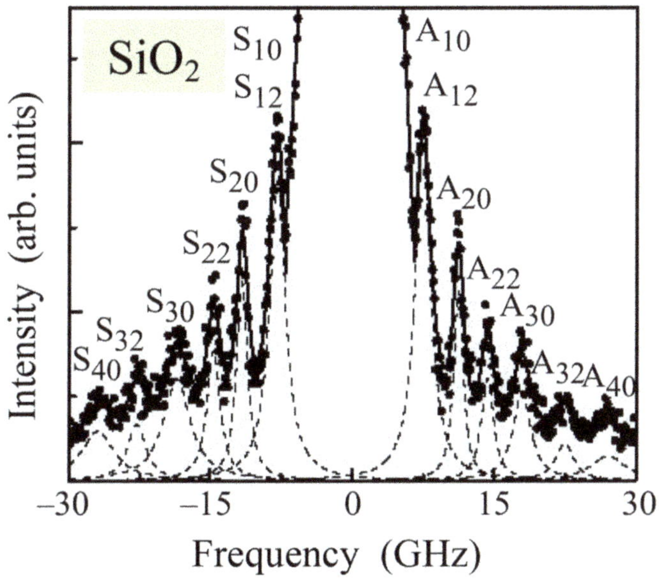

Figure 2.9. Brillouin spectrum showing light scattering from acoustic phonons in SiO_2 nanospheres with diameter 340 nm. The peaks are fitted with Lorentzian functions (dashed curves) and the resultant fitted spectrum (solid line) is compared with the experimental data. See the text for explanation of the labels for the confined acoustic phonon peaks. (Reproduced with permission from [25]. Copyright 2003 American Physical Society.)

obtained for a spherical sample of SiO_2, where the relevant theory (see section 2.3.2) indicates that the mode frequencies depend on integer quantum numbers n and ℓ with the latter being related to angular momentum. The acoustic phonon peaks are labelled here as $S_{n\ell}$ and $A_{n\ell}$ on the Stokes and anti-Stokes sides, respectively. These are all spheroidal modes, since the torsional modes are precluded by the selection rules of BLS (see [25]). For the same reason, the $\ell = 1$ spheroidal modes are not observed. Overall, the measured frequencies are in good agreement with experiment.

Next we show some results for the phonon dispersion relations obtained by the inelastic scattering of He atoms [35] at and just below the surface of a Bi (111) sample. In figure 2.10 the different symbols refer to the different phonon modes detected. As explained in section 1.3.2, the main result governing the dynamics is equation (1.34), which implies that a TOF spectrum gives a scan in terms of a parabolic path in the $\{\omega, q_{\|}\}$ space. By using a series of scans, a figure as shown here can be built up. The measurements are at a low sample temperatures T_S (as shown) for two different directions in the 2D hexagonal Brillouin zone, extending from the zone centre ($\bar{\Gamma}$) to the zone boundary $\bar{\text{K}}$ and $\bar{\text{M}}$ symmetry points (in a standard notation). According to convention, the phonon energies are expressed here in meV, where we note that 1 meV converts to approximately 242 GHz or 8.07 cm^{-1} in other

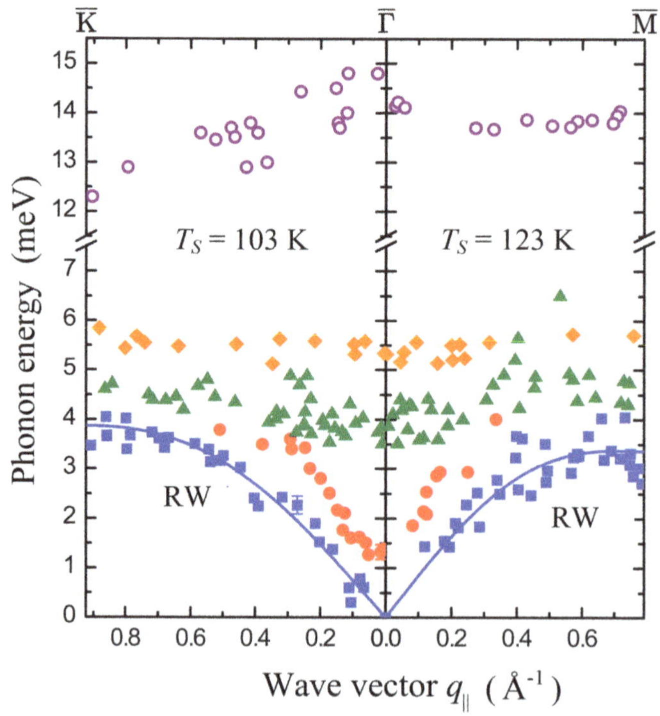

Figure 2.10. Results for phonon energy versus in-plane wave vector $q_{\|}$ for the inelastic scattering of He atoms from the surface of Bi (111). The different colour and symbols for the data points refer to different scan lines, as discussed in the text. The data include the Rayleigh wave (RW) as marked. (Reproduced with permission from [35]. Copyright 2013 American Physical Society.)

energy- and frequency-related units. The lowest excitation branches (indicated by the blue squares) labelled RW are for the Rayleigh wave extended to larger wave vector, and the solid line is lattice dynamics theory with adjusted force constants. It is seen that the agreement between experiment and theory is good. The other experimental points in figure 2.10 were compared with a lattice dynamics theory that employed density-functional perturbation theory (DFPT) with a software simulation package. It was found, for example, that the red circles and green triangles are different longitudinal phonons, the orange diamonds are localized phonons with shear vertical polarizations, and the open purple circles are a mixture of modes with shear vertical and longitudinal polarization. Again, agreement between experiment and the theory curves (not shown here) was generally good.

As a further example of particle scattering, we will illustrate inelastic electron scattering, which was pioneered as a technique for studying surface phonons by Lehwald *et al* [36]. A brief description of the methodology for electron energy loss spectroscopy (EELS) was given in section 1.3.2 (see also [37] for a detailed account). After the initial studies, which were mostly for clean surfaces, the technique has been applied to planar surfaces patterned with adsorbed atoms and with high-resolution EELS (or HREELS). A straightforward example of some earlier measurements are in figure 2.11, where the EELS spectra for phonons are shown for a Ni (110) surface with adsorbed H and D atoms in a (2×1) surface reconstruction depicted in the inset [38]. The main scattering peaks (at 640 cm^{-1} for H and at 510 cm^{-1} for D, as marked) are due to transitions between quantized vibrational states of the absorbed atoms. The ratio of these two frequencies is significantly less than the value $\sqrt{2}$ that would be expected for simple classical vibrations of D and H atoms, for which the

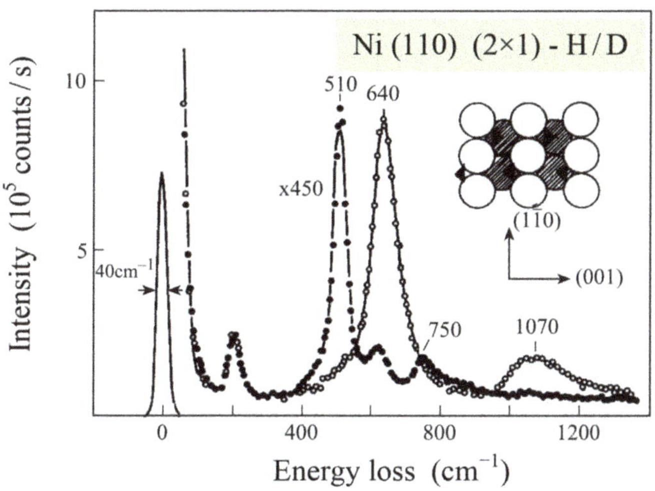

Figure 2.11. Electron energy loss scattering (EELS) from phonons for hydrogen H (open circles) and deuterium D (full circles) adsorbed on Ni (110). The instrumental resolution is 40 cm^{-1}. The surface geometry is shown in the inset. (Reproduced with permission from [38].)

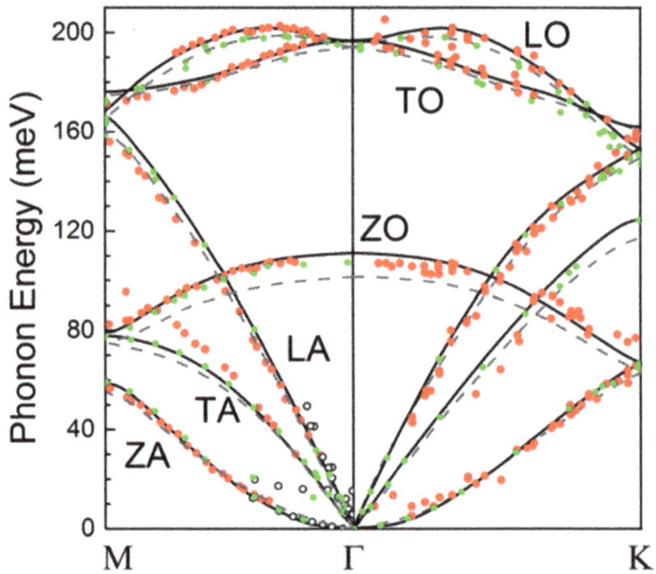

Figure 2.12. Phonon dispersion of monolayer graphene. Comparison of the theory (solid and dashed lines obtained from DFT) with experimental data compiled from several techniques: HREELS (red circles), INS (open circles) and inelastic x-ray scattering (green circles). The labelling of the branches is the same as in figure 2.5. See the text for details. (Reproduced with permission from [15].)

mass ratio is 2. The modified ratio was explained [38] in terms of the strong quantum-mechanical effects associated with the vibrational modes of these low-mass atoms.

Finally in this section we present some experimental studies for the dispersion of the six phonon branches in 2D graphene monolayers and their comparison with the DFT calculations mentioned in section 2.2. In figure 2.12 we show the comparison between theory (represented by black solid lines [11] and grey dashed lines [39]) and experiment using three techniques (high-resolution EELS (HREELS) [40, 41], inelastic neutron scattering (INS) [42] and inelastic x-ray scattering [43]). The agreement between all these results, which extend across the 2D Brillouin zone of wave vectors in the two different symmetry directions studied, is seen to be remarkably good.

2.5 Phonons in multilayers and superlattices

In the remaining parts of this chapter on phonons we consider some nanostructured arrays that have multiple planar interfaces between two or more materials. First in this section our attention is directed to the theoretical and experimental studies of phonons in multilayers, including, for example, periodic superlattices of the type depicted in figure 1.4(b). Next, in section 2.6, we turn to the case of lateral arrays, as occurring in 1D or 2D phononic crystals (see e.g. figure 1.4(c)).

2.5.1 Theoretical results

In the case of periodic superlattices the phonon modes may be studied by extending either the lattice dynamics approach or the elastic waves method of continuum acoustics, which were described in the earlier sections of this chapter. Both methods are considered here, but we start here with the latter one, since it is typically simpler to apply and is sufficient to explain many of the experimental studies. Also, we may follow a rather similar approach to that for our treatment of optical waves given in section 1.5.1. There we used a bulk-slab model for the system and we introduced the properties of the transfer matrix for the excitations.

Elastic waves method

We may employ the same geometry for a fully infinite two-component ...*ABABABAB*... superlattice as in figure 1.9, except that for acoustic phonons the layers will be labelled in terms of the corresponding density (ρ_A or ρ_B) and the elastic constant (C_A or C_B) instead of the dielectric constant. Within the layers the 1D wave equation is satisfied and it can be written as

$$\rho_i \frac{\partial^2 u}{\partial t^2} = C_i \frac{\partial^2 u}{\partial z^2}, \quad i = A, B. \tag{2.45}$$

Here the wave propagation is taken to be in the z-direction (perpendicular to the interfaces) with velocity $v_i = (C_i/\rho_i)^{1/2}$ and the displacement u may be either longitudinal (along z) or transverse (in the xy-plane). The applicable boundary conditions at each planar interface are that u and $C\, \partial u/\partial z$ are continuous (see sections 2.1 and 2.2). For simplicity, we are assuming no variations in the y- and z-directions in this calculation.

The theory now proceeds just as in the optical-wave example in section 1.5.1, except that u replaces an electric-field component and it is the acoustic phonon dispersion relations that replace the expressions from optics. Thus the Bloch wavenumber Q is introduced through the condition $u(z + D) = \exp(\mathrm{i}QD)\, u(z)$, which replaces equation (1.52) previously. Also, the wave-equation solutions for u at the upper and lower interfaces of either an A or B layer can be written down by analogy with equations (1.53) and (1.55) where now $q_{iz} = \omega/v_i$ (with $i = A, B$). Then, 2×2 matrices F_i and X_i can be written down as in equations (1.58) and (1.61), and the formal results related to the transfer matrix T in equations (1.62)–(1.68) still hold. Specifically, the implicit dispersion relation, which is analogous to equation (1.70), is obtained as

$$\cos(QD) = \cos(q_{Az}d_A)\cos(q_{Bz}d_B) - g\,\sin(q_{Az}d_B)\sin(q_{Bz}d_B). \tag{2.46}$$

This is the result attributed to Rytov [44], where we have

$$g = \frac{1}{2}\left(Z + \frac{1}{Z}\right), \tag{2.47}$$

and Z is the ratio of the acoustic impedances for the two media and is given by

$$Z = \frac{C_A q_{Az}}{C_B q_{Bz}} = \sqrt{\frac{\rho_A C_A}{\rho_B C_B}}. \tag{2.48}$$

We note that if the two media are impedance 'matched', meaning $Z = 1$, then the Rytov equation simplifies to $\cos(QD) = \cos(q_{Az}d_A + q_{Bz}d_B)$. Using the expressions for q_{iz} ($i = A, B$), the solutions of the Rytov equation in this case become

$$\omega\left(\frac{d_A}{v_A} + \frac{d_B}{v_B}\right) = \pm QD + n\pi, \quad (n = \text{integer}). \tag{2.49}$$

The interpretation of the above results is broadly similar to that given for figure 1.10 in section 1.5.1. Here there is a folding back of the acoustic phonon excitation in the reduced 1D Brillouin zone. In general, when $Z \neq 1$ we will have $g > 1$ in equation (2.46), and this leads to the opening up of band gaps (or stop bands) at the zone centre $Q = 0$ and zone edges $Q = \pm\pi/D$. This behaviour is illustrated in figure 2.13 where the dispersion branches are shown for $Z = 2$ (red solid lines) and $Z = 1$ (green broken lines). In the case of $Z \neq 1$ there are band gaps, as expected, and these are particularly evident in this example at the zone edge as labelled. No band gaps are seen when $Z = 1$ in accordance with equation (2.49). A consequence of the multiple branches, due to the folding back, is that several acoustic peaks (for a given polarization) may now be observed in a BLS experiment, as we discuss later.

There are two main respects in which the above type of theory has been generalized. One respect is to consider superlattice stacks that are either semi-infinite or finite in the z-direction, rather than infinite. Then the occurrence of localized superlattice modes becomes possible, in which there is a decay characteristic from one cell (of thickness D) with distance away from an external surface rather than adherence to Bloch's theorem. As expected, when surface modes exist they lie within the bands gaps. The second type of generalization is to include modes with a wave-vector component $\mathbf{q}_{\|} \neq 0$ in the xy-plane. In this latter case a distinction then arises between s- and p-polarizations, as discussed in simpler cases earlier with one or two interfaces.

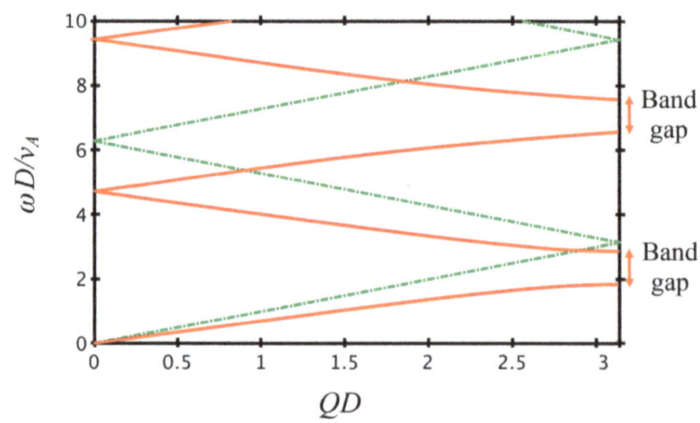

Figure 2.13. The folded acoustic phonon dispersion curves for a two-component superlattice calculated using equation (2.46) in the elastic wave approximation. The scaled frequency (in terms of $\omega D/v_A$ is plotted against QD across the first mini-Brillouin zone. The parameters are chosen to illustrate the cases of impedance matching (green broken lines) [$C_A = 2C_B$, $\rho_A = 0.5\rho_B$, implying $Z = 1$] and an impedance mismatch (red solid lines) [$C_A = 2C_B$, $\rho_A = 2\rho_B$, implying $Z = 2$]. We have taken $d_A = 2d_B = 2D/3$ in both cases.

Some calculations that explore these cases using elastic wave theory can be found in [45–47]. For example, a dispersion relation was obtained in [46] for the longitudinal and p-polarized transverse phonons in a semi-infinite Al/W superlattice, including results for both bulk and surface modes. When the mode frequencies are plotted in terms of $q_{\|}D$, the bulk modes found for $\mathbf{q}_{\|} \neq 0$ are much more complicated, and the bulk modes appear as a continuum. The properties of the surface modes are found to depend on whether it is an Al layer or a W layer that is located at the superlattice external surface.

2.5.1.1 Lattice dynamics method

Next we briefly discuss some of the results obtained for superlattice phonons using the lattice dynamics approach introduced in section 2.1. Much of the interest has been for the acoustic and optic phonons in structures involving III–V semiconductors, such as GaAs. For a 1D model (applicable when the in-plane wave vector $\mathbf{q}_{\|} = 0$) a simplified approach based on the linear chain structure in figure 2.14 may be utilized. Here there are sections of length d_A of material A (with alternating masses m_A and M) and sections of length d_B of material B (with masses m_B and M). The unit cell is of length $d_A + d_B$ as indicated. This approach, which involves a direct extension of earlier results in section 2.1.2 for diatomic chains combined with the transfer-matrix formalism, was first used in [48, 49] to study superlattice phonons (see also [19] for an outline derivation and further discussion). One of the main conclusions is that equation (2.46) obtained in the elastic wave approach still applies here, but with a redefinition of the factor g as

$$g = \frac{1}{2}\left\{\frac{\tan(q_{Az}a)}{\tan(q_{Bz}a)} + \frac{\tan(q_{Bz}a)}{\tan(q_{Az}a)}\right\}. \tag{2.50}$$

The values of the separation a between adjacent atoms along the chain and the force constant C are assumed, for convenience, to be the same for all nearest-neighbour atoms. In equations (2.46) and (2.50) the quantities q_{Az} and q_{Bz} represent effective wave-vector components along the z-direction in the two media. They are related to the angular frequency ω through the analogue of equation (2.15) with the appropriate masses substituted, so that for each material we have

$$\omega^2 = C\left(\frac{1}{M} + \frac{1}{m_i}\right) \pm C\left[\left(\frac{1}{M} + \frac{1}{m_i}\right)^2 - \frac{4\sin^2(q_{iz}a)}{Mm_i}\right]^{1/2} \quad (i = A,\, B). \tag{2.51}$$

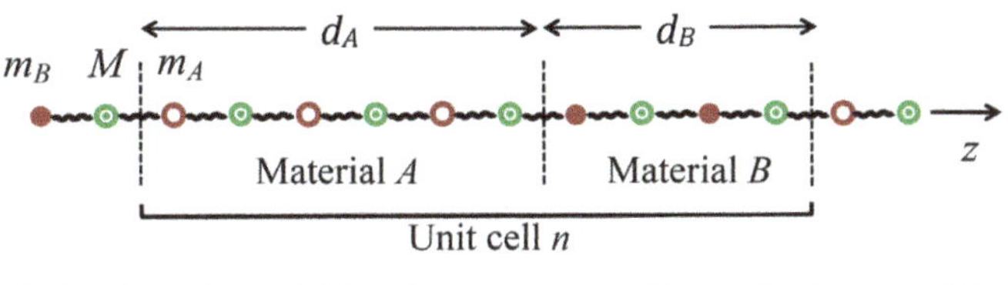

Figure 2.14. Simple lattice dynamics model for phonons in a periodic superlattice formed from alternating layers of materials A and B with layer thicknesses d_A and d_B, respectively, in each cell n. Material A is represented as a diatomic chain of atoms with masses m_A and M. The same applies for material B with masses m_B and M.

Several distinct cases can occur in the preceding theory depending on whether q_{Az} and q_{Bz} are real or complex (see [19]). For example, one possibility that arises is that q_{Az} may be real and q_{Bz} imaginary (or vice-versa) for certain values of ω. Then the superlattice modes may be wave-like in the z-direction in one medium, but may have a decaying behaviour in the other medium; in this case they are known as *confined phonon* modes. The above theory provides a simple description of GaAs/Al_xGa_{1-x}As superlattices, where $0 < x < 1$ for the Al concentration. Mass M would represent the As mass, while m_A and m_B represent the Ga mass and an average Al_xGa_{1-x} mass, respectively. Extensions of the above lattice dynamics theory to surface modes in semi-infinite superlattices and to other models in 3D have been carried out (see e.g. [50]).

Recently there have been many theoretical studies of the phonons in short-period superlattices using either lattice dynamical methods or *ab initio* techniques based on density-functional theory (DFT), or a combination of these. For example, Davydov *et al* [51] studied superlattices composed of alternating GaN and AlN layers that were just a few atomic layers in thickness. The theory led to predictions for the phonon frequencies and for their contributions to Raman spectra in a range of different polarizations. Comparisons were made with experimental results for their fabricated superlattice, and good agreement between theory and experiment was found in general.

2.5.1.2 Quasiperiodic superlattices

Another type of theoretical extension, which has been carried out using both the elastic continuum approach and the lattice dynamics approach, has been to superlattices that are quasiperiodic, rather than periodic, in their properties. Quasiperiodic multilayer structures were mentioned briefly in section 1.2.2. Some early accounts covering both theory and experiment for phonons were given by MacDonald [52] and Merlin [53], while a more comprehensive review that includes other excitations can be found in [54].

Here we will discuss mainly two-component superlattices, where there are two building blocks (taking the form of thin films or layers in the present context) labelled A and B as before. Instead of forming an alternating $ABABAB\ldots$ sequence, as in periodic case, the building blocks are ordered according to a mathematical growth rule such as that provided by the Fibonacci sequence mentioned earlier. By analogy with equation (1.30) for the Fibonacci numbers, we could define a structural growth rule in a recursive fashion such that the nth stage S_n is formed according to $S_n = S_{n-1}S_{n-2}$ for $n > 1$, starting with $S_0 \equiv B$ and $S_1 \equiv A$. Hence for $n > 1$ the first few generations are $S_2 = AB$, $S_3 = ABA$, $S_4 = ABAAB$, $S_5 = ABAABABA$, and so on. The number of individual blocks in the nth generation is just the Fibonacci number F_n given by equation (1.30). In general, for each generation, the number of B blocks is different from the number of A blocks: the ratio tends to the golden mean number $(1 + \sqrt{5})/2 \approx 1.618$ as $n \to \infty$. Another example of a quasiperiodic structure, which is generated recursively from the Thue–Morse sequence, corresponds to $S_n = S_{n-1}S_{n-1}^{+}$ and $S_n^{+} = S_{n-1}^{+}S_{n-1}$ for $n > 0$, taking $S_0 = A$ and $S_0^{+} = B$. The next few Thue–Morse generations are $S_1 = AB$, $S_2 = ABBA$, $S_3 = ABBABAAB$, and so on, which demonstrates the property that there are equal numbers of A and B

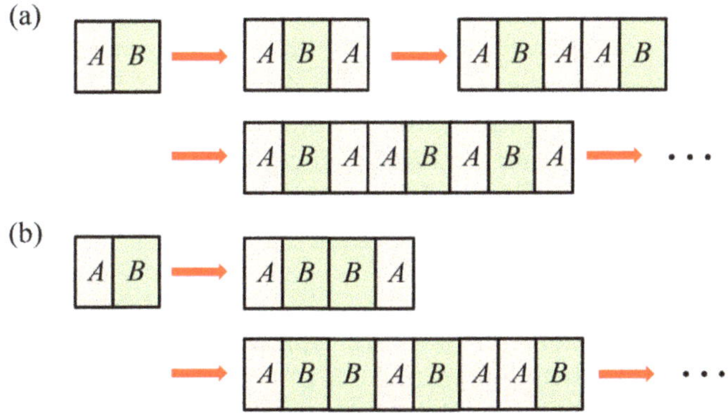

Figure 2.15. Successive generations of two different quasiperiodic superlattice structures using *A* and *B* building blocks. Each series starts with the juxtaposition *AB*. (a) Fibonacci and (b) Thue–Morse.

blocks in this case. The two quasiperiodic structures mentioned here are illustrated schematically in figure 2.15.

The modification of the transfer-matrix theory described earlier for periodic superlattices to apply to various quasiperiodic superlattices is described in detail in [54]. These quasiperiodic structures have been studied intensively because of their unusual localization effects and the multifractal properties of the excitations. Further theoretical calculations for the Lamb waves in Ta/Al Fibonacci superlattices were made, for example, by Gao *et al* [55] showing fragmentation of the bulk excitation bands.

2.5.2 Experimental results

The most successful experimental technique for studying the folded acoustic phonons and confined optical phonons in periodic superlattices has been inelastic light scattering, and we mention in particular the seminal work by Colvard *et al* [56, 57] and Jusserand *et al* [58]. A thorough review of these and other studies was given by Klein [59]. A focus of the experimental studies by Colvard *et al* was Raman scattering from a periodic GaAs/$Al_{0.3}Ga_{0.7}As$ superlattice. Spectra were obtained that corresponded to the (x, x) and (x, y) cases for the electric-field polarizations of the incident and scattered light beams. The selection rule, deduced from symmetry arguments, is that Raman scattering from phonons should occur only in 'diagonal' polarizations [60], and this was clearly verified in this work, where the phonon peaks were seen to occur in the (x, x) plot only. The multiple TA phonon peaks, occurring with Raman shifts between about 25 and 100 cm^{-1}, related to the mode folding described in the previous text. As mentioned earlier in this chapter, examples of Raman scattering measurements from vibrational modes in short-period GaN/AlN superlattices can be found in [51] and references therein.

Next we mention some experimental data and theory [61] for the acoustic phonons in two different quasiperiodic GaAs/AlAs superlattices (generated with Fibonacci and Thue–Morse sequences), along with a comparison made to a two-component superlattice where the GaAs and AlAs building blocks are chosen randomly. It was evident from these studies that the Raman spectra for the two types of quasiperiodic structures were significantly different from one another as regards the frequency shifts for the main peaks and the distribution of intensities. Also, these spectra for the quasiperiodic superlattices were found to be quite distinct in their features from the case of the randomly generated superlattice. In all cases the theoretical analysis led to fairly good agreement with the experimental results.

2.6 Phononic crystals

As explained earlier, phononic crystals are typically formed from *lateral* arrays of elements, mostly periodic with a repeating cell of order several nano- or micro-metres (see figure 1.4(c), for example). From the physics perspective, much the same effect can be achieved through a periodic patterning of a planar surface instead of an alternation of two different materials. A prototype phononic crystal of this type was originally made by Dutcher *et al* [62] by forming a holographic grating of grooves with separation $D = 0.25$ μm on a Si (001) surface. The system has 1D translational symmetry in the direction parallel to the grooves, and the mini-Brillouin-zone boundaries associated with the in-plane wave vector perpendicular to the grooves are at $m\pi/D$, where m is an integer. By analogy with the discussion in the previous section for superlattices, we can describe the phonon dispersion in a reduced-zone scheme in terms of folded modes with band gaps opening at the grating zone boundaries. The experimental data for the frequency of the Rayleigh surface mode (described for a planar surface in section 2.2.1) plotted versus the in-plane wave vector clearly demonstrated the mode folding that we discussed earlier and provided good consistency with a theory for the elastic modes at surfaces with 1D gratings [63].

In the last decade or so, there has been an intense interest in the bicomponent phononic crystals formed from lateral arrays (with either 1D or 2D periodicity) of two materials. For example, Gorishnyy *et al* [64] clearly demonstrated the effects of artificial band gaps opening up in a 2D hypersonic phononic crystal fabricated from a triangular-lattice arrangement of cylindrical holes in a 6 μm thick epoxy film. The holes were separated (centre-to-centre distance) by 1.36 μm and different hole radii were used to vary the sample porosity from close to zero up to about 40%. Subsequently, a particularly thorough investigation was carried out by Zhang *et al* [65] using samples in the specific alternating-stripe geometry of figure 1.4(c). For the materials they employed Fe and $Ni_{80}Fe_{20}$ (permalloy), which allowed them to study, by using BLS in different polarizations, both the phonon and magnon properties in the artificial crystal. Some of their data are shown in figures 2.16 and 2.17 for a sample with periodicity length $D = 500$ nm. In figure 2.16 the BLS spectra are displayed for two values, π/D and $2\pi/D$, of the 1D Bloch wave vector Q. These values correspond to the edges of the first and second Brillouin zone, respectively. The acoustic phonon results, which are in the upper panels of the figure, are seen in diagonal polarization (denoted here by p-p). Then the frequency versus Q plots for the phonon peaks are shown in figure 2.17, where there are band gaps at values of QD/π equal to 0, 1 and 2, as expected. The full lines (in blue) are theory for the phonons using numerical simulations with a finite-element approach. The magnon

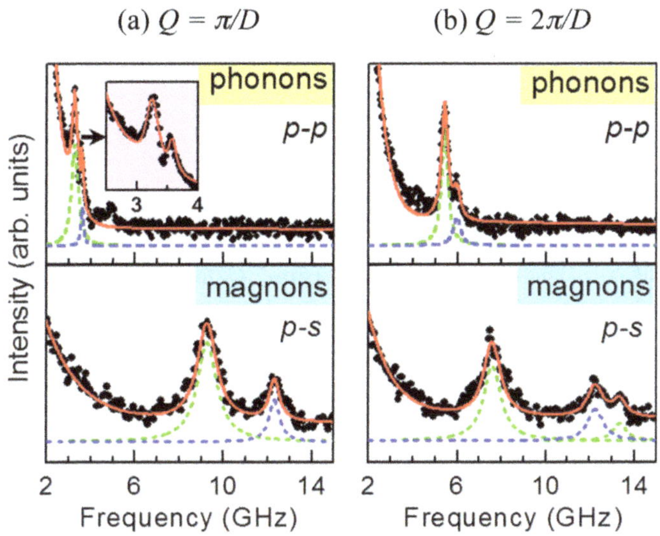

Figure 2.16. BLS spectra for scattering from phonons and magnons in different polarizations in a dual phononic/magnonic crystal composed of alternating Fe and $Ni_{80}Fe_{20}$ (permalloy) nanostripes. Results are shown with (a) $Q = \pi/D$ and (b) $Q = 2\pi/D$ for the in-plane Bloch wave vector. (Reproduced with permission from [65]. Copyright 2012 American Institute of Physics.)

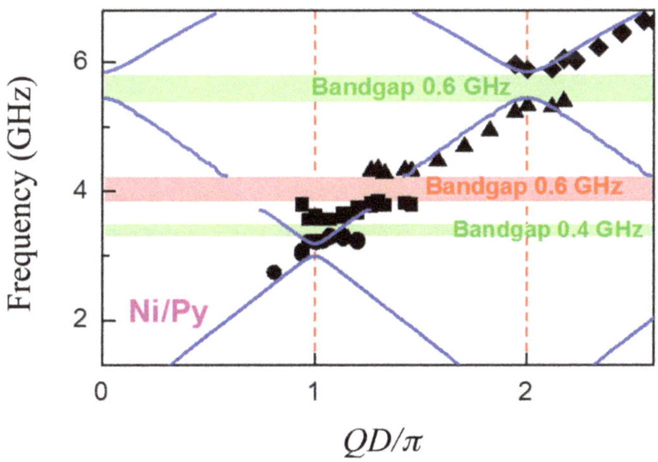

Figure 2.17. Plot of frequency versus reduced wave vector QD/π for acoustic phonons as deduced from BLS data (see also figure 2.16) for a dual phononic/magnonic crystal composed of alternating Fe and $Ni_{80}Fe_{20}$ (permalloy) nanostripes. (Reproduced with permission from [65]. Copyright 2012 American Institute of Physics.)

results will be discussed in chapter 4. The same group subsequently generalized their phonon studies to bicomponent structures with a chess (or checker) board array, i.e. a 2D phononic crystal [66].

Recently, thorough reviews of the different kinds of acoustic waves (localized and propagating) in 1D and 2D phononic materials have been given by Liu *et al* [67] and Cang *et al* [68].

References

[1] Cohen M L and Louie S G S 2016 *Fundamentals of Condensed Matter Physics* (Cambridge: Cambridge University Press)
[2] Ketterson J B 2016 *The Physics of Solids* (Oxford: Oxford University Press)
[3] Snoke D W 2020 *Solid State Physics: Essential Concepts* 2nd edn (Cambridge: Cambridge University Press)
[4] Ridley B K 2017 *Hybrid Phonons in Nanostructures* (Oxford: Oxford University Press)
[5] Srivastava G P 1990 *The Physics of Phonons* (New York: Taylor and Francis)
[6] Wallis R D 1957 *Phys. Rev.* **105** 540
[7] Benedek G and Toennies P 2018 *Atomic Scale Dynamics at Surfaces: Theory and Experimental Studies with Helium Atom Scattering* (Berlin: Springer)
[8] Esser N and Richter W 2000 *Light Scattering in Solids VIII* ed M Cardona and G Güntherodt (Berlin: Springer)
[9] Wang Y, Shang S-L, Fang H, Liu Z-K and Chen L-Q 2016 *npj Comput. Mater.* **2** 16006
[10] Grüneis A, Saito R, Kimura T, Cancado L G, Pimenta M A, Jorio A, Souza Filho A G, Dresselhaus G and Dresselhaus M S 2002 *Phys. Rev.* B **65** 155405
[11] Yan J A, Ruan W Y and Chou M Y 2008 *Phys. Rev.* B **77** 125401
[12] Sholl D S and Steckel J A 2009 *Density Functional Theory: A Practical Introduction* (Hoboken, NJ: Wiley)
[13] Balindin A A and Nika D L 2012 *Mater. Today* **15** 268
[14] Jiang J W, Wang B S, Wang J S and Park H S 2015 *J. Phys.: Condens. Matter* **27** 083001
[15] Al Taleb A and Farias D 2016 *J. Phys.: Condens. Matter* **28** 103005
[16] Kröger J, Néel N, Wehling T O and Brandsbyge M 2020 *Small Methods* **4** 1900817
[17] Landau L D and Lifshitz E M 1970 *Theory of Elasticity* (Oxford: Pergamon)
[18] Cottam M G and Haghshenasfard Z 2020 *Many-Body Theory of Condensed Matter Systems: An Introductory Course* (Cambridge: Cambridge University Press)
[19] Cottam M G and Tilley D R 2005 *Introduction to Surface and Superlattice Excitations* 2nd edn (Bristol: IOP Publishing)
[20] Albuquerque E L 1980 *J. Phys. C: Solid State Phys.* **13** 2623
[21] Albuquerque E L, Loudon R and Tilley D R 1980 *J. Phys. C: Solid State Phys.* **13** 1775
[22] Maradudin A A, Wallis R F, Mills D L and Ballard R L 1972 *Phys. Rev.* B **6** 1106
[23] Maradudin A A 1981 *Festkörperprobleme XXI* ed J Treusch (Braunschweig: Vieweg) 25
[24] Lamb H 1982 *Proc. London Math. Soc.* **13** 189
[25] Kuok M H, Lim H S, Ng S C, Liu N N and Wang Z K 2003 *Phys. Rev. Lett.* **90** 255502
[26] Arfken G B, Weber H J and Harris F E 2012 *Mathematical Methods for Physicists* 7th edn (Amsterdam: Elsevier)
[27] Hudson G E 1943 *Phys. Rev.* **63** 46
[28] Viktorov I A 1958 *Sov. Phys. Acoust.* **4** 131
[29] Rulf B 1969 *J. Acoust. Soc. Am.* **45** 493
[30] Speiser E, Plaikner J, Chandola S, Esser N, Halbig B and Geurts J 2019 *Phys. Status Solidi* **256** 1800341
[31] Hünermann M, Geurts J and Richter W 1991 *Phys. Rev. Lett.* **66** 640

[32] Sandercock J R 1978 *Solid State Commun.* **26** 547
[33] Wittkowski T, Cortina P, Jorzick J, Jung K and Hillebrands B 2000 *Diam. Relat. Mater.* **9** 1957
[34] Geilen M, Kohl F, Nicoloiu A, Müller A, Hillebrands B and Pirro P 2020 *Appl. Phys. Lett.* **117** 213501
[35] Tamtögl A, Kraus P, Mayrhofer-Reinhartshuber M, Campi D, Bemasconi M, Benedek G and Ernst W E 2013 *Phys. Rev.* B **87** 035410
[36] Lehwald S, Szeftel J M, Ibach H, Rahman T S and Mills D L 1983 *Phys. Rev. Lett.* **50** 518
[37] Egerton R F 2011 *Electron Energy-Loss Spectroscopy in the Electron Microscope* 3rd edn (Boston MA: Springer)
[38] Ibach H 1991 *Phys. Scr.* **T39** 323
[39] Maultzsch J, Reich S, Thomsen C, Requardt H and Ordejón P 2004 *Phys. Rev. Lett.* **92** 075501
[40] Oshima C, Aizawa T, Souda R, Ishizawa Y and Sumiyoshi Y 1988 *Solid State Commun.* **65** 1601
[41] Siebentritt S, Pues R, Rieder K H and Shikin A M 1997 *Phys. Rev.* B **55** 7927
[42] Nicklow R, Wakabayashi N and Smith H G 1972 *Phys. Rev.* B **5** 4951
[43] Mohr M, Maultzsch J, Dobardzić E, Reich S, Milosević I, Damnjanović M, Bosak A, Krisch M and Thomsen C 2007 *Phys. Rev.* B **76** 035439
[44] Rytov S M 1956 *Sov. Phys. Acoust.* **2** 68
[45] Camley R E, Djafari-Rouhani B, Dobrzynski L and Maradudin A A 1983 *Phys. Rev.* B **27** 7318
[46] Djafari-Rouhani B, Dobrzynski L, Hardouin-Duparc O, Camley R E and Maradudin A A 1983 *Phys. Rev.* B **28** 1711
[47] Babiker M, Tilley D R, Albuquerque E L and Gonçalves da Silva C E T 1983 *J. Phys. C: Solid State Phys.* **18** 1269
[48] Jusserand B, Paquet D, Kervarec J and Regreny A 1984 *J. Physique* **45** C5–154
[49] Djafari-Rouhani B, Sapriel J and Bonnouvrier F 1985 *Superlattices Microstruct.* **1** 29
[50] Molinari E, Baroni S, Giannozzi P and de Gironcoli S 1992 *Phys. Rev.* B **45** 4280
[51] Davydov V, Roginskii E, Kitaev Y, Smirnov A, Eliseyev I, Nechaev D, Jmerik V and Smirnov M 2021 *Nanomaterials* **11** 286
[52] MacDonald A H 1988 *Interfaces, Quantum Wells, and Superlattices* ed C R Leavens and R Taylor (New York: Plenum) p 347
[53] Merlin R 1989 *Light Scattering in Solids V* ed M Cardona and G Güntherodt (Berlin: Springer) p 214
[54] Albuquerque E L and Cottam M G 2003 *Phys. Rep.* **376** 225
[55] Gao J, Cheng J-C and Li B 2007 *Appl. Phys. Lett.* **90** 111908
[56] Colvard C, Merlin R, Klein M V and Gossard A C 1980 *Phys. Rev. Lett.* **45** 298
[57] Colvard C, Gant T A, Klein M V, Merlin R, Fischer R, Morkoc H and Gossard A C 1985 *Phys. Rev.* B **31** 2080
[58] Jusserand B, Paquet D and Regreny A 1984 *Superlattices Microstruct.* **1** 61
[59] Klein M V 1986 *IEEE J. Quantum Elect.* **QE-22** 1760
[60] Hayes W and Loudon R 1978 *Scattering of Light by Crystals* (New York: Wiley)
[61] Merlin R, Bajema K, Nagle J and Ploog K 1987 *J. Physique* **48** C5–503
[62] Dutcher J R, Lee S, Hillebrands B, McLaughlin G J, Nickel B G and Stegeman G I 1992 *Phys. Rev. Lett.* **68** 2464

[63] Glass N E and Maradudin A A 1983 *J. Appl. Phys.* **54** 796
[64] Gorishnyy T, Ullal C K, Maldovan M, Fytas G and Thomas E L 2005 *Phys. Rev. Lett.* **94** 115501
[65] Zhang V L, Ma F S, Pan H H, Lin C S, Lim H S, Ng S C, Kuok M H, Jain S and Adeyeye A O 2012 *Appl. Phys. Lett.* **100** 163118
[66] Zhang V L, Hou C G, Pan H H, Ma F S, Kuok M H, Lim H S, Ng S C, Cottam M G, Jamali M and Yang H 2012 *Appl. Phys. Lett.* **101** 053102
[67] Liu J, Guo H and Wang T 2020 *Crystals* **10** 305
[68] Cang Y, Jin Y, Djafari-Rouhani B and Fytas G 2022 *J. Phys. D: Appl. Phys.* **55** 193002

Chapter 3

Exchange-dominated magnetic excitations

In this chapter and the next one, we consider magnons, which are the quanta of spin waves associated with the fluctuating magnetization (or spins) in ordered magnetic materials, typically at temperatures well below their critical temperature. Just as in the phonon case, there are different approaches, depending on whether a continuum or discrete-lattice model is used for the theory. Also there can be competing types of magnetic interactions playing a role in the dynamics. Thus, an important consideration is the relative importance of the dynamical effects due to the quantum-mechanical exchange interactions and the magnetic dipole–dipole interactions of electromagnetism, among other factors. As explained later, these interactions are short range and long range, respectively, in their spatial behaviour, so their interplay gives rise to some interesting effects, particularly in the vicinity of surfaces and interfaces. The focus in chapter 3 will be on magnetic excitations for which the exchange effects are dominant. The conditions for this to be the case depend both on the type of magnetic material and the wave-vector regime being considered. In chapter 4 the additional role of the dipole–dipole interactions is included. For the most part, we will use the two terms *magnons* and *spin waves* interchangeably, as is commonly done in the literature.

Some general references on this topic are the books by Gurevich and Melkov [1], White [2] and Stancil and Prabhakar [3]. Additional references where the surface and/or interface properties of magnons are emphasized include [4–9]. Here we start with an overview of the interactions that are important for the static and dynamical behaviour of ordered magnetic materials, before proceeding with studies of the magnetic excitations in various lower-dimensional systems. We note that there have been exciting recent developments in the field of 2D and quasi-2D magnetic materials following the discovery of the so-called van der Waals ferromagnets and antiferromagnets in 2016, as well as renewed interest in the antisymmetric Dzyaloshinskii–Moriya exchange interactions at magnetic interfaces.

doi:10.1088/978-0-7503-3903-2ch3

3.1 Regimes of magnetization dynamics

The ordered magnetic materials of interest in this chapter are ferromagnets, ferrimagnets and antiferromagnets. Their basic characteristics are well established, and descriptions can be found in textbooks on solid-state physics (see e.g. [10–12]). Typically, ferromagnets consist of a lattice of identical magnetic moments or spins that are aligned parallel to one another at low temperatures. In contrast, in ferrimagnets and antiferromagnets there are two (or more) sublattices with some of the magnetic moments being aligned antiparallel. In a two-sublattice antiferromagnet the magnetic moments on each sublattice are equal in magnitude but opposite in direction (in the case of no externally applied field), and so the net magnetization is zero. In a ferrimagnet the magnetic moments of the sublattices do not cancel and hence there is a nonzero magnetization. The critical temperature, above which the long-range magnetic order essentially disappears, is known as the Curie temperature T_C in ferromagnets and ferrimagnets and the Néel temperature T_N in antiferromagnets. Sometimes the magnetic ordering scheme can be more complicated, e.g. in systems with spiral ordering or with competing interactions leading to spin frustration: these will be introduced as required later.

It is well known that the dominant magnetic interaction, which is responsible for the overall static magnetic order in these materials, is the quantum-mechanical *exchange interaction* (see e.g. [10–12]). In the simplest case this interaction arises from the overlap of the quantum-mechanical wave functions of two electrons on neighbouring atoms. The spin state for any electron pair may be either symmetric (where the spins are parallel) or antisymmetric (where the spins are antiparallel) with respect to an interchange. From the Pauli exclusion principle it follows that the corresponding spatial part of the wave function must be antisymmetric or symmetric, respectively, to ensure the antisymmetry of the total wave function. As a consequence of the Coulomb interaction between the electrons, it can be shown that the expectation value of the energy for the electron pair (e.g. as calculated in the Heitler–London theory) is different in the two states, with the difference usually expressed in the form $-J_{12}\mathbf{S}_1 \cdot \mathbf{S}_2$ characteristic of the Heisenberg Hamiltonian. Here $\mathbf{S}_1$ and $\mathbf{S}_2$ are the spin operators for the two electrons while the interaction strength J_{12}, which depends on the overlap of the electronic wave functions, is known as the exchange integral. Depending on the sign of the exchange, we may have either ferromagnetism (when $J_{12} > 0$) or antiferromagnetism (when $J_{12} < 0$).

The above arguments may be extended to other multi-electron situations and to provide an explanation of ferrimagnetism in more complex systems. As well, we shall be discussing other types of exchange interactions later in section 3.5.

The concept of spin waves (SWs) in bulk exchange-dominated ferromagnets was introduced by Bloch [13] using a semi-classical approach with spin vectors. He envisioned an excitation with one spin reversal being 'shared' over many spins throughout a crystal in terms of a collective wave-like disturbance with precessing spins. The amplitude of spin precession would be constant for all the equivalent spin sites in a bulk material, but the phase angles would be different. Nowadays a wide range of quantum-mechanical and semi-classical SW approaches are employed, as we discuss in the following sections.

By contrast with the short-range exchange interactions, we will also in general have energy contributions from the long-range magnetic dipole–dipole interactions [2, 3] proportional to

$$g^2\mu_B^2\left(\frac{\mathbf{S}_1 \cdot \mathbf{S}_2}{r_{12}^3} - \frac{3(\mathbf{S}_1 \cdot \mathbf{r}_{12})(\mathbf{S}_2 \cdot \mathbf{r}_{12})}{r_{12}^5}\right) \tag{3.1}$$

between spins 1 and 2 separated by vector $\mathbf{r}_{12}$. Here g and μ_B denote the Landé factor and Bohr magneton, respectively. Although the individual dipole–dipole interaction strengths may typically be much smaller than the nearest-neighbour exchange energy, the dipole–dipole effects will become important (and eventually will be dominant) at sufficiently small excitation wave vectors. This is because of their long-range character. As we will show later in this chapter, the exchange contribution to the energy of a ferromagnetic excitation with wave vector $\mathbf{q}$ is of order $SJ|\mathbf{q}|^2a^2$ for $|\mathbf{q}|a \ll 1$ (with J denoting the nearest-neighbour exchange and S the spin quantum number). By contrast, the dipolar contribution to the energy is of order $g\mu_B\mu_0 M_0$ (with M_0 denoting the static magnetization). Hence the two terms become comparable when

$$|\mathbf{q}|a \sim \sqrt{g\mu_B\mu_0 M_0/SJ} \ll 1. \tag{3.2}$$

The dipolar terms dominate in the dynamics below this magnitude of wave vector and the exchange terms dominate above. There is a 'crossover region' in the magnetic behaviour, where both terms need to be included, and this is known as the *dipole-exchange region*, as indicated in figure 3.1.

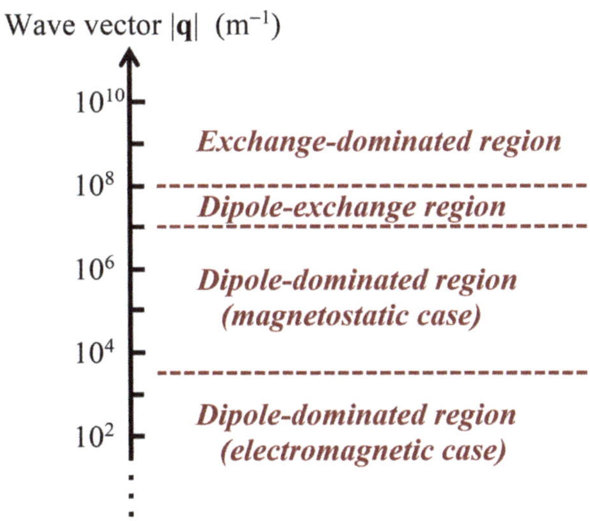

Figure 3.1. Schematic representation of the different regions of spin-wave (SW) behaviour for a ferromagnet in terms of the magnitude of its wave vector. Note that the value 10^{10} m^{-1} corresponds roughly to a Brillouin zone boundary wave vector.

Since the magnetic dipole–dipole interactions are usually important only for long wavelengths (compared to the lattice parameter a) they can often be treated in a continuous-medium representation using the magnetic field variables of electromagnetism together with Maxwell's equations, rather than the energy representation as in equation (3.1). We will use both approaches in this book, particularly for discussion of the dipole-exchange region in the next chapter, and there we will illustrate the connection between the methods. In the dipole-dominated region two cases arise as indicated. They correspond to either using a magnetostatic approximation for Maxwell's equations at larger wave vectors in this range or employing the electromagnetic equations with retardation effects included, as required at very small wave vectors.

As mentioned, we will focus in this chapter on the exchange-dominated magnons at larger wave vectors in the Brillouin zone. The dipole–dipole interactions can be largely ignored, except where their static contributions lead to a demagnetizing field in a finite magnetic system. The dipole-exchange and magnetostatic regions of excitations will be discussed later in chapter 4. The electromagnetic region is applicable to a situation where the magnons are strongly coupled to the photons to give mixed excitations known as *magnon-polaritons*. They will be treated along with other types of polaritons in chapter 6.

3.2 Heisenberg ferromagnetic films

We start by considering the magnons or SWs in ferromagnetic films and for the present we ignore the role of dipole–dipole interactions. Other types of magnetic behaviour and geometries will be introduced later. The Heisenberg spin Hamiltonian for the system may be expressed as

$$\mathcal{H} = -\frac{1}{2}\sum_{i,j} J_{ij}\mathbf{S}_i \cdot \mathbf{S}_j - g\mu_B B_0 \sum_i S_i^z, \tag{3.3}$$

where i and j label all the spin sites in the film, J_{ij} is the corresponding exchange interaction and the factor of $\frac{1}{2}$ before the first term is to avoid double counting in the summation. The second term represents the Zeeman energy of the spins in an applied magnetic field of magnitude B_0 in a fixed direction (taken as the z-axis). The film geometry is illustrated in figure 3.2, where we have assumed for simplicity an sc lattice structure with lattice parameter a. The parallel surfaces of the film are in the yz-plane and the N atomic layers are labelled by integer n (= 1, 2, 3, … , N) as shown. The film is infinite in the y- and z-directions. We adopt a simple model where the exchange interaction couples nearest neighbours only, having the value J_s between two spins in a surface layer and the bulk value J otherwise. The equilibrium orientation of the spins is assumed to be in-plane and along the z-direction of the applied magnetic field. A similar calculation could be carried for the case where the applied magnetic field and the magnetization direction are perpendicular to the surfaces. The only difference would be that B_0 has to be replaced by the internal field B_i, allowing for any static demagnetization effects.

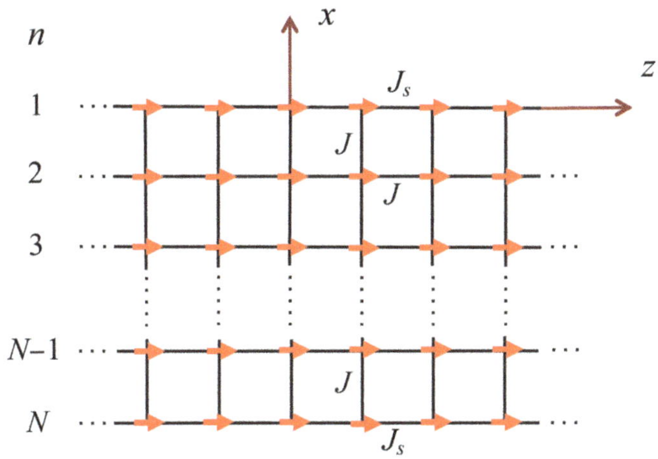

Figure 3.2. The geometry and coordinate axes for a Heisenberg ferromagnetic film. The integer n labels the atomic layers (in the yz-plane) parallel to the surfaces, while J and J_s are the bulk and surface nearest-neighbour exchange constants, respectively.

There are now some alternative ways in which we might proceed to calculate the magnon modes of a film. Mainly, we shall employ here a quantum-mechanical approach in terms of using the commutation properties of the spin operators directly (see section 3.2.1 below). Another method is based on the semi-classical torque equation of motion for the position-dependent magnetization vector, and for completeness we will also consider this approach in section 3.2.2.

3.2.1 Quantum-mechanical calculation

It is convenient at this stage to introduce the usual spin-raising S_j^+ and spin-lowering S_j^- operators at any site j by

$$S_j^{\pm} = S_j^x \pm iS_j^y. \tag{3.4}$$

The commutation properties of the Cartesian $\{x, y, z\}$ components of the spin operators are the same as those for the orbital angular momentum operators in quantum mechanics (see e.g. [14, 15]), so we have $[S_j^x, S_{j'}^y] = S_j^z \delta_{j,j'}$, etc. It follows from these results that the commutation relations satisfied by the set of operators S_j^+, S_j^- and S_j^z can be expressed as

$$[S_j^+, S_{j'}^-] = 2S_j^z\, \delta_{j,j'}, \quad [S_j^z, S_{j'}^{\pm}] = \pm S_i^{\pm}\, \delta_{j,j'}, \tag{3.5}$$

where we denote $[A, B] = AB - BA$ for any operators A and B, as before. Also, for convenience, we will employ units such that $\hbar = 1$.

The equation of motion [14, 15] for the time dependence of any operator A was given in equation (1.39) in terms of its commutator with the Hamiltonian $\mathcal{H}$. To study the magnons, which are excitations associated with the transverse spin components, it is natural that we choose to take $A = S_j^+$ (or $A = S_j^-$). Then using equations (3.3) and (3.5) we may obtain

$$\mathrm{i}\frac{\mathrm{d}}{\mathrm{d}t}S_j^+ = g\mu_B B_0 S_j^+ + \sum_i J_{ij}(S_i^z S_j^+ - S_j^z S_i^+). \tag{3.6}$$

This is a nonlinear equation in terms of the spin components, because of the products of operators under summation in the last term. It can be simplified (or linearized), however, by making the *random-phase approximation* (RPA) to replace the factors $S_i^z\, S_j^+$ and $S_j^z\, S_i^+$ by $\langle S_i^z\rangle S_j^+$ and $\langle S_j^z\rangle S_i^+$, respectively, where $\langle\ldots\rangle$ denotes a thermal average. This approximation involves 'decoupling' a product of operators by replacing each S^z operator by its scalar thermal average. The argument here is that, at low temperatures in an ordered magnetic material, the fluctuations in the z-components will be small (by comparison with the fluctuating components in the xy-plane), and so it is a good approximation to replace the z-components of spin by their average value (see also chapter 9 for further discussion). Critical reviews of decoupling approximations in general can be found in textbooks on many-body theory (e.g. [16–18]).

At each site we now make the low-temperature $T \ll T_C$ replacement $S_j^z \to \langle S_j^z\rangle \to S$ in equation (3.6) to find

$$\mathrm{i}\frac{\mathrm{d}}{\mathrm{d}t}S_j^+ = g\mu_B B_0 S_j^+ + S\sum_i J_{ij}(S_j^+ - S_i^+). \tag{3.7}$$

By analogy with the procedure used for phonons in chapter 2 we seek wave-like solutions of the form

$$S_j^+ = s_n \exp[\mathrm{i}(\mathbf{q}_\parallel \cdot \mathbf{r}_{\parallel,j} - \omega t)], \tag{3.8}$$

for any site j in layer n. Here the 2D vectors parallel to the film surfaces are $\mathbf{q}_\parallel = (q_y, q_z)$ and $\mathbf{r}_\parallel = (y, z)$. Equation (3.7) can be rewritten as a finite set of coupled equations in terms of the label n as

$$\begin{aligned}
&\left\{\omega - g\mu_B B_0 - SJ - 4SJ_s[1 - \gamma(\mathbf{q}_\parallel)]\right\}s_1 + SJs_2 = 0,\\
&\left\{\omega - g\mu_B B_0 - 2SJ - 4SJ[1 - \gamma(\mathbf{q}_\parallel)]\right\}s_n + SJ(s_{n-1} + s_{n+1}) = 0\\
&\quad (1 < n < N),\\
&\left\{\omega - g\mu_B B_0 - SJ - 4SJ_s[1 - \gamma(\mathbf{q}_\parallel)]\right\}s_N + SJs_{N-1} = 0,
\end{aligned} \tag{3.9}$$

where $\gamma(\mathbf{q}_\parallel)$ is a structure factor for the sc lattice defined by

$$\gamma(\mathbf{q}_\parallel) = \frac{1}{2}[\cos(q_y a) + \cos(q_z a)]. \tag{3.10}$$

The above finite-difference equations are formally similar to those encountered in the lattice-dynamics calculations in chapter 2, and so the same methods to obtain a solution can be employed. These methods include seeking solutions for $s_n(\mathbf{q}_{\|})$ formed by the superposition of two wave-like terms (one incident at a surface and the other one reflected), on the one hand, or using the TDM method (see section A.1 of the appendix) on the other hand. Following the former approach here for simplicity, we may write the x-dependence as

$$s_n = A\exp[-\mathrm{i}q_x(n-1)a] + B\exp[\mathrm{i}q_x(n-1)a], \tag{3.11}$$

where we note that the x-coordinate for layer n of the film is $(1-n)a$. It is now straightforward to verify by substitution that equation (3.11) is a solution of equation (3.9) provided $\omega = \omega_B(\mathbf{q})$, where $\mathbf{q} = (q_x, \mathbf{q}_{\|})$ is a 3D wave vector, and the bulk-magnon dispersion relation is found as

$$\omega_B(\mathbf{q}) = g\mu_B B_0 + 2SJ[3 - \cos(q_x a) - \cos(q_y a) - \cos(q_z a)]. \tag{3.12}$$

The first and last of the coupled equations in equation (3.9), i.e. those for $n = 1$ and $n = N$, respectively, constitute the boundary conditions. Each of them can be used to form an expression for the ratio B/A, and then the condition for consistency between them gives

$$\tan(q_x L) = \frac{\mathrm{i}[\Delta - \exp(\mathrm{i}q_x a)]^2 - \mathrm{i}[\Delta - \exp(-\mathrm{i}q_x a)]^2}{[\Delta - \exp(\mathrm{i}q_x a)]^2 + [\Delta - \exp(-\mathrm{i}q_x a)]^2}. \tag{3.13}$$

Here the film thickness is $L = (N-1)a$ and Δ is a surface-dependent parameter given by

$$\Delta = \left\{1 + \frac{4(J - J_s)}{J}\left[1 - \gamma(\mathbf{q}_{\|})\right]\right\}^{-1}. \tag{3.14}$$

Equation (3.13) is an implicit equation for q_x which can only be satisfied by a set of discrete values. Hence it can be interpreted as a quantization condition for the wave-vector component q_x and consequently a quantization of the bulk-magnon frequency ω_B when the film thickness is finite.

Similarly, by seeking localized (or attenuated) solutions for s_n, we may investigate the occurrence of surface magnons. We look for solutions of equation (3.9) of the form

$$s_n = C_1\exp(-\kappa n a) + C_2\exp(\kappa n a), \tag{3.15}$$

where Re $(\kappa) > 0$. For a thick film (in the semi-infinite limit where $N \to \infty$) it can easily be shown that the surface-magnon frequency consists of a single branch with $\omega = \omega_S(\mathbf{q}_{\|})$ where

$$\omega_S(\mathbf{q}_{\|}) = g\mu_B B_0 + 4SJ[1 - \gamma(\mathbf{q}_{\|})] - SJ\frac{(1-\Delta)^2}{\Delta}. \tag{3.16}$$

This solution only exists provides $|\Delta| < 1$ where Δ is defined in equation (3.14). In fact, the attenuation factor κ for the mode is found from $\exp(-\kappa a) = |\Delta|$ in this case. There are two ways in which the above existence condition can be satisfied.

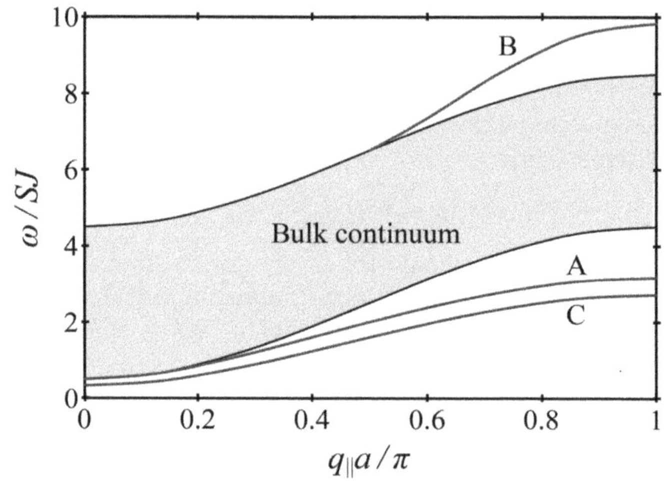

Figure 3.3. Plot of the magnon frequency (in terms of ω/SJ) against $|\mathbf{q}_{\|}a|/\pi$ for a semi-infinite Heisenberg ferromagnet with sc lattice structure, taking $\mathbf{q}_{\|} = (q_{\|}, 0)$. The bulk-magnon region is shown (shaded area) and the lines correspond to three examples of surface-magnon branches: acoustic mode A, $J_s/J = 0.5$ and $B_{AS} = 0$; optic mode B, $J_s/J = 2$ and $B_{AS} = 0$; and acoustic mode C, $J_s/J = 0.5$ and $g\mu_B B_{AS}/SJ = -0.5$. We have assumed an applied field such that $g\mu_B B_0/SJ = 0.5$.

One occurs when $J_s < J$, making $0 < \Delta < 1$ for all $\mathbf{q}_{\|}$, while the other occurs when $J_s > 1.25J$, making $-1 < \Delta < 0$ for at least some values of $\mathbf{q}_{\|}$. These are loosely referred to as acoustic and optic surface magnons, respectively, since they have frequencies below and above the bulk-magnon region. Some numerical examples are shown in figure 3.3 (see curves A and B). Curve C illustrates the effect of adding a surface anisotropy, acting as an additional field B_{AS} on the surface layer spins only, in splitting off an acoustic surface mode from the bulk-mode continuum.

The generalization of the above surface-magnon result to Heisenberg ferromagnetic films with finite thickness (the case of N finite) is given in [19]. For most values of the parameters each surface-magnon branch is split into two and the existence conditions (for localization) are modified. Typically, the mode splitting is smaller at larger wave vectors, as well as at larger N (see numerical examples given in [20]).

Within the quantum-mechanical approach, an alternative procedure would have been to define a transformation from the spin operators to boson creation and annihilation operators. The use of such a transformation becomes advantageous when the Hamiltonian is more complicated than in equation (3.3), as discussed in chapter 4 for dipole-exchange SWs, or when higher-order (nonlinear) SW properties are investigated, as in chapter 9.

3.2.2 Semi-classical torque equation of motion

To broaden our treatment of Heisenberg ferromagnetic films we remark that, in a long-wavelength limit where both $|\mathbf{q}a| \ll 1$ and $|\mathbf{q}_{\|}a| \ll 1$, the equivalent results for the dispersion relations may be derived using a continuum (or macroscopic) method in which the finite-difference terms are replaced by a continuous magnetic variable and its derivatives. Within such an approach we conventionally employ the torque equation of motion for the magnetization as

$$\frac{\mathrm{d}\mathbf{M}}{\mathrm{d}t} = \gamma(\mathbf{M} \times \mathbf{B}) = \gamma\mu_0(\mathbf{M} \times \mathbf{H}) \tag{3.17}$$

when the damping is ignored. This expression replaces the quantum-mechanical operator equation of motion (1.39) used earlier. In the above γ is the gyromagnetic ratio and $\mathbf{M}$ is the total magnetization, which can be written as

$$\mathbf{M} = M_0\hat{z} + \mathbf{m}(\mathbf{r}) \exp(-\mathrm{i}\omega t), \tag{3.18}$$

where M_0 is the static magnetization in the z-direction (specified by the unit vector $\hat{z}$) and $\mathbf{m}(\mathbf{r})$ is the fluctuating part of the magnetization. We are considering here the terms at angular frequency ω, so the total field has the form

$$\mu_0\mathbf{H} = B_0\hat{z} + \mu_0\mathbf{h}(\mathbf{r}) \exp(-\mathrm{i}\omega t). \tag{3.19}$$

When equations (3.18) and (3.19) are substituted into (3.17) and a linearization approximation of neglecting second-order small quantities is made, we obtain

$$-\mathrm{i}\omega\mathbf{m}(\mathbf{r}) = \gamma\hat{z} \times \left[\mu_0 M_0\mathbf{h}(\mathbf{r}) - B_0\mathbf{m}(\mathbf{r})\right], \tag{3.20}$$

In order to proceed we need an expression for the field $\mathbf{h}(\mathbf{r})$. In the present case of the exchange-dominated regime this may be deduced from the Heisenberg Hamiltonian (3.3), although we will later see that there can be additional dipolar contributions in general. The procedure adopted here is that the components of the total effective magnetic field at any site j can first be found from $\mathcal{H}$ by taking the functional derivative as

$$B_j^\alpha = -\frac{1}{g\mu_B}\frac{\delta\mathcal{H}}{\delta S_j^\alpha} \qquad (\alpha = x, y, z). \tag{3.21}$$

Then, on going over to continuous variables (by spatially averaging on the microscopic scale) and using a Taylor series expansion, it is found that (see e.g. [1, 3])

$$\mu_0\mathbf{h}(\mathbf{r}) = (a^2J/\gamma)\nabla^2\mathbf{m}(\mathbf{r}) \tag{3.22}$$

when $\mathbf{r}$ is in the interior of a magnetic material. Note that the first derivatives have cancelled out in the expansion, leaving a ∇^2 term. Also $g\mu_B$ has been equated with γ since we are using units such that $\hbar = 1$.

When equation (3.22) is substituted into equation (3.20) we arrive at the result that, for $\mathbf{r}$ in the interior of a Heisenberg ferromagnet, $\mathbf{m}(\mathbf{r})$ satisfies

$$\mathrm{i}\omega\mathbf{m}(\mathbf{r}) = \hat{z} \times [\gamma B_0 - D\nabla^2]\mathbf{m}(\mathbf{r}). \tag{3.23}$$

Here we used the relationships $D = SJa^2$ and $M_0 = S/a^3$ for an sc ferromagnet at low temperature $T \ll T_C$. Equation (3.23) has wave-like solutions for $\mathbf{m}$ of the form

$$\left[A \exp(\mathrm{i}q_x x) + B \exp(-\mathrm{i}q_x x)\right] \exp(\mathrm{i}\mathbf{q}_\parallel \cdot \mathbf{r}_\parallel) \tag{3.24}$$

yielding the bulk-magnon frequency as $\omega_B(\mathbf{q})$ with

$$\omega_B(\mathbf{q}) = g\mu_B B_0 + D\{q_{\parallel}^2 + q_x^2\}. \tag{3.25}$$

This is just the small wave-vector form of equation (3.12). Equation (3.25) holds for other cubic lattices (bcc and fcc) provided the constant D is redefined appropriately.

A result for the surface-magnon frequency can likewise be obtained. We go back to equation (3.21) to calculate $\mu_0\mathbf{h}(\mathbf{r})$ when $\mathbf{r}$ is at a free surface (corresponding, say, to the plane $x = 0$). The result to low orders in the spatial derivatives becomes [21]

$$\mu_0\mathbf{h}(\mathbf{r}) = \left[-\frac{aJ}{\gamma}\frac{\partial\mathbf{m}(\mathbf{r})}{\partial x} + \frac{a^2J}{2\gamma}\frac{\partial^2\mathbf{m}(\mathbf{r})}{\partial y^2} + \frac{a^2J_s}{\gamma}\left(\frac{\partial^2\mathbf{m}(\mathbf{r})}{\partial y^2} + \frac{\partial\mathbf{m}(\mathbf{r})}{\partial y^2}\right)\right] \tag{3.26}$$

$$(\text{at } x = 0).$$

The analogue at $x = 0$ of equation (3.23) becomes

$$\mathrm{i}\omega\mathbf{m}(\mathbf{r}) = \hat{z}\times\left[\gamma(B_0 + B_{\mathrm{AS}}) + \frac{aJ}{\gamma}\frac{\partial}{\partial x} - \frac{a^2J}{2\gamma}\frac{\partial^2}{\partial y^2} - \frac{a^2J_s}{\gamma}\left(\frac{\partial^2}{\partial y^2} + \frac{\partial}{\partial y^2}\right)\right]\mathbf{m}(\mathbf{r}), \tag{3.27}$$

where we have included the effect of a surface anisotropy field B_{AS}, as before. The solution for any surface (localized) mode will have the form $[a\exp(\lambda x) + b\exp(-\lambda x)]\exp(\mathrm{i}\mathbf{q}_{\parallel}\cdot\mathbf{r}_{\parallel})$. Assuming $\mathrm{Re}(\lambda) > 0$ and $|\lambda|a \ll 1$ (consistent with the continuum approximation), we arrive at $\omega = \omega_S(\mathbf{q}_{\parallel})$ for the surface mode frequency where

$$\omega_S(\mathbf{q}_{\parallel}) = g\mu_B B_0 + D\{q_{\parallel}^2 - \xi^2\}. \tag{3.28}$$

Here ξ is a surface-related parameter (with dimensions of an inverse length) defined by

$$\xi = \frac{g\mu_B B_{\mathrm{AS}}}{SJa} + \left(1 - \frac{J_s}{J}\right)q_{\parallel}^2 a, \tag{3.29}$$

and we have $\lambda = \xi$ for the decay constant. The localization condition $\mathrm{Re}(\lambda) > 0$ becomes simply $\xi > 0$, which can be satisfied for any $J_s < J$ in the absence of an anisotropy field. This is the condition obtained earlier in section 3.2.1 for an acoustic SW mode. There are no optic SWs in this model because the requirement $|\lambda|a \ll 1$ would preclude them.

As a final result in this section, we note that the above equations (3.26)–(3.29) imply at the $x = 0$ surface that

$$\frac{\partial\mathbf{m}(\mathbf{r})}{\partial x} - \xi\,\mathbf{m}(\mathbf{r}) = 0 \qquad (x = 0). \tag{3.30}$$

This form of boundary condition is often proposed phenomenologically, and ξ is referred to as a *pinning parameter*, because if ξ becomes large it effectively requires $\mathbf{m}(\mathbf{r})$ to become small (or 'pinned') at that surface. The boundary condition was first derived by Rado and Weertman [22]. The connection between the pinning parameter and the parameter Δ in the microscopic theory is $\exp(-\xi a) = \Delta$.

There are some direct consequences of pinning in the context of experiments on thin films using techniques such as BLS and spin wave resonance (SWR). The effect can be seen by solving for the x-dependence of the wave-like amplitudes for bulk SWs, e.g. using equation (3.27). Then, in the limits of either large or small pinning at each surface of a film it can be shown that there is an approximate quantization of q_x values as being equal to $n\pi/L$ with $n = 1, 2, 3, \ldots$, which is consistent with fitting an integer number of half-wavelengths into the film thickness L (see e.g. [19]). The cases of large or small pinning would correspond to a standing wave in the x-direction with nodes or antinodes, respectively, at the film surfaces. This general behaviour can be illustrated by the theory lines in figure 3.4 for the SW frequencies plotted against film thickness L and by the corresponding BLS data points for a Fe film. The bulk-mode frequencies become large for small L because the quantized wave vectors q_x (with the values as mentioned above) become larger. The surface-mode branch (labelled ω_S here) depends strongly on the dipole–dipole interactions in this case, and so the discussion of it will be deferred until chapter 4.

It is useful to mention here that the torque equation of motion for the magnetization is sometimes supplemented by the inclusion of a phenomenological damping term. In principle, this could done in various ways, but typically the previous equation (3.17) is generalized to have an extra term with the form (see e.g. [1, 2]) written as

$$\frac{\mathrm{d}\mathbf{M}}{\mathrm{d}t} = \gamma\mu_0(\mathbf{M} \times \mathbf{H}) - \alpha\mathbf{M} \times (\mathbf{M} \times \mathbf{H}). \tag{3.31}$$

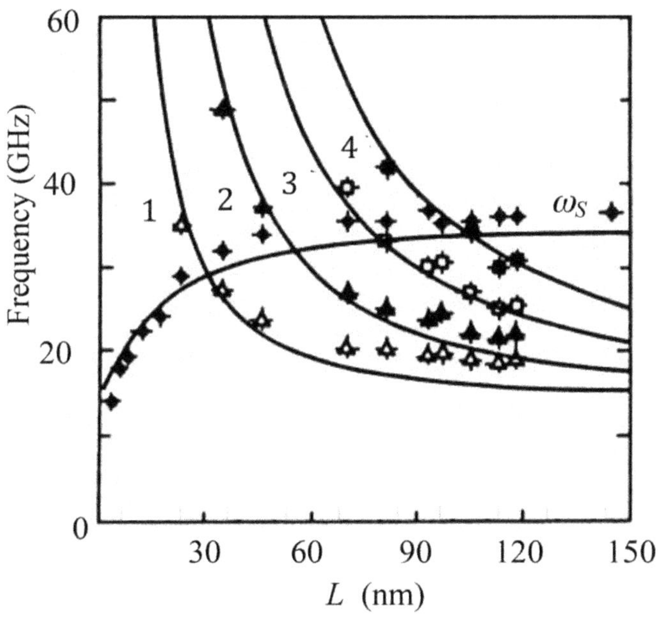

Figure 3.4. The dependence of the SW frequencies on the ferromagnetic film thickness L. The experimental data points derived from BLS measurements on Fe films are shown, as well as calculated values from dipole-exchange theory (full lines). Results for the lowest-frequency standing bulk modes (labelled with their n values) are given, along with the surface mode (labelled as ω_S). (Reproduced with permission from [23]. Copyright 1982 American Institute of Physics.)

In a classical picture where the magnetization vector is regarded as precessing about the net field direction, the extra term on the right-hand side represents a vector that is directed inwards, pointing to the axis of precession, and thus causing an inwards spiralling motion of precessional decay. Here α is a damping constant that has a value depending on the magnetic material. The result in equation (3.31) is commonly known as the Landau–Lifshitz–Gilbert (or LLG) equation.

3.3 Heisenberg antiferromagnetic films

As mentioned earlier, antiferromagnets generally have two interpenetrating sublattices with equal and opposite magnetic moments, so they cancel out to give no net static magnetization when the applied magnetic field is zero. The Heisenberg Hamiltonian can typically be written in the form

$$\mathcal{H} = \sum_{i,j} J_{ij}\mathbf{S}_i \cdot \mathbf{S}_j - g\mu_B B_0\left[\sum_i S_i^z + \sum_j S_j^z\right] - g\mu_B B_A \sum_i S_i^z + g\mu_B B_A \sum_j S_j^z, \quad (3.32)$$

where i and j are chosen here to label the sites on sublattice 1 (spin-up) and sublattice 2 (spin down), respectively, and $J_{ij} > 0$ denotes the nearest-neighbour exchange interaction between sites on opposite sublattices. The second term represents the Zeeman energy due to an applied field B_0 along the z-direction, while the last two terms account for the anisotropy energy expressed in terms of an effective anisotropy field B_A that has the same magnitude but opposite sign on the two sublattices. The anisotropy is known to play an important role in stabilizing the antiferromagnetic order and it typically arises due to crystal-field effects that lead to *single-ion anisotropy* (see e.g. [2, 3]).

It is well known, according to standard textbooks on solid-state physics, that there are two magnon branches for infinite antiferromagnets. This is essentially because there are two spin sites per unit cell, so the situation is broadly analogous to that encountered when calculating the lattice dynamics of diatomic chains (see section 2.1.2) where two phonon excitation branches were obtained. In the antiferromagnetic case it will be seen later that the two bulk-magnon branches are degenerate in magnitude in the case of an effectively infinite material with zero applied magnetic field ($B_0 = 0$).

When a planar surface is introduced to form a semi-infinite structure the effects are more subtle for antiferromagnets than was the case for ferromagnets. This is because the surface may, at least in some cases, remove the symmetry between the up and down sublattices, leading to important consequences for the localized surface excitations in particular. To illustrate this point we will compare two lattice structures, namely, body-centred tetragonal (bct, as in the rutile-structure antiferromagnets MnF_2 and FeF_2) and simple cubic (sc, as in the perovskite structure antiferromagnets $RbMnF_3$ and $KNiF_3$), taking (100) surfaces in both cases. These are shown schematically in figure 3.5. It is evident that in case (a) each layer parallel to the surface (at the top) contains magnetic sites belonging to one sublattice only in an alternating fashion. Moreover the nearest-neighbour intersublattice exchange

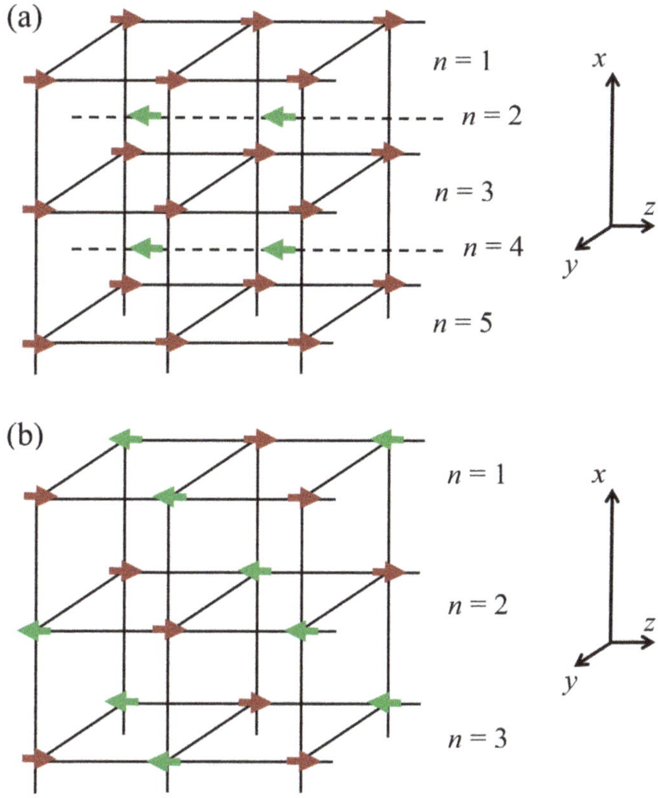

Figure 3.5. Schematic illustrations of (a) body-centred tetragonal antiferromagnets and (b) simple-cubic antiferromagnets with (100) surfaces. The numbering of the atomic layers away from the surface (at the top) is shown.

connects sites on adjacent layers only. By contrast, in case (b) there are equal numbers of sites from both sublattices in all layers and the exchange coupling connects sites in the same as well as adjacent layers. After comparing results for the bulk and surface SWs in these two structures, as obtained next in sections 3.3.1 and 3.3.2 for $B_0 = 0$, we proceed to discuss the dependences of the SW frequencies on applied magnetic field in the following section 3.3.3.

3.3.1 Body-centred tetragonal antiferromagnets

We begin with the bulk and surface SWs or magnons in a semi-infinite bct antiferromagnet, as represented in figure 3.5(a). The surface layer $n = 1$ and all other odd-numbered layers have only spin-up sites (i.e. their spin projections are along the z-direction), and the even layers have only spin-down sites (spin projections along the $-z$-direction). We may now form a set of coupled-finite-difference equations for the spin operators, as was done in the ferromagnetic case, except that we employ equation (3.32) for the Hamiltonian and we take the sublattice structure into account. A linearization approximation is made for the operator equations of motion, and the analogous results to those in equation (3.9) are found to be

$$\begin{aligned}
&(\omega - g\mu_B B_A - 4SJ)s_1 - 4SJ\gamma_0(\mathbf{q}_\parallel)s_2 = 0, \\
&(\omega + g\mu_B B_A + 8SJ)s_n + 4SJ\gamma_0(\mathbf{q}_\parallel)(s_{n-1} + s_{n+1}) = 0 \ (n > 1,\ \ n \text{ even}), \\
&(\omega - g\mu_B B_A - 8SJ)s_n - 4SJ\gamma_0(\mathbf{q}_\parallel)(s_{n-1} + s_{n+1}) = 0 \ (n > 1,\ \ n \text{ odd}),
\end{aligned} \tag{3.33}$$

where J is the nearest-neighbour exchange interaction and the spin amplitudes $s_n(\mathbf{q}_\parallel))$ are defined as in equation (3.8). Also we denote

$$\gamma_0(\mathbf{q}_\parallel) = \cos(q_y a/2)\cos(q_z c/2), \tag{3.34}$$

with c being the lattice parameter corresponding to the tetragonal (z) axis in the bct lattice. For simplicity we have assumed here that the applied magnetic field B_0 is zero and that the spin averages $\langle S^z \rangle$ may be approximated by the values S and $-S$ for sites on sublattices 1 and 2, respectively.

The subsequent analysis is made easier if one notes that the odd-layer equations can be employed to eliminate all of the odd-layer spin amplitudes from the remaining equations, so that a new set of finite-difference equations is obtained that involves only the even-layer spin amplitudes. The solutions for these new coupled equations can be investigated to obtain the bulk and surface magnons in the same manner as for the ferromagnetic case in section 3.2 (for details see e.g. [24, 25]), and therefore we will quote only the final results here. In the case of the bulk magnons, wave-like solutions are found with $\omega = \pm\omega_B(\mathbf{q})$ where

$$\omega_B(\mathbf{q}) = \left[(g\mu_B B_A + 8SJ)^2 - \{8SJ\cos(q_x a/2)\gamma_0(\mathbf{q}_\parallel)\}^2\right]^{1/2}. \tag{3.35}$$

For the special case of wave vector $\mathbf{q} = 0$ this result can be written as

$$\omega_B(0) = [\omega_A(2\omega_E + \omega_A)]^{1/2}, \tag{3.36}$$

where we have defined $\omega_A = g\mu_B B_A$ and $\omega_E = 8SJ$ as the bulk anisotropy frequency and exchange frequency, respectively, for this lattice structure. Equation (3.36) then has the form of the well-known antiferromagnetic resonance (AFMR) frequency in zero applied field [26]. It is also found that the surface magnon consists of a single branch at $\omega = \omega_S(\mathbf{q}_\parallel)$ where (see e.g. [27])

$$\begin{aligned}
\omega_S(\mathbf{q}_\parallel) = &-(\omega_E/4)[1 - \gamma_0^2(\mathbf{q}_\parallel)] \\
&+ \left[(\omega_E + \omega_A)\{\omega_A + (\omega_E/2)[1 - \gamma_0^2(\mathbf{q}_\parallel)]\} + (\omega_E/4)^2[1 - \gamma_0^2(\mathbf{q}_\parallel)]^2\right]^{1/2}.
\end{aligned} \tag{3.37}$$

This surface branch occurs at frequencies below the band of bulk magnons, i.e. it is an acoustic-type surface mode. It is interesting to note that its zero wave-vector limit is

$$\omega_S(0) = [\omega_A(\omega_E + \omega_A)]^{1/2}, \tag{3.38}$$

which may be compared with equation (3.36). In many bct antiferromagnets the ratio ω_A/ω_E is small (e.g. it is about 0.015 in MnF_2), leading to the conclusion that the ratio $\omega_S(0)/\omega_B(0)$ is approximately $\sqrt{0.5}$ ($\simeq 0.71$).

3.3.2 Simple-cubic antiferromagnets

The theory for a semi-infinite sc antiferromagnet, with the structure depicted in figure 3.5(b), is more complicated than the case just described. In part, this is because there is now exchange coupling between sites in the same layer, as well as to adjacent layers, and also each layer contains sites from both sublattices. Therefore there are twice as many equations of motion for the spin amplitudes and the coupling scheme is more intricate. Details of the modified theory are described in [21, 24], so we will quote just the main results for comparison. In the case of the bulk modes it is found that $\omega = \pm\omega_B(\mathbf{q})$, where

$$\omega_B(\mathbf{q}) = \left[(g\mu_B B_A + 6SJ)^2 - 4S^2J^2\{\cos(q_x a) + 2\gamma(\mathbf{q}_\parallel)\}^2\right]^{1/2} \tag{3.39}$$

and $\gamma(\mathbf{q}_\parallel)$ was defined in equation (3.10). At $\mathbf{q} = 0$ the bulk AFMR frequency takes the same form as equation (3.36) provided we use the appropriate expression that $\omega_E = 6SJ$ to define the exchange frequency. If all nearest-neighbour exchange interactions are taken to have the same value as the bulk constant J, the result for the surface magnon is $\omega = \omega_S(\mathbf{q}_\parallel)$ where

$$\omega_S(\mathbf{q}_\parallel) = \left[\frac{(\omega_A + \omega_E)\{4\omega_E^2[1 - \gamma(\mathbf{q}_\parallel)^2] + 12\omega_A\omega_E + 9\omega_A^2\}}{3(2\omega_A + 3\omega_E)}\right]^{1/2}, \tag{3.40}$$

which is again an acoustic-type surface branch. However, if $\mathbf{q}_\parallel = 0$ and if it is assumed that $\omega_A/\omega_E \ll 1$, the above result leads to

$$\omega_S(0) \simeq [\omega_A(2\omega_E + \omega_A/2)]^{1/2}. \tag{3.41}$$

It follows that this frequency for the surface magnon is very close to, but slightly less than, the bulk AFMR frequency in equation (3.36). The situation is therefore quite different from the result in equation (3.38) for the surface AFMR frequency in the bct case.

We remark that all these results for semi-infinite antiferromagnets become modified if other factors are taken into account, e.g. if the exchange parameters near the surface are varied, if there is a surface anisotropy field that is different from the bulk value B_A, or if an applied magnetic field is included. The last of these possibilities is considered next.

3.3.3 Field dependence of antiferromagnetic magnons

When $B_0 \neq 0$ the calculations outlined for bct and sc antiferromagnetic films need to be generalized by including the extra terms due to the Zeeman energy (see the Hamiltonian $\mathcal{H}$ in equation (3.32) in the operator equations of motion). There is also the complication that the spin averages now have the form $\langle S^z\rangle_1 = \langle S^z\rangle_0 + \chi_{\parallel} B_0$ for the spin-up sublattice 1 and $\langle S^z\rangle_2 = -\langle S^z\rangle_0 + \chi_{\parallel} B_0$ for the spin-down sublattice 2. Here $\langle S^z\rangle_0$ is the magnitude of the spin average when $B_0 = 0$ and $\chi_{\parallel}$ is the magnetic susceptibility along the field direction.

Fortunately, the results simplify at low temperatures $T \ll T_N$ where typically $\chi_{\parallel}$ can be neglected and we may approximate by taking $\langle S^z\rangle_0 \simeq S$. In this case, the generalization of the previous results is simple: we find that that the bulk-magnon frequencies are shifted to occur at $\omega_B(\mathbf{q}) \pm g\mu_B B_0$, where $\omega_B(\mathbf{q})$ denotes the zero-field result quoted in equation (3.35) or (3.39) for the two structures. In particular, putting $\mathbf{q} = 0$ for the AFMR frequencies, we have

$$\omega^{\pm}_{\text{AFMR}} = [\omega_A(2\omega_E + \omega_A)]^{1/2} \pm g\mu_B B_0. \tag{3.42}$$

We now see that one branch (ω^{+}_{AFMR}) increases in frequency with applied field, while the other (ω^{-}_{AFMR}) decreases in magnitude, and the splitting between the branches is equal to $2g\mu_B B_0$. This persists as B_0 is increased until the frequency of the lower branch reaches zero, signalling the onset of a phase transition known as the spin-flop (SF) transition. It occurs when $B_0 = B_{\text{SF}}$, where we define $g\mu_B B_{\text{SF}} = [\omega_A(2\omega_E + \omega_A)]^{1/2}$. In the SF phase the sublattice spins have reoriented as depicted in figure 3.6(b) and are inclined at an angle θ to the field direction. If the field is increased further in the SF phase, the angle θ gradually decreases and reaches zero at a second critical field value (which we denote by B_{FP}). For $B_0 > B_{\text{FP}}$ the spin ordering is paramagnetic (or ferromagnetic in the low-temperature limit). For an infinite bulk antiferromagnet we have $g\mu_B B_{\text{FP}} \sim \omega_E$, so typically $B_{\text{SF}} \ll B_{\text{FP}}$ depending on the anisotropy term. The magnons in all three of the above magnetic phases have been thoroughly investigated for bulk magnetic materials, e.g. by inelastic light scattering (see [28]).

In the film geometry it is interesting to recall our result for $\omega_S(0)$ in equation (3.38) for bct antiferromagnets. The fact that this surface-mode frequency is significantly smaller than the bulk $\omega_B(0)$ means that the SF transition will occur for the surface regions in bct antiferromagnets at a lower applied field than the bulk value given

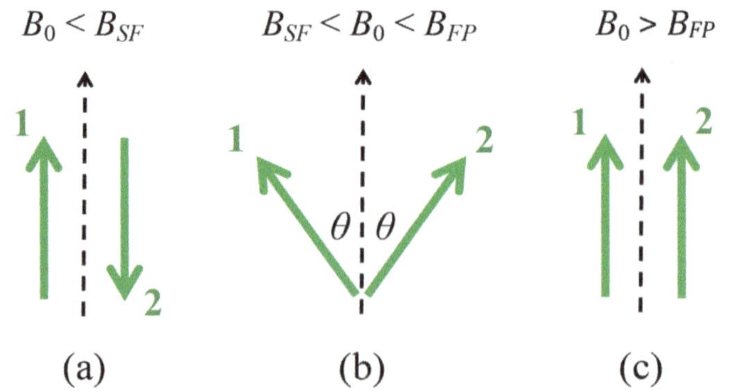

Figure 3.6. Schematic illustration of the magnetic ordering phases for a typical antiferromagnet (with sublattice spins labelled as 1 and 2) as the field B_0 is varied: (a) the antiferromagnetic phase, (b) the spin-flop phase and (c) the paramagnetic (ferromagnetic) phase. See the text for details.

above. This is in contrast with the situation for sc antiferromagnets, where there is essentially no difference.

3.4 One- and two-dimensional magnetic systems

For magnetic systems in 1D or 2D at nonzero temperatures, the quantum fluctuations in the spins become larger and the spontaneous long-range magnetic order may be inhibited. The constraint is usually summed up in terms of the well-known theorem due to Mermin and Wagner [29]. Their rigorous result essentially establishes that, for an infinitely extended Heisenberg magnetic system in 1D or 2D having finite-range exchange interactions, there can be no spontaneous long-range magnetic order at any finite temperature. The crucial properties here are the continuous rotational symmetry of a Heisenberg system and the finite range of the exchange interactions.

As a simple illustration of the formal difficulties that would be encountered in the analysis for lower-dimensional systems if a spontaneous order is assumed, we note that linear SW theory would imply a dispersion relation $\omega(\mathbf{q}) = Dq^2$. This is applicable for a Heisenberg ferromagnet with small wave vector of magnitude $q = |\mathbf{q}|$ in zero applied field, e.g. by analogy with equation (3.28). It then follows that the deviation ΔM in the magnetization due to the thermally excited SWs at temperature T would be given by an expression of the form (see [1–3])

$$\Delta M(T) \propto \int_0^{q_{\max}} \frac{q^{n-1}\mathrm{d}q}{\exp(Dq^2/k_BT) - 1}, \tag{3.43}$$

where n is an integer denoting the dimensionality of the system. The quantity $q_{\max}$ is a wave-vector cut-off whose value can be chosen according to the size of the Brillouin zone. In a 3D ferromagnet (putting $n = 3$) the above integral is well-behaved (including near $q \approx 0$), and it can be evaluated to give the usual result for ΔM proportional to $T^{3/2}$, as found in the solid-state physics textbooks cited earlier. In contrast, if $n = 1$ or 2 the integral becomes infinite due to the divergent behaviour of the integrand when $q \approx 0$, and the simple theory breaks down for the reasons mentioned earlier.

Long-range magnetic order in 1D and 2D systems, however, does become allowable if *either* there is a magnetic anisotropy that introduces an energy gap in the SW dispersion relation *or* there are magnetic interactions of lower symmetry than the isotropic Heisenberg interaction. The magnetic anisotropy could arise, for example, from single-ion anisotropy or from anisotropic exchange terms. The lower-symmetry interactions could be due to the long-range magnetic dipole–dipole interactions. We will consider specific forms for the anisotropic Hamiltonian in the following sections and in chapter 4.

3.4.1 Two-dimensional van der Waals magnets

The search for novel magnetically ordered materials that are truly two-dimensional, in the same sense that nonmagnetic graphene forms in 2D atomic monolayers, has for a long time been elusive. However, a few *quasi-2D magnets* had been identified since around the 1980s and were thoroughly studied then and subsequently. A notable example is the two-sublattice antiferromagnet $MnPS_3$, which has a layered structure with a honeycomb (hexagonal) 2D lattice of Mn^{2+} ions in each layer. The layers are weakly held together to form the 3D solid via the interlayer van der Waals forces. The spins in this $S = 5/2$ material order antiferromagnetically in the directions perpendicular to the layers at temperatures below the Néel temperature $T_N = 78$ K. The material has well-defined SW excitations, which have been studied by inelastic neutron scattering (INS) and interpreted theoretically by several groups (see e.g. [30, 31]). The findings are consistent with a 2D Heisenberg antiferromagnet for the honeycomb layers with a weak intralayer anisotropy as well as a weak (mainly dipole–dipole) interaction between the layers. It is the combination of the latter two effects in this compound that facilitate the stable long-range antiferromagnetic ordering and is sufficient to suppress the large quantum fluctuations.

This situation changed in 2016 with the discovery that individual magnetic monolayers could be extracted from several of the quasi-2D magnets, or van der Waals (vdW) magnets, often by a simple exfoliation process using adhesive tape or in other cases by more sophisticated techniques such as molecular beam epitaxy or chemical vapour deposition. The existence of the long-range magnetic order (as well as the associated magnetic excitations) was demonstrated soon afterwards, for example, by inelastic light scattering [32, 33]. Already there is a rapidly growing list of these 2D vdW antiferromagnets and ferromagnets that have been reported in the literature (see [34–37] for reviews). They can be fabricated either as atomic monolayers and/or with just a few atomic layers. Some examples of these materials, which generally have hexagonal (honeycomb) or triangular 2D lattices in the atomic planes, are $MnPS_3$, $FePSe_3$, $NiPS_3$ (all antiferromagnets), and $Cr_2Ge_2Te_6$, Fe_3GeTe, $CrBr_3$, CrI_3 (all ferromagnets). Depending on the material and the scheme for the anisotropies, the magnetization may either lie in the plane or be perpendicular to it.

On the experimental side, ferromagnetic CrI_3, where the magnetization is perpendicular to the 2D atomic plane, has been investigated in detail. Jin *et al* [38] used micro-Raman techniques to show that there are the two SW branches occurring in monolayers. Also, magnetic resonance techniques were used by Kapoor *et al* [39] to detect standing SWs excited across few-layer films of CrI_3 and the SW modes in antiferromagnetically-coupled ferromagnetic bilayers of CrI_3 were detected using optical pumping and Kerr magneto-optical techniques [40]. Detailed observations of the magnetic field dependence of the acoustic and optic SWs were reported by Cenker *et al* [41].

Theoretically, the 2D monolayer vdW magnets are often represented by a simplified spin Hamiltonian of the general form

$$\mathcal{H} = \sum_{\langle i,j \rangle} \left\{ J_{ij}^T \left(S_i^x S_j^x + S_i^y S_j^y \right) + J_{ij}^L S_i^z S_j^z \right\} - \sum_i K_i (S_i^z)^2 - g\mu_B \sum_i \mathbf{B}_0 \cdot \mathbf{S}_i. \quad (3.44)$$

Here the easy axis of magnetization (or sublattice magnetization for an antiferromagnet) is selected as the z-axis (about which there is assumed to be uniaxial anisotropy), and we use the notation $\langle i, j \rangle$ to indicate summations over all distinct pairs of sites i and j in the 2D lattice. The exchange terms in the anisotropic interaction are denoted by transverse J_{ij}^T and longitudinal J_{ij}^L coefficients, allowing us to encompass several special cases. Thus, if $J_{ij}^T = J_{ij}^L$ we have the isotropic Heisenberg model assumed previously, whereas if $J_{ij}^T = 0$ and $J_{ij}^L \neq 0$ we have the *Ising model*, and if $J_{ij}^T \neq 0$ and $J_{ij}^L = 0$ we have the *XY model*. Also, K_i is a coefficient of uniaxial single-ion anisotropy, which can be positive or negative, at site i. The final term in equation (3.44) represents the Zeeman energy due to an applied field $\mathbf{B}_0$. In terms of the above Hamiltonian, the necessary gap in the SW spectrum at zero wave vector can arise from combinations of the anisotropic exchange (when $J_{ij}^T \neq J_{ij}^L$) and the single-ion anisotropy (when $K_i \neq 0$).

If the lattice structure, the easy direction (z) of magnetization, and the Hamiltonian parameters are all known, the SW spectrum can, in principle, be evaluated from equation (3.44) using the standard SW methods [1–3], which include a direct extension of the spin operator equation-of-motion methods as presented in sections 3.2 and 3.3. Some SW calculations of this type using either spin operators or boson operators have recently been reported by several authors [40, 42, 43]. For some of the 2D vdW materials it may be relevant to add other anisotropic interaction terms to the Hamiltonian for a realistic description of the magnetic properties. These terms might represent the biquadratic exchange and/or the Dzyaloshinskii–Moriya exchange; these exchange interactions will both be discussed in the upcoming section 3.5. Additional effects in vdW magnetic materials could also come from the magnetic dipole–dipole interactions, which will be described in chapter 4.

3.4.2 One-dimensional chain magnets

Examples of quasi-1D materials, which exhibit anisotropic interactions and a spontaneous magnetic ordering along a line (or chain) of atoms at low temperatures, have been known since around the 1970s. Among these, the compound hexagonal perovskite materials of the type $ACoX_3$ with A as a monovalent cation, such as Cs or Rb and X = (Cl, Br, I), have attracted renewed interest as Ising-dominated chain magnets exhibiting unusual magnetic phases when the temperature is varied.

One of the systems most thoroughly investigated to date is the spin $S = \frac{1}{2}$ antiferromagnet $RbCoCl_3$. From Raman scattering and other experimental studies (see e.g. [44]) it is known that there are two distinct magnetic ordering temperatures for this material, which occur at $T_{N1} \simeq 28$ K and $T_{N2} \simeq 12$ K. Below T_{N1} the spins along any chain order antiferromagnetically and there is a strong 1D characteristic, whereas below T_{N2} the role of the interchain interactions becomes more significant. The dominant interaction between the Co^{2+} ions along the magnetic chains is a mainly Ising exchange interaction (with a small admixture of XY exchange interaction) for the antiferromagnetic coupling between nearest neighbours. More recent work has served to emphasize the crucial stabilizing role played by the weaker interactions, likely due to interchain coupling and the intrachain next-nearest-neighbour exchange interactions [45–47]. The spin Hamiltonian for the magnetic interactions can be written in the form (see [45])

$$\mathcal{H} = \sum_j \Big\{ J_1\big[S_j^z S_{j+1}^z + \alpha_1(S_j^x S_{j+1}^x + S_j^y S_{j+1}^y)\big] + (-1)^j h \\ + J_2\big[S_j^z S_{j+2}^z + \alpha_2(S_j^x S_{j+2}^x + S_j^y S_{j+2}^y)\big]\Big\}. \tag{3.45}$$

Here the Co spins along the chain are numbered by the index j, while J_1 and J_2 are the exchange interactions between nearest and next-nearest neighbours, respectively. The small ratios between the XY and Ising exchange terms are denoted by α_1 and α_2, and the term in h represents the so-called staggered anisotropy field along the length of the chain. The inclusion of this term (with $h > 0$) is consistent with the antiferromagnetic ordering, and the magnitude of h is a measure of interchain effects. The recent theoretical and experimental studies [45, 46], which were mainly applied to inelastic neutron scattering (INS), facilitated in determining the magnetic parameters in equation (3.45). It was deduced that $\alpha_1 \sim 0.1$ and $|J_2|/J_1 \sim 0.1$, thus confirming the dominance of the 1D magnetic chain behaviour with strong Ising-like nearest-neighbour interactions.

Interpretations of the magnetic excitations in $S = \frac{1}{2}$ antiferromagnetic 1D (chain) materials have generally been based on ideas proposed by Villain [48] and later extended by others [49, 50]. In simple terms, Villain's original description, in which the interchain effects were neglected, was in terms of a single spin flip (reversal) producing magnetic domain walls, which could then move apart along the chain. The XY model terms of the Hamiltonian were treated as a perturbation (compared to the dominant Ising exchange terms), leading to a soliton-pair type of description and the prediction of a broadened (continuum) band for the spectrum of magnetic excitations. Subsequently, Shiba [49] and Matsubara *et al* [50] introduced the staggered-field term in h, mentioned above, to account for the role of exchange coupling due to the neighbouring spin chains, treated in a mean-field approximation. Remarkably, it was shown that employing the hierarchy of possible nonzero values for h, in what is known as a 'Zeeman ladder' approximation, resulted in Villain's continuum band becoming quantized with a series of discrete peaks (representing the SWs) becoming superimposed on the continuum.

In the case of $RbCoCl_3$ with its two different Néel temperatures, as mentioned earlier, it has been established that when $T < T_{N1}$ the ordering along any chain is antiferromagnetic but the interchain coupling may only have a small effect. In contrast, at very low temperatures when $T < T_{N2}$ the magnetic coupling between the neighbouring chains is more significant since thermal effects are reduced. These considerations have a bearing on the values to be adopted for h, which can be deduced (in principle) using statistical arguments. This leads to a spectrum of magnetic excitations consisting of discrete peaks across the low-temperature region below T_{N1}. These theoretical interpretations are all found to be consistent with the experimental observations from Raman scattering [44, 47] and from inelastic neutron scattering [45, 46]. Another similar example of an antiferromagnetic chain material is $CsCoBr_3$, which has also been studied by inelastic neutron scattering and optical techniques (see e.g. [51] and references therein). There is evidence that its magnetic phase behaviour is more complicated than in $RbCoCl_3$.

We note that soliton excitations, which were mentioned in this subsection, will be described more generally in chapters 8 and 9 of this book. Chapter 9 also includes further discussion involving magnetic chain materials in the context of models for magnetic solitons.

3.5 More types of exchange interactions

A simplified account for the origin of exchange interactions was previously given in section 3.1 in terms of considering two electrons on adjacent atoms. Then, if the electrons have overlapping wave functions and the requirements of the Pauli exclusion principle are taken into account, we obtain a quantum-mechanical exchange interaction. This approach led us to adopt (as the simplest choice) a dependence of the magnetic energy on a combination of the spin operators proportional to $\mathbf{S}_1 \cdot \mathbf{S}_2$ at magnetic sites labelled as 1 and 2.

This type of exchange, sometimes referred to as *bilinear exchange*, provides the basis of the Heisenberg model, which embodies the Ising model and XY model as special cases. It is important to note, however, that other types of exchange interactions may play a role, in particular when the symmetry is reduced, as in some low-dimensional magnetic materials or close to magnetic interfaces.

3.5.1 RKKY and biquadratic exchange interactions

A typical situation involving exchange-coupled magnetic bilayers is when two ferromagnetic layers are weakly coupled by the exchange interaction across a very thin nonmagnetic spacer layer, as depicted in figure 3.7. Typically the spacer layer may be a metal such as Cr with a thickness of a few nanometres or less. A different exchange mechanism comes into play here, in contrast with the mechanism mentioned in section 3.1 that involved localized spins and their overlapping wave functions. Instead, for the coupled bilayers, the polarization effects in the 'cloud' of conduction electrons in the Cr metallic layer provide an additional exchange interaction that may extend over greater distances, up to a few nanometres.

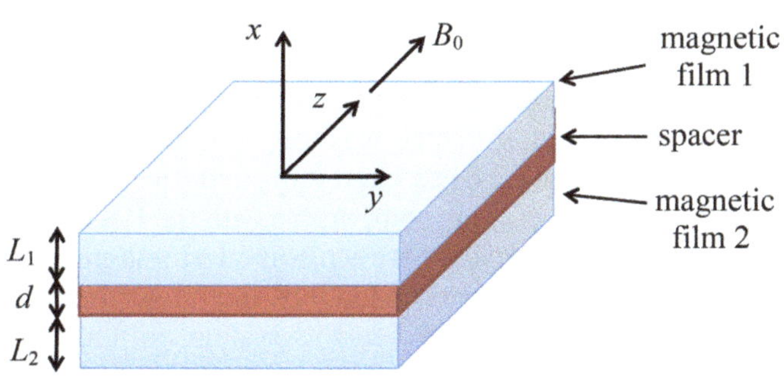

Figure 3.7. The assumed geometry and choice of coordinate axes for the consideration of interfacial exchange coupling between two magnetic films separated by a thin nonmagnetic spacer layer.

As in the case of the Heisenberg Hamiltonian, the interaction typically has a bilinear form proportional to $\mathbf{S}_1 \cdot \mathbf{S}_2$. This is the Ruderman–Kittel–Kasuya–Yosida (or RKKY) interaction, which was originally proposed as an indirect exchange mechanism that takes place between the 4*f* ion cores in rare-earth metals via the surrounding s electrons. A remarkable property, in addition to the longer-range nature (over distances up to about 2 or 3 nm typically), is that it can oscillate in sign with respect to the distance apart of the ions or with the thickness d of the nonmagnetic metallic spacer in the present context [52]. The oscillations mean that, depending on the spacer thickness, the interface coupling may give a preference for either parallel or antiparallel magnetic ordering between the adjacent magnetic films. The RKKY magnetic coupling is thus quite different from the long-range magnetic dipole–dipole coupling across a thin spacer. The latter type of coupling will be considered in the next chapter, but we shall see that it has a considerably longer range than the RKKY interaction (with dynamical effects extending over tens of nanometres or more) and it does not have an oscillatory character.

In work carried out in the late 1980s that eventually led to them being awarded a Nobel Prize, Grünberg and Fert separately investigated trilayer systems such as Fe/Cr/Fe in which the Cr thickness was chosen such that the Fe layers were weakly coupled antiferromagnetically (see e.g. [53, 54]). By studying the magnetization-switching behaviour of this system in an applied magnetic field using optical and electrical techniques, they discovered the phenomenon of *giant magnetoresistance* (GMR), which has become of immense technological importance in recording heads and for information storage, etc. Parkin *et al* [55], in related experiments that also used Fe/Cr/Fe trilayers, made a careful characterization of the RKKY coupling across the Cr spacer, confirming that it oscillates with increasing thickness d before its magnitude eventually decays. Review accounts of this topic can be found in, e.g. [56, 57].

This bilinear exchange, as just discussed, involves the spin operators in second order, but other dependences are possible when higher orders of perturbation are taken into account for the interaction energy between a pair of electrons. In particular, it is found that there may be exchange contributions proportional to $(\mathbf{S}_1 \cdot \mathbf{S}_2)^2$ known as *biquadratic exchange.* The total exchange energy acting between two interface spins 1 and 2 across the spacer can then be written as

$$-J\,(\mathbf{S}_1 \cdot \mathbf{S}_2) - J_{bq}(\mathbf{S}_1 \cdot \mathbf{S}_2)^2, \tag{3.46}$$

where J and J_{bq} denote the bilinear and biquadratic exchange constants, respectively. The phenomenon of biquadratic exchange was long known to occur as a very small effect in some bulk materials, but renewed interest arose in the 1990s with the discovery that biquadratic exchange could become comparable with the RKKY bilinear exchange interaction across metallic spacer layers between some ferromagnets.

This behaviour was clearly demonstrated for the Fe/Cr/Fe trilayer system in experiments by Rührig *et al* [58]. Since J in this system oscillates between positive and negative values as the spacer thickness d is varied, there are some circumstances arising where $J \approx 0$ and the second term in equation (3.46) dominates. Then if

$J_{bq} < 0$, which is usually the case for biquadratic exchange, the minimum energy will occur when $\mathbf{S}_1$ and $\mathbf{S}_2$ are perpendicular to one another. This behaviour was confirmed for Fe/Cr/Fe systems in studies where d was varied. Various explanations have been put forward for the enhanced role of biquadratic exchange at interfaces, including interface roughness and the presence of interface impurities (see e.g. [59]). Further discussion of the bilinear RKKY and biquadratic exchange interactions will come later in chapter 4 in the context of including the additional role of the dipole–dipole interactions in trilayers and other multilayered structures.

3.5.2 Dzyaloshinskii–Moriya antisymmetric exchange interaction

As a final topic in this section, we will discuss antisymmetric exchange interactions. We start by noting that the general form of a bilinear interaction between two spins 1 and 2 can be expressed in terms of combinations of products of spin operators having the form $S_1^{\alpha} S_2^{\alpha'}$, where α and α' denote Cartesian components x, y or z. In a crystal lattice where there is a high degree of symmetry, some terms will be found to vanish. Those that remain can be grouped into combinations that are either symmetric or antisymmetric with respect to the interchange $1 \rightleftarrows 2$ of the site labels. The Heisenberg, Ising, XY and RKKY models are all examples of the former case. The possible role of antisymmetric exchange combinations was proposed by Dzyaloshinskii [60] to account for a small canting of the spins in haematite ($\alpha - Fe_2O_3$), and shortly afterwards Moriya [61] explained the microscopic origin of the interaction. This type of antisymmetric exchange is now known as the Dzyaloshinskii–Moriya (or DM) interaction. Its contribution to the spin Hamiltonian can be written in the vector form

$$\mathbf{J}_{\mathrm{DM}} \cdot (\mathbf{S}_1 \times \mathbf{S}_2), \tag{3.47}$$

where $\mathbf{J}_{\mathrm{DM}}$ is a vector with its direction specifying an axis and its magnitude defining the strength of the DM exchange interaction.

Although the initial interest in DM interactions was directed to spin canting in ferrimagnets and other multi-sublattice magnetic compounds, it was first pointed out by Melcher [62] that the DM interactions could lead to an overall linear dependence on the wave vector for the SW dispersion relation of ferromagnets. This is a property that has been exploited in recent years, particularly in the context of interface exchange effects and low-dimensional magnetic materials. Typically, the occurrence of DM interactions can be associated with *either* the lack of inversion symmetry in the crystal lattices of some bulk magnetic materials (such as MnSi and FeCoSi) *or* effects occurring at the interfaces between materials. For example, in the latter case of interfacial DM interactions, it is observed that a bilayer formed between a ferromagnetic metal and a heavy metal (e.g. such as Mn/W or Fe/Ir) might typically be employed to achieve a high value of the interfacial $|\mathbf{J}_{\mathrm{DM}}|$ interaction. Recent reviews of this topic can be found in [63, 64].

As a simple illustration of the SW behaviour at an interface in the presence of both symmetric Heisenberg and antisymmetric DM exchange interactions, we consider a 1D array of spins as depicted schematically in figure 3.8. Here the

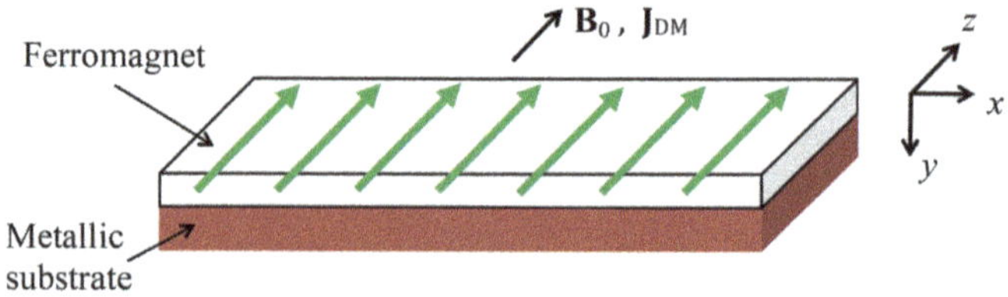

Figure 3.8. Simple geometry with a 1D array of spins, as employed to illustrate interfacial DM exchange interactions for a ferromagnetic layer on a symmetry-breaking metallic substrate. The applied magnetic field $\mathbf{B}_0$ and the axial DM vector $\mathbf{J}_{DM}$ are both taken to be along the z-direction and the propagation direction for the SW is along the x-axis. See the text for details.

interface is in the xz-plane with the spin vectors and the applied magnetic field vector $\mathbf{B}_0$ oriented in the z-direction. The system is assumed to be effectively infinite along the x-axis, and so we may anticipate a wave-vector component q_x associated with SW propagation along this direction. The model spin Hamiltonian in this example can be expressed as

$$\mathcal{H} = -\sum_j J\, \mathbf{S}_j \cdot \mathbf{S}_{j+1} + \sum_j \mathbf{J}_{DM} \cdot (\mathbf{S}_j \times \mathbf{S}_{j+1}) - g\mu_B \sum_j \mathbf{B}_0 \cdot \mathbf{S}_j. \tag{3.48}$$

The first term in this expression represents a bilinear Heisenberg ferromagnetic contribution (with $J > 0$ for the nearest-neighbour exchange interaction only), the second term gives the nearest-neighbour DM interaction, and finally there is a term for the Zeeman energy due to the applied field. Specifically, we shall assume here that the axial DM vector and magnetic field vector lie along the z-axis and have magnitudes J_{DM} and B_0, respectively. The antisymmetric DM term then simplifies to become $J_{DM}(S_j^x S_{j+1}^y - S_j^y S_{j+1}^z)$ with there being a summation over site label j.

The SW calculation now proceeds by generalizing the steps employed in section 3.2.1 for a Heisenberg ferromagnet. First, we form the commutator of the spin operator S_j^+ with the Hamiltonian in equation (3.48) to form its operator equation of motion, and a low-temperature decoupling approximation in which we take $S_j^z \to \langle S_j^z \rangle \to S$ can be made as before. Wave-like solutions for the spin amplitudes are then sought that have the form $S_j^+ = s_j \exp(\mathrm{i}q_x x - \mathrm{i}\omega t)$, which is just the analogue of equation (3.8). After some straightforward algebra we find solutions with angular frequency $\omega = \omega_B(q_x)$, where the SW dispersion relation is

$$\omega_B(q_x) = g\mu_B B_0 + 2SJ_{DM} \sin(q_x a) + 2SJ[1 - \cos(q_x a)]. \tag{3.49}$$

It follows that at small wave vectors (such that $|q_x a| \ll 1$) the approximate form of the dispersion relation becomes $\omega_B(q_x) \simeq g\mu_B B_0 + 2SJ_{DM}\,(q_x a) + SJ(q_x a)^2$. From this analysis we conclude that the FMR frequency (at zero wave vector) is unchanged due to the DM interaction, but significantly when $q_x \neq 0$ a term linear in q_x appears in the dispersion relation, in contrast with the case involving a Heisenberg exchange interaction only. At sufficiently small $q_x a$, this linear term will provide the dominant wave-vector dependence, even when we have $J_{DM} \ll J$. Also, it is interesting to remark that we have encountered here an example of *nonreciprocal* wave propagation. The significance is that an SW propagating in the $+x$-direction will have a different frequency from that of an SW with the same magnitude of wave vector propagating in the $-x$-direction, i.e. we have $\omega_B(-q_x) \neq \omega_B(q_x)$. Nonreciprocal SW propagation due to DM interactions was also recently predicted at the lateral edges in nanoribbons (or stripes) of vdW antiferromagnets [65].

We will see in the next chapter that nonreciprocity of magnetic waves may also occur in different ways when the magnetic dipole–dipole interactions are taken into account in low-dimensional structures. A general account of the conditions that may lead to nonreciprocal wave propagation at surfaces and interfaces in magnetic and other related materials is given in [66].

3.6 Magnetic superlattices and magnonic crystals

We now turn to the consideration of SWs in multilayer magnetic systems (superlattices or magnonic crystals) in the exchange-dominated case. This topic will be treated rather briefly at this stage, with the main discussion being deferred until chapter 4. This is because most studies of these magnetic systems have involved dipole–dipole interactions. Nevertheless, it is relevant to examine some cases where the exchange interactions, both within the magnetic layers and across the magnetic interfaces, provide the dominant role. In particular, we may consider superlattices with exchange-coupled layers (films or slabs) that follow either a periodic or quasiperiodic growth rule. The magnonic crystals (MCs) formed as lateral arrays of finite-width magnetic stripes with direct exchange coupling are also of interest. There are several excellent review accounts that give theoretical and experimental details of this topic; see, for example, [21, 67–70] for mainly periodic structures and [71] for quasiperiodic structures.

As a preliminary example, we consider a case where the exchange interactions provide a direct coupling across the interface between two exchange-dominated ferromagnets is the Gd/Fe system. This was modelled by Camley [72] as a thin film of Gd on a substrate of bulk Fe. Both materials are ferromagnetic, having an exchange ratio $J_{\mathrm{Fe}}/J_{\mathrm{Gd}} \sim 6$, but they couple antiferromagnetically across the planar interface with an exchange J_I comparable in magnitude (but opposite in sign) to J_{Fe}. Thus there are competing exchange interactions along with other competing effects if a magnetic field is applied. At low values of magnetic field, applied parallel to the interface, the magnetizations in the two materials are uniform and oppositely directed. At larger field values, however, it was shown that the Fe and Gd spins rotate in the plane of the interface, giving a 'twisted' state for the magnetizations that has some broad similarities to the spin-flop phase mentioned earlier in simple antiferromagnets.

Generalizing next to a periodic superlattice with multiple interfaces, we consider a discrete-lattice model where alternating media A and B are both sc ferromagnets (assumed for simplicity to have the same lattice constant a), as depicted in figure 3.9. The magnetization directions in each material are parallel to the (010) interfaces. The nearest-neighbour exchange interactions and spin quantum numbers are J_A and

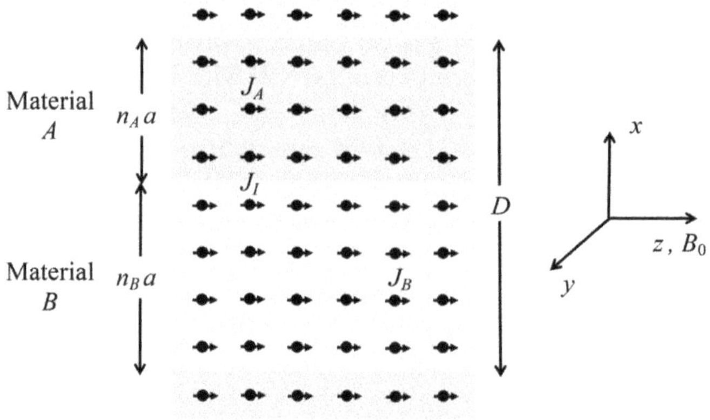

Figure 3.9. Schematic model for an exchange-dominated magnetic superlattice of alternating layers of two ferromagnetic materials A and B with nearest-neighbour exchange interactions J_A and J_B, respectively, and interface exchange J_I. The layers are stacked in the x-direction, and the periodic length D is given by $D = (n_A + n_B)a$ where a is the lattice constant.

S_A, respectively, in material A, and likewise J_B and S_B in material B. Nearest neighbours directly across an interface are assumed to be coupled by an exchange interaction J_I.

The calculations for an infinitely extended superlattice can now proceed using transfer-matrix methods by direct analogy with the cases described in section 1.5 (for photonics bands in a bulk-slab model) and section 2.4 (for acoustic phonons or elastic waves in superlattices). The equations of motion for the spin operators S_j^+ at each atomic site j are written down and then simplified using the random-phase approximation (RPA) just as in section 3.2.1. Expressions for each spin operator are written in terms of forward- and backward-travelling waves (in the x-direction here) by analogy with equations (1.53) and (1.55), and a 2×2 matrix representation can be introduced. The expressions become quite complicated, even for relatively modest values of n_A and n_B. Details can be found in, e.g. [69, 71]. The general behaviour can be anticipated by analogy with the folding of acoustic phonon modes in chapter 2 (see figure 2.13 and discussion in the text in section 2.5.1). Here the bulk SWs associated with the constituent films are folded in the reduced Brillouin zone with its edge occurring at $Q = \pi/D$.

For waves propagating in the direction perpendicular to the interface, the modes of the individual films have frequencies extending from $g\mu_B B_0$ to $g\mu_B B_0 + 4S_A J_A$ in material A, and have a similar range in material B. This formalism has been modified by Barnás [69] in two respects: first, he extended the derivations to include waves with in-plane wave vector $\mathbf{q}_{\|} = (q_y, q_z) \neq 0$ and, second, he considered semi-infinite superlattices with an outer surface at the terminated end of the stack. In the latter case there is the possibility of surface SW modes of the superlattice (with an envelope function decaying with distance from the external surface). This mode may

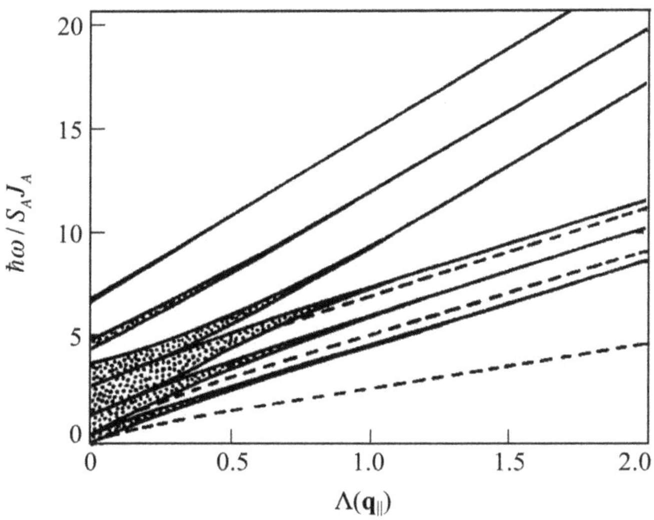

Figure 3.10. Bulk (shaded regions) and surface (dashed lines) exchange-dominated SWs for a semi-infinite superlattice of alternating ferromagnetic materials A and B with nearest-neighbour exchange interactions. The mode frequency (as the dimensionless quantity $\hbar\omega/S_AJ_A$) is plotted in terms of $\Lambda(\mathbf{q}_{\parallel}) \equiv [2 - \cos(q_y a) - \cos(q_z a)]/2$. The assumed parameters are $n_A = n_B = 3$, $J_B/J_A = 2$, $J_I/J_A = 1.4$, $S_B = S_A$ and $B_0 = 0$. There is a B layer assumed at the surface with $J_{BS} = \frac{1}{2}J_A$ for the surface exchange. (Reproduced with permission from [69]. Copyright 1994 World Scientific.)

occur when the appropriate modification to Bloch's theorem is made to allow for localized modes. A calculation for the dispersion relations of the bands of superlattice bulk SWs (shaded regions) and discrete surface SWs (dashed lines) is given in figure 3.10 for a semi-infinite superlattice. The existence (or otherwise) of localized surface modes is controlled, in part, through a modification to the exchange interaction at the external surface of the superlattice.

Various extensions of the work for the exchange-dominated superlattices have been made. For example, in some systems there may be a more complicated magnetic ground states for the spin ordering than was considered in the above case. Thus, Hinchey and Mills [73] considered superlattices with alternating ferromagnetic and antiferromagnetic films, while Camley and Tilley [74] analysed Fe/Gd superlattices where the spins in successive layers have their equilibrium orientations rotated in the plane parallel to the interfaces, leading to a 'twisted phase' of the superlattice (generalizing [72]). Other calculations using transfer-matrix methods have been made for the exchange-dominated SWs in superlattices and MCs when a quasiperiodic, instead of periodic, growth rule is followed for the layers (see e.g. [75]). On the experimental side, inelastic neutron scattering was used by Schreyer *et al* [76] to probe the SWs in short-period Dy/Y magnetic superlattices. Also, as mentioned, we will return to this topic of SWs in magnetic superlattices and MCs in more detail in chapter 4 using other models of magnetic behaviour.

References

[1] Gurevich A G and Melkov G A 1996 *Magnetization Oscillations and Waves* (Boca Raton, FL: CRC Press)

[2] White R M 2007 *Quantum Theory of Magnetism: Magnetic Properties of Materials* 3rd edn (Heidelberg: Springer)

[3] Stancil D D and Prabhakar A 2009 *Spin Waves: Theory and Application* (Heidelberg: Springer)

[4] Hillebrands B and Ounadjela K (ed) 2001 *Spin Dynamics in Confined Magnetic Structures* (Berlin: Springer)

[5] Hillebrands B and Ounadjela K (ed) 2003 *Spin Dynamics in Confined Magnetic Structures II* (Berlin: Springer)

[6] Hillebrands B and Thiaville A (ed) 2006 *Spin Dynamics in Confined Magnetic Structures III* (Berlin: Springer)

[7] Barman A and Haldar A 2014 *Solid State Phys.* **65** 1

[8] Demokritov S O (ed) 2017 *Spin Wave Confinement: Propagating Waves* 2nd edn (Singapore: Pan Stanford)

[9] Gubbiotti G (ed) 2019 *Three-Dimensional Magnonics* (Singapore: Jenny Stanford)

[10] Cohen M L and Louie S G S 2016 *Fundamentals of Condensed Matter Physics* (Cambridge: Cambridge University Press)

[11] Ketterson J B 2016 *The Physics of Solids* (Oxford: Oxford University Press)

[12] Snoke D W 2020 *Solid State Physics: Essential Concepts* 2nd edn (Cambridge: Cambridge University Press)

[13] Bloch F 1930 *Z. Phys.* **61** 206

[14] Griffiths D J and Schroeter 2018 *Introduction to Quantum Mechanics* 3rd edn (Cambridge: Cambridge University Press)

[15] Sakurai J J and Napolitano J 2020 *Modern Quantum Mechanics* 3rd edn (Cambridge: Cambridge University Press)

[16] Nolting W 2009 *Fundamentals of Many-Body Physics: Principles and Methods* (Berlin: Springer)

[17] Coleman P 2015 *Introduction to Many-Body Physics* (Cambridge: Cambridge University Press)

[18] Cottam M G and Haghshenasfard Z 2020 *Many-Body Theory of Condensed Matter Systems: An Introductory Course* (Cambridge: Cambridge University Press)

[19] Cottam M G and Kontos D E 1980 *J. Phys.* C **13** 2945

[20] Costa Filho R N, Cottam M G and Farias G A 2000 *Phys. Rev.* B **62** 6545

[21] Cottam M G and Tilley D R 2005 *Introduction to Surface and Superlattice Excitations* 2nd edn (Bristol: IOP Publishing)

[22] Rado G T and Weertman J R 1959 *J. Phys. Chem. Solids* **11** 315

[23] Grünberg P, Cottam M G, Vach W, Mayr C M and Camley R E 1982 *J. Appl. Phys.* **53** 2078

[24] Wolfram T and Dewames R E 1972 *Prog. Surf. Sci.* **2** 233

[25] Mills D L 1984 *Surface Excitations* ed V M Agranovich and R Loudon (Amsterdam: North-Holland) 379

[26] Kittel C 2004 *Introduction to Solid State Physics* 8th edn (New York: Wiley)

[27] Mills D L and Saslow W M 1968 *Phys. Rev.* **171** 488

[28] Cottam M G and Lockwood D J 1986 *Light Scattering in Magnetic Solids* (New York: Wiley)

[29] Mermin N D and Wagner H 1966 *Phys. Rev. Lett.* **17** 1133
[30] Wildes A R, Roessli B, Lebech B and Godfrey K W 1998 *J. Phys.: Condens. Matter* **10** 6417
[31] Goossens D J 2010 *Eur. Phys J.* B **8** 305
[32] Lee S, Choi K-Y, Lee S, Park B H and Park J-G 2016 *APL Mater.* **4** 086108
[33] Lee J-U, Lee S, Ryoo J H, Kang S, Kim T Y, Kim P, Park C-H, Park J-G and Cheong H 2016 *Nano Lett.* **16** 7433
[34] Burch K S, Mandrus D and Park J G 2018 *Nature* **563** 47
[35] Gong C and Zhang X 2019 *Science* **363** 706
[36] Wang M C *et al* 2020 *Ann. Phys. (Berlin)* **532** 1900452
[37] Blei M, Lado J L, Song Q, Dey D, Erten O, Pardo V, Comin R, Tongay S and Botana A S 2021 *Appl. Phys. Rev.* **8** 021301
[38] Jin W *et al* 2018 *Nat. Commun.* **9** 5122
[39] Kapoor L N, Mandal S, Adak P C, Patankar M, Manni S, Thamizhavel A and Deshmukh M M 2021 *Adv. Mater.* **33** 2005105
[40] Zhang X X, Li L, Weber D, Goldberger J, Mak K F and Shan J 2020 *Nat. Mater.* **19** 838
[41] Cenker J *et al* 2021 *Nat. Phys.* **17** 20
[42] Kim H H *et al* 2019 *Proc. Natl Acad. Sci.* **116** 11131
[43] Mkhitaryan V V and Ke L 2021 *Phys. Rev.* B **104** 064435
[44] Lockwood D J, Johnstone I W, Labbe H J and Briat B 1983 *J. Phys. C: Solid State Phys.* **16** 6451
[45] Mena M *et al* 2020 *Phys. Rev. Lett.* **124** 257201
[46] Hänni N P *et al* 2021 *Phys. Rev.* B **103** 094424
[47] Cottam M G and Lockwood D J 2022 *Phys. Rev.* B **105** 064411
[48] Villain J 1975 *Physica B+C* **79** 1
[49] Shiba H 1980 *Prog. Theor. Phys.* **64** 466
[50] Matsubara F, Inawashiro S and Ohhara H 1991 *J. Phys.: Condens. Matter* **3** 1815
[51] van Duijn J, Gaulin B D, Lumsden M A, Castellan J P and Buyers W J L 2004 *Phys. Rev. Lett.* **92** 077202
[52] Hathaway K B 1994 *Ultrathin Magnetic Structures II* ed B Heinrich and J A C Bland (Berlin: Springer) p 45
[53] Grünberg P, Schreiber R, Pang Y, Brodsky M B and Sowers H 1986 *Phys. Rev. Lett.* **57** 2442
[54] Baibich M N, Broto J M, Fert A, Nguyen Van Dau F, Petroff F, Etienne P, Creuzet G, Friederich A and Chazelas J 1988 *Phys. Rev. Lett.* **61** 2472
[55] Parkin S S P, More N and Roche K P 1990 *Phys. Rev. Lett.* **64** 2304
[56] Heinrich B and Bland J A C (ed) 1994 *Ultrathin Magnetic Structures II* (Berlin: Springer)
[57] Hartmann U (ed) 2000 *Magnetic Multilayers and Giant Magnetoresistance* (Berlin: Springer)
[58] Rührig M, Schäfer R, Hubert A, Mosler R, Wolf J A, Demokritov S and Grünberg P 1991 *Phys. Stat. Solidi* a **125** 635
[59] Slonczewski J C 1993 *J. Appl. Phys.* **73** 5957
[60] Dzyaloshinskii I E 1958 *J. Phys. Chem. Solids* **4** 241
[61] Moriya T 1960 *Phys. Rev.* **160** 91
[62] Melcher R L 1973 *Phys. Rev. Lett.* **30** 125
[63] Moon J H, Seo S M, Lee K J, Kim K W, Ryu J, Lee H W, McMichael R D and Stiles M D 2013 *Phys. Rev.* B **88** 184404
[64] Gallardo R A, Cortés-Ortuño D, Troncoso R E and Landeros P 2019 *Three-Dimensional Magnonics* ed G Gubbiotti (Singapore: Jenny Stanford) p 121

[65] Ghader D and Khater A 2019 *Nat. Sci. Rep.* **9** 15220
[66] Camley R E 1987 *Surf. Sci. Rep.* **7** 103
[67] Grünberg P 1989 *Light Scattering in Solids V* ed M Cardona and G Güntherodt (Berlin: Springer) p 303
[68] Mills D L 2000 *Light Scattering in Solids V* ed M Cardona and G Güntherodt (Berlin: Springer) 13
[69] Barnás J 1994 *Linear and Nonlinear Spin Waves in Magnetic Films and Superlattices* ed M G Cottam (Singapore: World Scientific) p 157
[70] Hillebrands B 2000 *Light Scattering in Solids VII* ed M Cardona and G Güntherodt (Berlin: Springer) p 174
[71] Albuquerque E L and Cottam M G 2003 *Phys. Rep.* **376** 225
[72] Camley R E 1987 *Phys. Rev.* B **35** 3608
[73] Hinchey L L and Mills D L 1986 *Phys. Rev.* B **33** 3329
[74] Camley R E and Tilley D R 1988 *Phys. Rev.* B **37** 3413
[75] Costa C H O and Vasconcelos M S 2013 *J. Phys.: Condens. Matter* **25** 266002
[76] Schreyer A, Schmitte T, Siebrecht R, Bödecker P, Zabel H, Lee S H, Erwin R W, Majkrzak C F, Kwo J and Hong M 2000 *J. Appl. Phys.* **87** 5443

Chapter 4

Dipole-exchange magnetic excitations

In this chapter we continue with considering the magnetic excitations (spin waves or magnons) associated with the fluctuating magnetization in ordered magnetic materials. In contrast with chapter 3, where we focused on materials and/or wave-vector regimes where the short-range exchange interactions dominated, we now incorporate the novel effects due to the long-range magnetic dipole–dipole interactions. Previously, in section 3.1, we discussed in general terms the relative importance of the dynamical effects due to the quantum-mechanical exchange interactions on the one hand and the magnetic dipole–dipole interactions on the other, and different regimes of magnetic behaviour were identified. We concluded that, among other factors, the role of the wave vector of the excitations was important, as was illustrated in figure 3.1. In this chapter we shall be concerned with the so-called dipole-exchange region, where dipolar and exchange effects may play a comparable role in the dynamics, and with the magnetostatic case of the dipole-dominated region (for which the dynamical effects of the exchange are neglected). Eventually, at even smaller wave vectors, the magnetostatic approximation breaks down and we will have the full electromagnetic case where retardation effects have to be included, as discussed for magnetic and nonmagnetic excitations in chapter 6 on polaritons.

As general references for spin waves (SWs) in the context of the magnetic dipole–dipole interactions, we mention the textbooks by Gurevich and Melkov [1] and Stancil and Prabhakar [2]. Some additional references that place an emphasis on the surface and/or interface properties of the magnetic modes are [3–8], among others.

4.1 Dipole-dominated spin waves in films

We first consider the magnetic excitations for ferromagnetic and antiferromagnetic thin films in the wave-vector regime associated with the dipole-dominated SWs or

doi:10.1088/978-0-7503-3903-2ch4

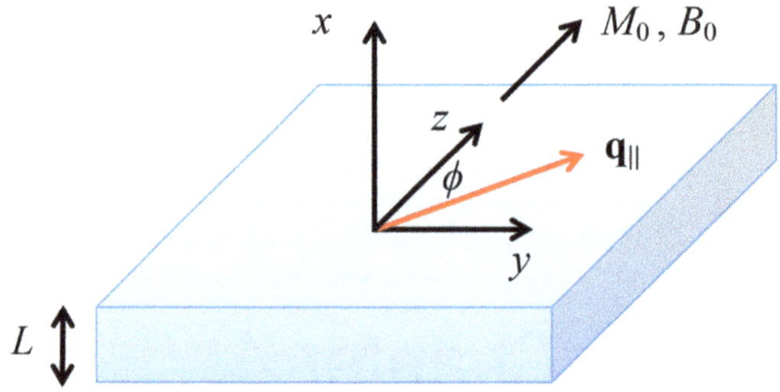

Figure 4.1. Assumed geometry and choice of coordinate axes for the dipole-dominated (magnetostatic) and dipole-exchange spin waves in a ferromagnetic film.

magnetostatic modes (see figure 3.1). The more complicated 'mixed' region of dipole-exchange waves, where both dipolar and exchange interactions influence the magnetization dynamics to a comparable extent, will be discussed in section 4.2.

The usual configuration for considering localized dipole-dominated waves in a ferromagnetic film is shown in figure 4.1, where the static magnetization M_0 is along the z-direction in the yz-plane parallel to the film surfaces. Since the relevant wavelengths are large compared to any lattice constant, a macroscopic or continuum approach is usually followed. In particular, this means that the dipole–dipole interactions can be treated in terms of Maxwell's equations of electromagnetism for the field variables. The first detailed treatment for a ferromagnetic thin film or slab with planar surfaces was due to Damon and Eshbach [9]. They predicted a magnetostatic surface wave with unusual propagation characteristics: this wave is now generally known as the Damon–Eshbach (or DE) wave (see e.g. [1, 3–5] for reviews). After first describing the ferromagnetic case, we will give an account of the analogous magnetostatic theory for antiferromagnets.

4.1.1 The case of ferromagnetic film

As a preliminary step to the mode calculation we need to need to establish a dynamic magnetic susceptibility relationship between the fluctuating magnetization $\mathbf{m}(\mathbf{r})$ and the fluctuating field $\mathbf{h}(\mathbf{r})$. This can be achieved by referring back to the vector equation (3.20), which was deduced using the torque equation of motion for the total magnetization. Suppose we now write it out in component form. Within the linearization approximation used for its derivation, this equation implies $m_z = 0$ while the x- and y-components of the fluctuating fields are related by

$$\begin{pmatrix} m_x \\ m_y \end{pmatrix} = \begin{pmatrix} \chi_a & \mathrm{i}\chi_b \\ -\mathrm{i}\chi_b & \chi_a \end{pmatrix} \begin{pmatrix} h_x \\ h_y \end{pmatrix}. \tag{4.1}$$

Here we have introduced the magnetic susceptibility factors

$$\chi_a = \frac{\omega_M \omega_0}{\omega_0^2 - \omega^2}, \qquad \chi_b = \frac{\omega_M \omega}{\omega_0^2 - \omega^2}, \tag{4.2}$$

where angular frequencies related to the applied field and static magnetization have been defined by

$$\omega_0 = \gamma B_0, \qquad \omega_M = \gamma \mu_0 M_0. \tag{4.3}$$

We note that both of the susceptibility terms in equation (4.2) have poles for the frequency at the ferromagnetic resonance (FMR) frequency ω_0. The 2×2 matrix appearing in equation (4.1) is often referred to as the *gyromagnetic susceptibility tensor* (or matrix).

The fundamental relationships between $\mathbf{m}(\mathbf{r})$ and $\mathbf{h}(\mathbf{r})$, when exchange interactions are ignored, is provided by Maxwell's equations. If the electromagnetic retardation effects are negligible, as assumed here, the relevant equations for these field variables are [10, 11]

$$\nabla \times \mathbf{m}(\mathbf{r}) = 0, \qquad \nabla \cdot [\mathbf{h}(\mathbf{r}) + \mathbf{m}(\mathbf{r})] = 0. \tag{4.4}$$

These are the *magnetostatic equations*, the first of which means that a magnetostatic scalar potential $\psi(\mathbf{r})$ can be introduced with the property that

$$\mathbf{h}(\mathbf{r}) = \nabla\psi(\mathbf{r}). \tag{4.5}$$

Then, using equation (4.1) and the second condition in equation (4.4), we have at all position vectors within the magnetic material

$$(1 + \chi_a)\left(\frac{\partial^2\psi}{\partial x^2} + \frac{\partial^2\psi}{\partial y^2}\right) + \frac{\partial^2\psi}{\partial z^2} = 0. \tag{4.6}$$

Outside the magnetic material (where $\chi_a = 0$) we have simply $\nabla^2\psi = 0$.

Assuming now that the magnetic film occupies the region $0 < x < -L$ and a nonmagnetic material fills the regions outside, we seek solutions for $\psi(\mathbf{r})$ that have the form $\psi_1(x)\exp(\mathrm{i}\mathbf{q}_\parallel \cdot \mathbf{r}_\parallel)$ in accordance with Bloch's theorem. Here we have denoted $\mathbf{r}_\parallel = (y, z)$ and $\mathbf{q}_\parallel = (q_y, q_z) = q_\parallel(\sin\phi, \cos\phi)$ for the propagation angle ϕ as indicated in figure 4.1. The general solutions for the x-dependence in the three spatial regions (choosing the terms that vanish at $x = \pm\infty$ as required) are

$$\psi_1(x) = \begin{cases} a_1 \exp(-q_\parallel x) & \text{for } x > 0, \\ a_2 \exp(\mathrm{i}q_x x) + a_3 \exp(-\mathrm{i}q_x x) & \text{for } 0 > x > -L, \\ a_4 \exp(q_\parallel x) & \text{for } x < -L. \end{cases} \tag{4.7}$$

By using equation (4.6) inside the film it is seen that the quantity q_x, which may be real or imaginary, must satisfy

$$(1 + \chi_a)(q_x^2 + q_\parallel^2) - \chi_a q_z^2 = 0. \tag{4.8}$$

The amplitudes in equation (4.7) can now be found by applying the usual electromagnetic boundary conditions at the film surfaces $x = 0$ and $x = -L$. This step will determine the physical nature of the modes as bulk-like or localized at the surface(s). In the magnetostatic regime the boundary conditions are, first, that ψ

must be continuous across the film surfaces and, second, that $(h_x + m_x)$ must be continuous across the film surfaces. The four conditions allow us to solve for the amplitudes and hence to obtain a consistency condition that takes the form

$$q_\parallel^2 + 2q_\parallel q_x(1 + \chi_a)\cot(q_x L) - q_x^2(1 + \chi_a)^2 - q_y^2\chi_b^2 = 0. \tag{4.9}$$

Taken together with equation (4.8) for q_x, the above result provides an implicit dispersion relation for the modes.

The results are particularly interesting, as well as having a simple form, for the case when the propagation angle $\phi = \pi/2$ relative to the magnetization direction, giving $q_z = 0$. This is known as the *Voigt configuration*. When $q_z = 0$ it follows from equation (4.6) that either we have $(1 + \chi_a) = 0$ or we have $(q_x^2 + q_\parallel^2) = 0$. The first condition holds for all real values of q_x provided that $\chi_a = -1$, which gives solutions $\omega = \pm\omega_B$ where

$$\omega_B(q_x, \mathbf{q}_\parallel) = [\omega_0(\omega_0 + \omega_M)]^{1/2}. \tag{4.10}$$

This equation gives the frequency of the bulk magnetostatic modes in the Voigt configuration, and they are seen to be independent of the wave-vector components. The second condition obtained above corresponds to q_x being imaginary with $q_x = \pm \mathrm{i}q_\parallel$. From the form of the solution for the magnetostatic potential given by equation (4.7) in the region $0 > x > -L$, this case must correspond to a localized mode near one or other of the surfaces. The solution for the frequency is obtained as $\omega = \omega_S(\mathbf{q}_\parallel)$ where

$$\omega_S(\mathbf{q}_\parallel) = \left[(\omega_0 + \omega_M/2)^2 - (\omega_M/2)^2\exp(-2q_\parallel L)\right]^{1/2}. \tag{4.11}$$

This is the Damon–Eshbach (DE) mode, for which the frequency increases monotonically as a function of $q_\parallel L$. Its limiting values correspond to

$$\omega_S(\mathbf{q}_\parallel) = \begin{cases} [\omega_0(\omega_0 + \omega_M)]^{1/2} & (q_\parallel L \ll 1), \\ (\omega_0 + \omega_M/2) & (q_\parallel L \gg 1). \end{cases} \tag{4.12}$$

Except at $q_\parallel = 0$, it follows that we have $\omega_S > \omega_B$ corresponding to an optical-type surface branch, unlike the situation in the exchange-dominated region. If $q_y > 0$, a calculation of the mode amplitudes shows that the two cases of $q_x = -\mathrm{i}q_\parallel$ or $+\mathrm{i}q_\parallel$ correspond to the surface magnetostatic wave being localized near the upper $(x = 0)$ or the lower $(x = -L)$ surface, respectively. It is vice-versa if $q_y < 0$, but the main point to emphasize is that the surface wave is localized near just one of the two surfaces, instead of near both surfaces as in the Heisenberg exchange-dominated case. This remarkable property is another example of *nonreciprocal* wave propagation. It arises here due to the dipole–dipole interactions, contrasting with the case in section 3.5.2 involving antisymmetric DM exchange interactions.

The dispersion relations represented by equations (4.11) and (4.12) in the Voigt geometry have been amply verified using BLS experiments by several groups.

For example, the dispersion relation plot (the mode frequency versus L at a constant $q_{\|}$) for an Fe films discussed earlier in figure 3.4 includes a surface branch labelled ω_S. It is seen that there is a good fit between experiment and theory. Further comparisons for a thick Ni and Fe films ($q_{\|}L \gg 1$) were made by Sandercock and Wettling [12]. They found that, when the bulk and surface-mode frequencies were plotted as a function of the applied magnetic field B_0, there was excellent agreement with the theory just described. The nonreciprocal nature of the surface (DE) mode in ferromagnetic films is also evident from BLS experiments in terms of an asymmetry between the Stokes and anti-Stokes parts of the spectra. This behaviour is illustrated in figure 4.2, where we refer in particular to the spectrum for angle $\phi = 90°$ (the Voigt geometry).

The theory may also be extended to apply to other values of the propagation angle, apart from the $\phi = 90°$ (or $\pi/2$) case considered in the preceding analysis. In particular, when $\phi = 0$ which corresponds to the propagation being parallel or antiparallel to the magnetization direction, it can be shown that there are solutions of equations (4.6) and (4.9) only when q_x is real. Hence there are no surface modes in this case, and the equations for the bulk modes can be separated into two types for modes that are either symmetric or antisymmetric with respect to the middle of the film:

$$
\begin{aligned}
q_x \cot(q_x L/2) &= -\, q_{\|} \quad \text{(symmetric modes)},\\
q_x \tan(q_x L/2) &= q_{\|} \quad \text{(antisymmetric modes)}.
\end{aligned}
\tag{4.13}
$$

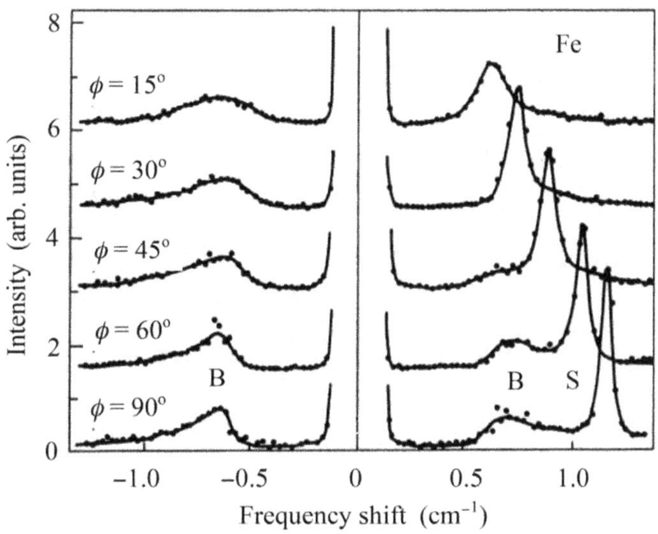

Figure 4.2. Spectra for Brillouin light scattering from the surface of a thick Fe sample with applied field B_0 = 0.19 T for different propagation angles ϕ. The bulk (B) and surface (S) SW peaks are shown. (Reproduced with permission from [12]. Copyright 1979 American Institute of Physics.)

The discrete real solutions for q_x correspond to a quantization of the bulk modes, whose frequencies are now given by

$$\omega_B(q_x, \mathbf{q}_{\|}) = \left[\omega_0\left(\omega_0 + \frac{\omega_M q_x^2}{(q_x^2 + q_{\|}^2)}\right)\right]^{1/2}. \tag{4.14}$$

It is clear from the above expression that the frequency $\omega_B(q_x, \mathbf{q}_{\|})$ decreases with increasing $q_{\|}$, and so the modes have a negative group velocity $\partial\omega_B/\partial q_{\|}$ parallel to the surface. Hence they are often called *magnetostatic backward bulk modes*. In contrast the group velocity of the bulk modes in the Voigt configuration is zero.

The behaviour for intermediate values of ϕ between 0 and $\pi/2$ is more complicated than for the special cases just described. The main property is that a DE surface mode still exists, having a modified frequency, provided ϕ exceeds a critical value ϕ_c that is given by

$$\phi_c = \left[\frac{\omega_0}{(\omega_0 + \omega_M)}\right]^{1/2}. \tag{4.15}$$

In the physical range for ϕ, the frequency of the modified DE mode in a thick film (such that $q_{\|}L \gg 1$) is approximately

$$\omega_S(\mathbf{q}_{\|}) = \frac{1}{2}\left\{\frac{\omega_0}{\sin\phi} + (\omega_0 + \omega_M)\sin\phi\right\}, \tag{4.16}$$

which steadily decreases in value as ϕ is reduced from $\pi/2$ to the critical value. Some BLS spectra are shown in figure 4.2 for several different values of the propagation angle ϕ. When $\phi = 90°$ the surface peak S is a strong feature on one side of the spectrum (as a consequence of nonreciprocity), but as the angle is reduced the surface peak shifts to lower frequency, weakens, and eventually disappears in accordance with theory. By using equation (4.15) and the parameters for Fe, we obtain a reasonable estimate of $\phi_c \simeq 20°$.

In other BLS studies, the integrated intensity (which is proportional to the area under the scattering peak) of the DE surface mode in thin ferromagnetic films was investigated. Figure 4.3 shows the theoretical and experimental BLS integrated intensity for the surface SW mode as a function of L for a ferromagnetic film with two different choices of substrate material. Although both substrates are nonmagnetic they are found to have a large influence on the intensity due to modifications of the SW reflection properties at the Fe/substrate interface. The theory curves were obtained from a magnetostatic Green's function (GF) analysis [13].

The case of perpendicular magnetization

If similar calculations are made for the magnetostatic modes of a ferromagnetic film with its magnetization perpendicular to the surface [3, 5], it is concluded that there are no surface modes and that the bulk-mode frequencies satisfy

$$\omega_B(\mathbf{q}_{\|}, q_z) = \left[\omega_0\left(\omega_0 + \omega_M - \frac{\omega_M q_z^2}{(q_{\|}^2 + q_z^2)}\right)\right]^{1/2}, \tag{4.17}$$

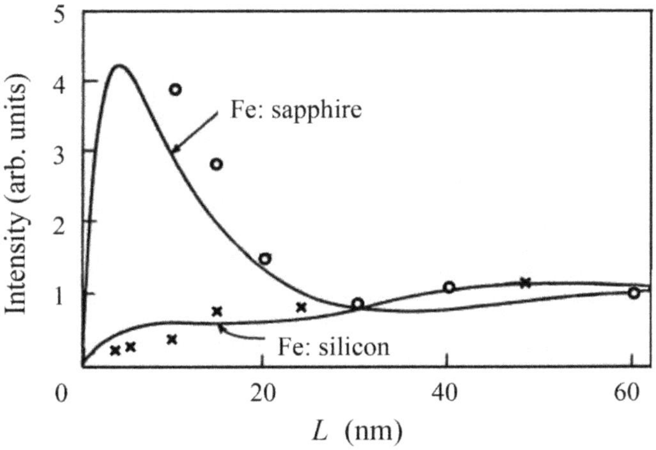

Figure 4.3. BLS intensity for the anti-Stokes scattering from surface SWs in Fe films as a function of film thickness L for two choices of substrate material as indicated. The solid lines are theory and the experimental points (circles and crosses) are due to P Grünberg and co-workers. (Reproduced with permission from [13].)

where the definition of ω_0 is modified to include the static demagnetizing field. Note that the z-axis is now perpendicular to the surfaces (since this is the direction of M_0) and so $\mathbf{q}_{\|} = (q_x, q_y)$. The quantized values for the perpendicular wave-vector component q_z are obtained from

$$\begin{aligned} q_z \tan(q_L/2) &= q_{\|} \quad \text{(symmetric modes)}, \\ q_z \cot(q_z L/2) &= -q_{\|} \quad \text{(antisymmetric modes)}. \end{aligned} \tag{4.18}$$

The group velocity $\partial\omega_B/\partial q_{\|}$ parallel to the surface is positive for these modes, and they are often called *magnetostatic forward bulk modes.*

The case of a double-layer ferromagnetic system

For a double-layer ferromagnetic system we consider the same geometry as in figure 3.7, but with two ferromagnetic films that are now dipolar coupled to one another across a nonmagnetic spacer. For simplicity, we take the two ferromagnets to be of the same material and to have their static magnetizations parallel to each other and to the surfaces. Also we assume the special case of the Voigt configuration mentioned earlier. The surfaces of the two films can be taken at $x = 0$ and $x = -L_1$ for the top film and at $x = -(L_1 + d)$ and $x = -(L_1 + L_2 + d)$ for the lower film. The magnetostatic scalar potential satisfies the same equations as before inside and outside the ferromagnets. Thus a set of equations for the x-dependence of this potential can be written down by analogy with equation (4.7) and the magnetostatic boundary conditions can be applied (see [14] for details). The mode frequencies can eventually be written in the form

$$\omega = [(\omega_0 + \omega_M/2)^2 - (\omega_M/2)^2\alpha)]^{1/2}. \tag{4.19}$$

Here we have $\alpha = 1$ for a bulk mode, giving the same result as in equation (4.10), whereas in the case of a surface mode α satisfies the following quadratic equation:

$$\begin{aligned}&[\alpha - \exp(-2q_{\|}L_1)][\alpha - \exp(-2q_{\|}L_2)] \\ &\quad - \alpha \exp(-2q_{\|}d)[1 - \exp(-2q_{\|}L_1)][1 - \exp(-2q_{\|}L_2)] = 0.\end{aligned} \tag{4.20}$$

The two solutions for α correspond to the coupled surface magnetostatic modes of the double ferromagnetic film. It is easily verified that in the limit of large separation ($d \to \infty$) the solutions become $\alpha = \exp(-2q_{\|}L_1)$ and $\alpha = \exp(-2q_{\|}L_2)$, as expected, corresponding to surface modes of the single isolated films. Extensions of this theory have been made to cases where the magnetization directions of the films are not parallel [15] and where there is a direct interface (the case of $d = 0$) between two different magnetic materials [16].

On the experimental side, BLS measurements were reported by Grünberg *et al* [17] for Fe films separated by a quartz spacer of variable thickness up to about 100 nm. The observed behaviour for the coupled modes was found to be in good agreement with the type of theoretical analysis presented here, where the coupling is entirely due to the long-range magnetic dipole–dipole interactions across the spacer. As discussed in chapter 3, however, for certain metallic spacers there may be further types of coupling due to RKKY exchange interactions, sometimes occurring together with biquadratic exchange. We mentioned earlier that this phenomenon is extensively utilized in devices that depend on giant magnetoresistance (GMR). The coupled SW modes can be studied experimentally using BLS and/or FMR techniques (see e.g. [18, 19] for studies of Fe/Cu/Fe and $Ni_{80}Fe_{20}$/Ru/$Ni_{80}Fe_{20}$ magnetic bilayer films).

4.1.2 The case of antiferromagnetic film

To enable an extension of the magnetostatic theory to be made to antiferromagnets, it is first necessary to obtain the equivalent expressions to equations (4.1) and (4.2) for the gyromagnetic susceptibility. The derivation for a two-sublattice antiferromagnet is given, for example, in [4, 20]; it is a direct extension of what is done for a ferromagnet, starting from the torque equation of motion given in equation (3.17). Briefly, the torque equation for each sublattice individually can be written down and linearized, as before. A substitution for the total effective magnetization and field can be made, as in equations (3.18) and (3.19), including here the static exchange and anisotropy fields. Then the total fluctuating magnetizations from each sublattice are added together and their response to a fluctuating field is deduced. Eventually, it is found that equation (4.1) still applies as a dynamic susceptibility result, but the expressions for χ_a and χ_b are modified to

$$\chi_a = \frac{1}{2}(\chi^+ + \chi^-), \qquad \chi_b = \frac{1}{2}(\chi^+ - \chi^-), \tag{4.21}$$

where

$$\chi^{\pm} = \frac{2\omega_A\omega_M}{\omega_A(2\omega_E + \omega_A) - (\omega \mp \omega_0)^2}. \tag{4.22}$$

The frequencies ω_A, ω_E and ω were all defined previously in section 3.3, and the definition of ω_M is unchanged from equation (4.3) provided we now take M_0 to refer to the sublattice magnetization. Note that it is only the *static* part of the exchange that is being included through ω_E. We see that, in the absence of a static applied field ($\omega_0 = 0$), we will have $\chi_b = 0$, so the susceptibility matrix then takes a diagonal form. Also, we note that the poles of χ_a occur at the AFMR frequency, which is given by $[\omega_A(2\omega_E + \omega_A)]^{1/2}$.

Next, the magnetostatic theory for an antiferromagnetic film can be carried out in a similar manner to the ferromagnetic case, provided equations (4.21) and (4.22) are used and the two-sublattice structure is taken into account [21, 22]. We will quote the results only for the special case of a semi-infinite antiferromagnet in zero applied field. Then the two branches for the bulk modes are at frequencies $\omega_B^{\pm}(\mathbf{q})$ where

$$\begin{aligned} \omega_B^{+}(\mathbf{q}) &= [\omega_A(2\omega_E + \omega_A) + 2\omega_A\omega_M \sin^2\theta]^{1/2}, \\ \omega_B^{-}(\mathbf{q}) &= [\omega_A(2\omega_E + \omega_A)]^{1/2}. \end{aligned} \tag{4.23}$$

Here θ is the angle between the 3D wave vector $\mathbf{q} = (q_x, \mathbf{q}_{\|})$ and the in-plane z-axis. At first sight it is surprising that the two bulk magnetostatic frequencies are different, even in zero applied field, unlike the case for the exchange-dominated bulk magnons in antiferromagnets. This property is a consequence of including the lower-symmetry dipolar interactions, which shifts one of the branches (ω_B^{+}) to a higher frequency than that for the AFMR frequency. The frequency of the surface magnetostatic mode is found to be (see [21])

$$\omega_S(\mathbf{q}_{\|}) = \left[\omega_A(2\omega_E + \omega_A) + \frac{2\omega_A\omega_M \sin^2\phi}{1 + \sin^2\phi}\right]^{1/2}, \tag{4.24}$$

where ϕ is the angle between the 2D wave vector $\mathbf{q}_{\|}$ and the z-axis. It can easily be verified that this expression satisfies $\omega_B^{+}(\mathbf{q}) > \omega_S(\mathbf{q}_{\|}) > \omega_B^{-}(\mathbf{q})$, except when angle $\phi = 0$. The surface branch exists as a localized mode for all nonzero values of ϕ and it satisfies $\omega_S(\mathbf{q}_{\|}) = \omega_S(-\mathbf{q}_{\|})$. The latter property holds only when the applied field is zero, and it contrasts with the nonreciprocal propagation behaviour mentioned earlier for surface magnetostatic modes in ferromagnetic films.

4.2 Dipole-dominated spin waves in wires and tubes

In long wires there is still translational invariance symmetry in 1D along the length axis (taken as the z-direction), so the modes will be characterized by a longitudinal propagation wave number q_z, as well as typically by spatial variations in the transverse xy-plane. Here we will employ magnetostatic theory to consider cases where the cross section of the wire is circular, and analytic solutions of the magnetostatic equations are available. The generalization to cylindrical tubes will also be given. In many practical applications the lengths involved might be of the order of tens of micrometres and the lateral dimensions less than about 400 nm. Examples to illustrate the theory will be discussed later. The SW modes in wires with other cross sections (such as rectangular, corresponding to magnetic stripes or ribbons) will be discussed later in this chapter in the context of dipole-exchange theory.

We begin by considering the magnetostatic (dipole-dominated) theory for a long cylindrical wire of a ferromagnetic material with circular cross section of radius R. The calculations are given in detail in [23], so here we will just summarize the main steps. Taking the applied magnetic field B_0 and the magnetization M_0 to be along the z-axis, which is also the longitudinal axis of the cylinder, we introduce a scalar potential ψ as was done in section 4.1.1. This must again satisfy equation (4.6) inside the material and $\nabla^2\psi = 0$ outside in the nonmagnetic region.

We now transform from Cartesian (x, y, z) to the more convenient cylindrical polar coordinates (r, θ, z) and seek separable solutions for ψ of the form $f(r)\exp(im\theta)\exp(iq_z z)$. Here m must be an integer for ψ to be single-valued, and the z-dependence has been chosen in accordance with the 1D Bloch's theorem. It follows that the radial part $f(r)$ inside the ferromagnet (where $r < R$) must satisfy

$$(1 + \chi_a)\left(\frac{\mathrm{d}^2 f}{\mathrm{d}r^2} + \frac{1}{r}\frac{\mathrm{d}f}{\mathrm{d}r} - \frac{m^2}{r^2}f\right) - q_z^2 f = 0, \tag{4.25}$$

where χ_a is the susceptibility term given by equation (4.2). The radial solutions that are well-behaved (non-divergent) at $r = 0$ are $f(r) = a\, J_m(kr)$, where a is a constant and J_m is a Bessel function of the first kind. The quantity k, which may be complex in general, satisfies

$$[\omega_0(\omega_0 + \omega_M) - \omega^2](k^2 + q_z^2) - \omega_0\omega_M q_z^2 = 0. \tag{4.26}$$

Outside the cylinder (where $r > R$) the solution that is well-behaved as $r \to \infty$ is $f(r) = b\, K_m(kr)$ where K_m is a modified Bessel function of the second kind. The magnetostatic boundary conditions at $r = R$, which have the same form as discussed in section 4.1.1, can be used to solve for the coefficients a and b.

The results for the magnetostatic modes can be summarized as follows. First, if the quantity k in equation (4.26) is real, the expression can be re-arranged to give the continuum of bulk modes, which extend from ω_0 up to $[\omega_0(\omega_0 + \omega_M)]^{1/2}$, just as previously for a film. The quantity k now plays the role of a radial wave vector. The surface magnetostatic modes correspond to k in equation (4.26) taking discrete imaginary values that can be deduced on applying the magnetostatic boundary conditions at $r = R$. There are multiple surface branches, characterized by $m = 1, 2, \ldots$ and having complicated dispersion relations. At long wavelengths such that $|q_z|R \ll 1$ the surface-mode frequencies simplify to give [23]

$$\omega_S(q_z) = \begin{cases} \omega_0 + \frac{1}{2}\omega_M - \frac{1}{4}\omega_M(q_z R)^2[0.366 - \ln(q_z R)] & (m = 1), \\ \omega_0 + \frac{1}{2}\omega_M - \frac{1}{4}\omega_M(q_z R)^2[m^2 - 1]^{-1} & (m > 1) \end{cases} \tag{4.27}$$

We note that these surface modes all correspond to frequencies that are above the bulk-mode region. The surface modes become degenerate in frequency at

$\omega_S = \omega_0 + \frac{1}{2}\omega_M$ when $|q_z|R \to 0$. Also they have a negative group velocity at nonzero q_z since it can be shown that $\partial\omega_S/\partial q_z < 0$ for all modes.

The above magnetostatic theory may straightforwardly be extended to apply to long cylindrical tubes with inner and outer radii R_1 and R_2, respectively [24]. Again the applied magnetic field B_0 and the magnetization M_0 are taken to be along the longitudinal z-axis. Cylindrical polar coordinates (r, θ, z) are introduced as before, and separable solutions are found in the physical regions that correspond to $R_1 < r < R_2$ for the tube filled with the magnetic material and the two regions $r < R_1$ and $r > R_2$ external to the tube. The magnetostatic boundary conditions are applied at the surfaces $r = R_1$ and $r = R_2$. The bulk modes still occupy the frequency range between ω_0 to $[\omega_0(\omega_0 + \omega_M)]^{1/2}$, but the surface-mode frequencies are modified. These modes occur in a frequency range above the bulk modes, but there are now two non-degenerate branches of surface modes for each value of m, as might be expected since there are two non-equivalent surfaces. The modes for different m are no longer degenerate at zero wave vector $q_z = 0$ and satisfy $\omega_S < \omega_0 + \frac{1}{2}\omega_M$.

4.3 Dipole-exchange waves in films

The dipole-exchange region for SWs, as identified in figure 3.1, is much more challenging theoretically than the magnetic cases treated up to now, where one of the interactions (either the exchange or the dipole–dipole interactions) was dominant over the other in the dynamics. Now we allow the two interactions to be comparable. In part, the extra complexity in the theory is because of the interplay between the long-range nature of the dipole–dipole interactions and the short-range exchange, but perhaps more significantly the boundary conditions at the surfaces and interfaces are more complicated to implement.

Two approaches will be presented here, where the first is based on extensions of the macroscopic method used in section 3.2.2 and section 4.1. The second is a microscopic (or Hamiltonian-based) method in which the dipole–dipole interactions are represented by a Hamiltonian as in equation (3.1) instead of using Maxwell's equations of electromagnetism. The calculations will initially be applied to the thin-film geometry for ferromagnets (as shown in figure 4.1).

4.3.1 Macroscopic dipole-exchange theories

Here we start by summarizing the formalism as developed initially by Wolfram and Dewames (see [3]). It involves taking into account both the dipolar and exchange terms in the previous torque equation of motion for the magnetization (3.17), leading to a sixth-order differential equation for the magnetostatic scalar potential. A different macroscopic technique was later introduced by Kalinikos and Slavin (see [25]), based on a tensorial Green's function (GF) method in which the matrix elements of the dipole–dipole interaction are treated perturbatively. While both methods are quite complicated mathematically and typically require numerical evaluations, the latter approach has some advantages in yielding approximate analytical results in films with an arbitrary magnetization direction for applications to inelastic light scattering and SWR measurements. Both methods are outlined in the following text.

4.3.1.1 Wolfram–Dewames theory

The expression governing the magnetization dynamics can be formed from equations (3.17)–(3.20) by including the exchange contribution to the total **H** field, as in equation (3.23), as well as the dipolar contribution (denoted by $\mathbf{h}_d$) using the susceptibility relationship in equation (4.1). The result is that the expression used in the dipole-dominated case becomes augmented as

$$-\mathrm{i}\omega\mathbf{m}(\mathbf{r}) = \hat{z} \times [\omega_M\mathbf{h}_d - (\omega_0 - D\nabla^2)\mathbf{m}(\mathbf{r})], \tag{4.28}$$

where the SW stiffness constant D is related to the exchange constant as defined earlier in chapter 3. Following now the same procedure as for the magnetostatic case and adopting the same film geometry as in figure 4.1, we may rewrite equation (4.28) in terms of the scalar potential ψ as

$$(\hat{w}^2 + \omega_M\hat{w} - \omega^2)\left(\frac{\partial^2\psi}{\partial x^2} + \frac{\partial^2\psi}{\partial x^2}\right) + (\hat{w}^2 - \omega^2)\frac{\partial^2\psi}{\partial z^2} = 0, \tag{4.29}$$

where we define the differential operator $\hat{w}$ by $\hat{w} = \omega_0 - D\nabla^2$. The above result generalizes equation (4.6) for the purely dipolar case. It replaces a second-order differential equation by a sixth-order one, so the change is profound. In the regions outside the film, where $x > 0$ or $x < -L$, we have simply $\nabla^2\psi = 0$ as before.

On putting $\psi(\mathbf{r}) = \psi_1(x)\exp(\mathrm{i}\mathbf{q}_\parallel \cdot \mathbf{r}_\parallel)$ in accordance with Bloch's theorem, we find that the x-dependence of $\psi_1(x)$ for $x > 0$ and $x < -L$ has just the same form as in equation (4.7) whereas inside the film there is now a linear superposition of terms, giving

$$\psi_1(x) = \sum_{j=1}^{3}\left[a_{2j}\exp(\mathrm{i}q_{xj}x) + a_{3j}\exp(-\mathrm{i}q_{xj}x)\right](0 > x > -L). \tag{4.30}$$

Here the quantities q_{xj} (with $j = 1, 2, 3$) are the roots of

$$(q_x^2 + q_\parallel^2)\left[(\omega_0 + Dq_\parallel^2 + Dq_x^2)^2 + \omega_M(\omega_0 + Dq_\parallel^2 + Dq_x^2) - \omega^2\right] = 0, \tag{4.31}$$

where we have simplified the analysis by taking the special case of the Voigt geometry (with $q_z = 0$ implying $\mathbf{q}_\parallel \perp M_0$). The roots of this equation are given by

$$\begin{aligned} q_{x1}^2 &= -q_\parallel^2, \\ q_{x2}^2 &= \left[\left(\omega^2 + \frac{1}{4}\omega_M^2\right)^{1/2} - \frac{1}{2}\omega_M - \omega_0 - Dq_\parallel^2\right]/D, \\ q_{x3}^2 &= -\left[\left(\omega^2 + \frac{1}{4}\omega_M^2\right)^{1/2} + \frac{1}{2}\omega_M + \omega_0 + Dq_\parallel^2\right]/D. \end{aligned} \tag{4.32}$$

It follows that q_{x1} is pure imaginary with the same value as found for the surface magnetostatic (or DE) mode in section 4.1.1. We see that q_{x3} is also pure imaginary, corresponding to another localized mode that has no analogue in the dipole-dominated case. By contrast, q_{x2} is real for most frequencies of interest, and so it corresponds to a bulk-like mode. After some rearrangement, its frequency can be expressed as

$$\omega_B(q_{x2}, q_{\|}) = \left[\omega_0 + D(q_{x2}^2 + q_{\|}^2)\right]^{1/2} \left[\omega_0 + \omega_M + D(q_{x2}^2 + q_{\|}^2)\right]^{1/2}. \tag{4.33}$$

This expression generalizes equation (4.10) through the addition of wave-vector-dependent terms that are proportional to D.

The statistical weighting (or degree of 'mixing') of these three fundamental modes is determined by the boundary conditions, which must include the magnetostatic boundary conditions described in section 4.1.1. It readily becomes apparent that these are insufficient to provide a complete solution for the sets of a_{2j} and a_{3j} coefficients (with $j = 1, 2, 3$) appearing in equation (4.30), and so *additional boundary conditions* (or ABCs) are required. These can be deduced from the linearized torque equation of motion for the spins at the surface of the ferromagnetic film, as was done in section 3.2.2 for a Heisenberg ferromagnet (see also [1]). The important result was equation (3.30), enabling us to adopt the so-called 'exchange' boundary conditions which are written here as

$$\left[\frac{\partial m^{\mu}(\mathbf{r})}{\partial x} - \xi m^{\mu}(\mathbf{r})\right]_{x=0} = 0, \tag{4.34}$$

for the surface at $x = 0$, with $\mu = y, z$ for a transverse component of the fluctuating magnetization. There is an analogous condition that holds for the surface at $x = -L$. We note that extensions of the theory can be made to generalize the exchange boundary conditions to cases where the magnetization is not in the plane of the surfaces (see e.g. [1]). Here ξ is an effective surface anisotropy (or 'pinning') coefficient, for which a simple form was deduced in section 3.2.2. A more general discussion of pinning, including the concept of dynamic pinning, was given in [26]. We comment that the calculations presented in figure 3.4 for Fe films employed this type of macroscopic dipole-exchange theory for the surface and quantized bulk SWs and gave good agreement with BLS measurements.

The macroscopic dipole-exchange theory has been extended to antiferromagnetic films [27] by using a two-sublattice model appropriate to body-centred tetragonal (bct) materials such as MnF_2. It was found that, instead of a cubic equation in q_x^2 being required as in the ferromagnetic case, a fifth-order polynomial in q_x^2 is obtained instead. This implies a mixing of five waves, so the analysis becomes significantly more complicated.

4.3.1.2 Tensorial Green's function theory

As mentioned, this approach was pioneered by Kalinikos and Slavin [25], and is reviewed by one of these authors in [28]. The method is particularly adaptable to films in which the external magnetic field is applied in an arbitrary direction, causing the static magnetization vector to be tilted out of the plane of the surfaces. We will continue to specify the film geometry in terms of coordinates (x, y, z) as before, but note that B_0 and M_0 will no longer be parallel to one another or directed along the z-direction, in general. The in-plane wave vector $\mathbf{q}_{\|}$ for the SW propagation is taken to be along a direction defined by the coordinate $r_{\|}$ in the xz-plane. Thus we may express $\mathbf{m}(\mathbf{r}) = \mathbf{m}_1(x)\exp(\mathrm{i}\mathbf{q}_{\|} \cdot \mathbf{r}_{\|} - \mathrm{i}\omega t)$ and similarly $\mathbf{h}_d(\mathbf{r}) = \mathbf{h}_{d1}(x)\exp(\mathrm{i}\mathbf{q}_{\|} \cdot \mathbf{r}_{\|} - \mathrm{i}\omega t)$ for the dipolar part of the fluctuating field.

Then, instead of forming a sixth-order differential equation for the fluctuating magnetization (as in the Wolfram–Dewames formalism just described), the methodology is based on the relationship

$$\mathbf{h}_{d1}(x) = \int_{-L}^{0} \hat{\mathbf{G}}(x, x')\, \mathbf{m}(x')\, \mathrm{d}x', \tag{4.35}$$

where we are continuing to use x for the spatial coordinate perpendicular to the film surfaces (at $x = 0$ and $-L$, as before). The quantity $\hat{\mathbf{G}}(x, x')$ is the so-called tensorial Green's function (GF), and it has a form given explicitly in [25, 28]. The details are complicated, but the methodology enables an *approximate* dispersion equation for the dipole-exchange SWs to be obtained. For example, for the case of unpinned surface spins, meaning that the parameter ξ in equation (4.34) is zero at both film surfaces, the result is

$$\omega_{n,q_{\|}} = \left[\{\omega_0 + D(n\pi/L)^2 + Dq_{\|}^2\}\{\omega_0 + D(n\pi/L)^2 + Dq_{\|}^2 + \omega_M F_{nn}\}\right]^{1/2}. \tag{4.36}$$

Here the quantity F_{nn} has a complicated form that depends on the orientation of the film magnetization (as specified by a polar angle θ with respect to the x-axis and an azimuthal angle ϕ in the yz-plane) [25]. We may write

$$F_{nn} = (1 - P_{nn}\cos^2\phi)\sin^2\theta + P_{nn}\left\{\cos^2\theta + \frac{\omega_M}{\omega_{n,q_{\|}}}(1 - P_{nn})\sin^2\theta\sin^2\phi\right\}, \tag{4.37}$$

where $0 < P_{nn}(q_{\|}L) < 1$ is a matrix element of the dipole–dipole interaction deduced using equation (4.35) and the tensorial GF. For the lowest SW mode, which corresponds to $n = 0$, we have

$$P_{00}(q_{\|}L) = 1 - \frac{\{1 - \exp(-q_{\|}L)\}}{q_{\|}L}, \tag{4.38}$$

which has the limiting values 0 and 1 as $q_{\|}L \to 0$ and ∞, respectively. It was shown in [25] how these results can lead to surface modes (e.g. when F_{nn} exceeds unity for certain values of the angles θ and ϕ) and to the quantized, or standing, bulk modes that arise in spin wave resonance (SWR).

4.3.2 Microscopic dipole-exchange theory

While a macroscopic approach generally provides a good approximation for the dipole-exchange SWs, a microscopic (or Hamiltonian-based) approach becomes advantageous if the wave vectors are larger and/or if the film thickness is very small corresponding to just a few atomic layers. Following its initial formulation for ultrathin films by Benson and Mills [29], the technique was extended by several authors for films of ferromagnetic (see e.g. [30, 31]) and antiferromagnetic (see e.g. [32]) materials.

We start with the dipole-exchange spin Hamiltonian, which can be written as

$$\mathcal{H} = -\frac{1}{2}\sum_{i,j} J_{ij}\mathbf{S}_i \cdot \mathbf{S}_j - g\mu_B B_0 \sum_i S_i^z + \frac{1}{2}g^2\mu_B^2 \sum_{i,j}\sum_{\alpha,\beta} D_{i,j}^{\alpha,\beta} S_i^\alpha S_j^\beta, \tag{4.39}$$

where we have combined contributions from equation (3.1) with (3.3). The final term describes the dipole–dipole interactions, where α and β denote x, y or z, and the dipolar coefficient is

$$D_{i,j}^{\alpha,\beta} = \left[\left|\mathbf{r}_j - \mathbf{r}_i\right|^2 \delta_{\alpha\beta} - 3(r_j^\alpha - r_i^\alpha)(r_j^\beta - r_i^\beta)\right] / |\mathbf{r}_j - \mathbf{r}_i|^5. \tag{4.40}$$

One way to proceed would be to write down the operator equation of motion for the spin component S_j^+, as was done in equation (3.7) for the exchange-dominated case. However, we find that the right-hand side involves terms with S^- operators as well as other S^+ operators. Then, in turn, we could write down the equation of motion satisfied by S_j^-, which would involve both S^- and S^+ operators after linearization. This approach is discussed in [20]. An alternative method, which will be useful also when presenting generalizations to *nonlinear* SWs in chapter 9, is to rewrite the spin operators in terms of boson operators. This is motivated by the property that SWs at low temperatures behave like weakly interacting boson particles. The connection can be made in various formal ways, but the Holstein–Primakoff (or HP) transformation [33] is commonly used. This is defined by

$$\begin{aligned} S_j^+ &= (2S)^{1/2}\left[1 - (a_j^\dagger a_j/2S)\right]^{1/2} a_j, \\ S_j^+ &= (2S)^{1/2} a_j^\dagger \left[1 - (a_j^\dagger a_j/2S)\right]^{1/2}, \\ S_j^z &= S - a_j^\dagger a_j. \end{aligned} \tag{4.41}$$

The operators for creating and annihilating a boson at site j are written as $a_j^\dagger$ and a_j, respectively. These operators satisfy the standard quantum-mechanical commutation relations [34, 35] that (see also the analogous discussion in section 2.1.1)

$$[a_i, a_j^\dagger] = \delta_{ij}, \; [a_i^\dagger, a_j^\dagger] = [a_i, a_j] = 0. \tag{4.42}$$

It is a simple exercise, using equations (4.41) and (4.43), to check that the spin commutation relations quoted in equation (3.5) are recovered.

The HP transformation is commonly employed in an approximate form for low temperatures $T \ll T_C$ on the basis that $S^z \simeq S$ which implies that $a_j^\dagger a_j/2S \ll 1$. This property allows us to use the binomial expansion for the square roots in equation (4.41) to give

$$S_j^+ = (2S)^{1/2}\left[a_j - (a_j^\dagger a_j a_j/4S) + \cdots\right] \tag{4.43}$$

and a similar result for S_j^-. Then the dipole-exchange Hamiltonian in equation (4.39) can be expanded in products of the boson operators to yield

$$\mathcal{H} = \text{constant} + \mathcal{H}^{(2)} + \mathcal{H}^{(3)} + \mathcal{H}^{(4)} + \cdots. \tag{4.44}$$

Apart from the constant, which is a contribution to the ground state energy of the ferromagnetic film, each term $\mathcal{H}^{(m)}$ involves a product of m operators. There is no $\mathcal{H}^{(1)}$ term since it vanishes by symmetry in most crystal structures (including cubic), so the lowest-order effect is provided by $\mathcal{H}^{(2)}$ and this will be our focus in the following paragraphs. The higher-order terms, starting with $\mathcal{H}^{(3)}$ and $\mathcal{H}^{(4)}$, describe nonlinear SW processes and will be discussed in chapter 9.

If we now make a 2D Fourier transform from $a_j^\dagger$ in the site representation to $a_n^\dagger(\mathbf{q}_\parallel)$ in a mixed representation with layer number n and in-plane wave vector $\mathbf{q}_\parallel$ (just as in section 3.2.1 for spin operators), we find a result for $\mathcal{H}^{(2)}$ of the form

$$\mathcal{H}^{(2)} = \sum_{\mathbf{q}_\parallel, n, m} \left[A_{n,m}^{(2)}(\mathbf{q}_\parallel) a_n^\dagger(\mathbf{q}_\parallel) a_m(\mathbf{q}_\parallel) + \{ B_{n,m}^{(2)}(\mathbf{q}_\parallel) a_n^\dagger(\mathbf{q}_\parallel) a_m^\dagger(-\mathbf{q}_\parallel) + \text{H.c.} \} \right]. \tag{4.45}$$

Here $A^{(2)}$ and $B^{(2)}$ depend on the exchange and dipolar interactions, as well as on the applied magnetic field, and H.c. indicates the Hermitian conjugate. For a model of a thin film with N atomic layers, the explicit form of the coefficients is given in [30]. It is important to note that, in the absence of dipolar coupling, we have $B^{(2)} = 0$ and so only the first term remains in the above equation. The $A^{(2)}$ coefficients simplify in this special case to give a formal consistency with equation (3.9).

Next a linear transformation can be made from the $a^\dagger$ and a operators to a new set of boson operators, which we denote by $b^\dagger$ and b, satisfying boson commutation relations like those in equation (4.43). The object of making such a transformation (as detailed in [30]) is to 'diagonalize' the linearized Hamiltonian $\mathcal{H}^{(2)}$ so that it can be re-expressed as

$$\mathcal{H}^{(2)} = \sum_{\mathbf{q}_\parallel, \ell} \omega_\ell(\mathbf{q}_\parallel) b_\ell^\dagger(\mathbf{q}_\parallel) b_\ell(\mathbf{q}_\parallel). \tag{4.46}$$

The interpretation of this result is that ℓ (= 1, 2, … , N) labels the N discrete SW modes which have frequencies $\omega_\ell(\mathbf{q}_\parallel)$. If we denote $\mathbf{A}^{(2)}(\mathbf{q}_\parallel)$ and $\mathbf{B}^{(2)}(\mathbf{q}_\parallel)$ as the $N \times N$ matrices whose elements are the $A^{(2)}$ and $B^{(2)}$ coefficients appearing in equation (4.45), it is found that the SW frequencies are given by the positive solutions (for ω) of

$$\det \begin{pmatrix} \mathbf{A}^{(2)}(\mathbf{q}_\parallel) - \omega \mathbf{I}_N & 2\tilde{\mathbf{B}}^{(2)}(\mathbf{q}_\parallel) \\ 2\mathbf{B}^{(2)}(\mathbf{q}_\parallel) & \mathbf{A}^{(2)}(\mathbf{q}_\parallel) + \omega \mathbf{I}_N \end{pmatrix} = 0. \tag{4.47}$$

Here the tilde denotes a matrix transpose and $\mathbf{I}_N$ is the $N \times N$ unit matrix. The SW frequencies can be obtained numerically from equation (4.47). An example obtained for an fcc ferromagnetic film is shown in figure 4.4, where the calculated dependence of the frequency on applied magnetic field is shown for the lowest microscopic dipole-

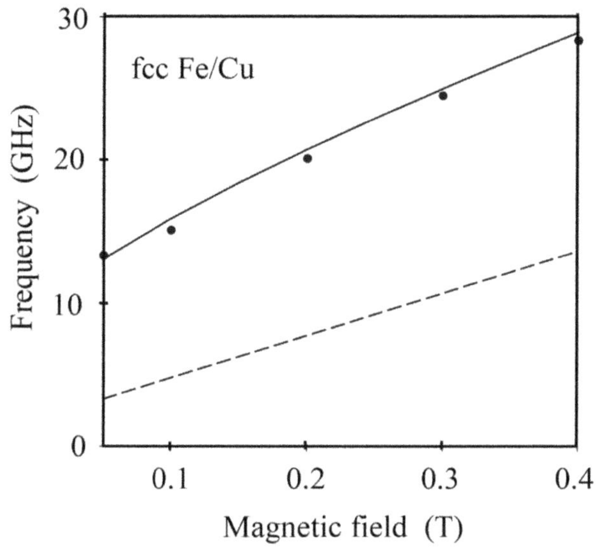

Figure 4.4. Frequency of the lowest SW mode plotted against magnetic field for a 3.8 nm thick film of fcc Fe on a Cu substrate. The in-plane wave vector corresponds to $q_y = 0$ and $q_z \simeq 0.017$ nm^{-1}. The solid line represents dipole-exchange theory for an fcc film with SW propagation parallel to $\mathbf{B}_0$ and the circles are BLS experimental data [36]. For comparison the dashed line shows the theory for the exchange limit. (Reproduced with permission from [37].)

exchange mode (solid line), giving close agreement with BLS measurements. The dashed line illustrates the effect of ignoring dipole–dipole interactions in the film.

4.4 Magnetic nanowires and stripes

Our focus here will be on quasi-1D magnetic wires that are effectively infinite along their longitudinal axis but have finite dimensions for their cross section in the transverse plane. The possibility arises of spatial quantization of the modes in the transverse plane, depending on the shape of the cross section. We already considered cases of cylindrical symmetry (longitudinally magnetized wires and tubes) in the magnetostatic limit in section 4.2, but now we want to include exchange effects through a dipole-exchange SW analysis. Another case of special interest is the magnetic stripe (or ribbon) geometry, where the cross section of the wire is rectangular.

Magnetic chains (single lines of interacting spins) represent a limiting case They have been a topic of intense interest, mainly in the exchange-dominated regime, where we already mentioned quasi-1D chain materials such as $RbCoCl_3$ in section 3.4.2. There will be discussion of other chain materials in chapter 9 in the context of magnetic solitons. Apart from this, magnetic chains have attracted attention for many years as model theoretical systems for studying quantum phase transitions, since various schemes of competing anisotropic exchange interactions may be envisaged. Detailed review accounts have been given, for example, by Mikeska and Kolezhuk [38] and Parkinson and Farnell [39].

4.4.1 Cylindrical wires and tubes

The incorporation of exchange effects into the previous magnetostatic SW calculation, outlined in section 4.2, for a long cylindrical wire was made by Arias and Mills [40]. They employed a macroscopic dipole-exchange theory developed using the Wolfram–Dewames type of approach described in section 4.3.1, in which the magnetostatic boundary conditions were supplemented by exchange-dependent additional boundary conditions (ABCs) for the film geometry. A modified form of the above theory was employed by Wang *et al* [41] to compare with their BLS measurements on cylindrical Ni nanowires with radius R ranging from 12.5 to 27.5 nm. They deduced that the BLS spectrum should be dominated by standing bulk modes with quantization of the radial wave vector. The approximate values of the mode frequencies, if it is assumed that there is small 'pinning' of the modes at the wire surface, correspond to

$$\omega = \left\{ D\left(\frac{a_m}{R}\right)^2 \left[D\left(\frac{a_m}{R}\right)^2 + \omega_M \right] \right\}^{1/2}, \; (m = 1, 2, 3, \ldots). \tag{4.48}$$

The values of the numerical constants a_m are related to properties of Bessel functions, enabling their values to be deduced [41]. Equation (4.48) holds when $B_0 = 0$ provided $q_z^2 \ll (a_m/R)^2$, which was well satisfied in the experiments. The above simple equation provides excellent agreement with the BLS experimental data, as illustrated in figure 4.5.

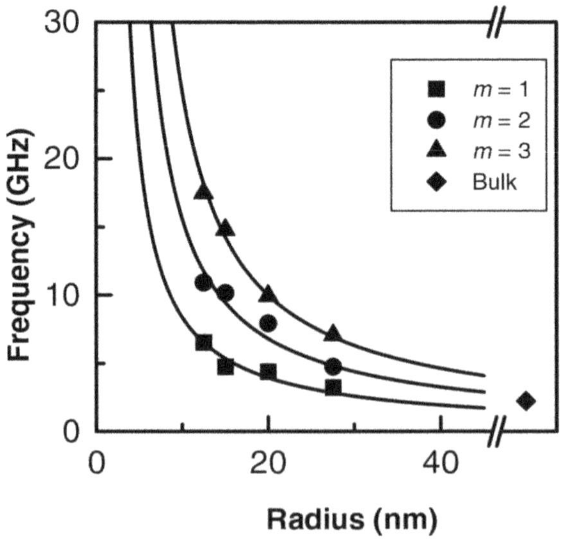

Figure 4.5. Bulk SW frequencies plotted versus radius for Ni cylindrical nanowires in zero applied field. The solid lines are obtained from macroscopic dipole-exchange theory using equation (4.48) for three values of m and the experimental points are deduced from BLS measurements. (Reproduced with permission from [41]. Copyright 2002 American Physical Society.)

Generalizations of the cylindrical wire results to tube geometries (where there is an inner and outer radius) were made using the macroscopic dipole-exchange theory. We mention, in particular, some BLS experimental and theoretical work reported in [42] on Ni nanorings with outer radii of about 50 nm and wall thickness of about 15 nm. The quantized SWs were detected and studied as a function of a longitudinal applied magnetic field. The accompanying dipole-exchange theory was supplemented using the OOMMF micromagnetic simulation package, which was mentioned in chapter 1.

The microscopic dipole-exchange theory has also been applied to describe the spin dynamics of cylindrical nanowires and is particularly useful when the saturation magnetization is spatially inhomogeneous (e.g. due to edge/interface effects or competing interactions). As an example, we mention theoretical and BLS experimental studies on permalloy ($Ni_{80}Fe_{20}$) cylindrical nanowires [43]. It was found that the presence of a weak single-ion anisotropy in these samples modified the SW properties in a way that contrasted with the behaviour for the Ni studies cited earlier in this subsection. In applying the microscopic theory, it is convenient and customary to divide the sample up into very small cells (e.g. cubes with sides of length a), which might contain many atomic spins. Then, if a is sufficiently small compared with the length scale associated with the fluctuations in the dynamical magnetization, all the atomic spins in the cell will behave as one effective spin and a plays the role of an effective lattice parameter. The usual criterion employed is that a should be chosen to be smaller than a characteristic parameter of the material known as the *exchange length*, or more specifically the exchange correlation length, l_{ex}. This is given by

$$l_{ex} = \sqrt{D/\gamma\mu_0 M_0} \tag{4.49}$$

in terms of the static magnetization M_0 and the SW exchange stiffness parameter D. Physically l_{ex} is the SW wavelength at which the dipolar and exchange contributions in the bulk dispersion relation are comparable in magnitude (see e.g. equation (4.33)). In permalloy and Ni, for example, we have $l_{ex} \simeq 5$ and $6 - 7$ nm, respectively.

4.4.2 Rectangular wires

Rectangular wire samples are typically fabricated by starting with a complete thin metallic film of thickness L on a nonmagnetic substrate. Then, by lithography and etching, long wires of width W can be formed in arrays with a chosen separation, so that the wires are either independent of one another (if well separated laterally) or otherwise are magnetically coupled. Consideration of the latter case will be deferred until section 4.6 on magnonic crystals, whereas we focus here on individual wires. Typically, the thickness L is a few tens of nanometres while the lateral width W may be of order 1 μm. Thus we may have an aspect ratio such that $W/L \gg 1$, and these systems are often referred to as 'stripes' or 'ribbons'.

The pioneering work in studying these ferromagnetic nanowires (often fabricated from a thin permalloy layer on a substrate of Si) was carried out by Jorzick *et al* [44], among others, using mainly BLS as a probe for the SWs. This topic was

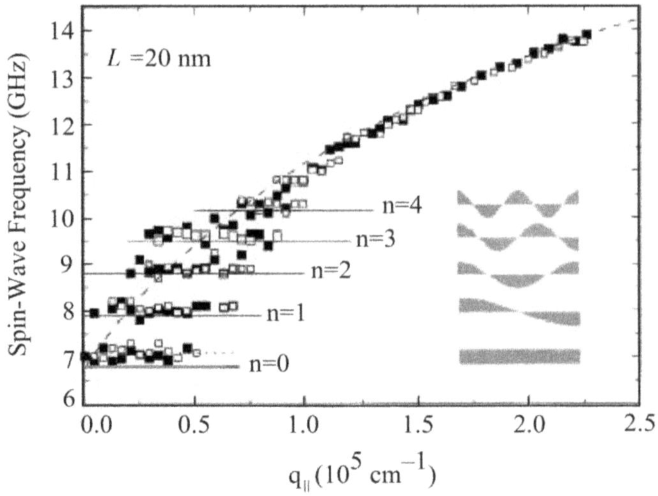

Figure 4.6. SW dispersion for a permalloy nanowire (stripe) of thickness $L = 20$ nm and width $W = 1.8$ μm. The open and closed squares are experimental data points from BLS in the Voigt geometry. The solid horizontal lines are theory, and for comparison the DE curve for a complete film with the same L is shown as a dashed line. The mode profiles are depicted as insets alongside the corresponding values of n (see the text). (Reproduced with permission from [44]. Copyright 1999 American Physical Society.)

comprehensively reviewed by Demokritov and co-workers [45, 46]. Some typical data for the SW frequencies as measured by BLS at different values of the in-plane wave vector are illustrated in figure 4.6. The light scattering geometry is such that $\mathbf{q}_{||}$ is in the plane of the original film and perpendicular to the longitudinal axis of the stripe, whereas the magnetization is along the longitudinal axis of the stripe. Hence this is the analogue of the Voigt configuration for a film, but there is no translational symmetry in the selected direction of $\mathbf{q}_{||}$. As a consequence, it was argued in [44] that the in-plane wave vector of the DE surface mode, appearing in equation (4.11) for a complete film, would effectively become quantized as

$$q_{||}^{(n)} = n\pi/W \quad (n = 0, 1, 2, \ldots), \tag{4.50}$$

which corresponds to fitting an integer number n of half-wavelengths into the stripe width W. The effect of this quantization for the SW frequency is clearly evident at small n in figure 4.6, but it is less obvious at larger n because the discrete frequencies become closer together. The experimental points lie to a good approximation along the dispersion curve for the DE mode of a complete film (dashed line). The spread of $q_{||}$ values for the SWs with low n was estimated in [44] through mode intensity calculations. Subsequent work focused on improving the exchange-type boundary conditions, analogous to equation (4.34), but applied now at the lateral edges of the stripe, e.g. through the concept of dynamical pinning to replace the previous ξ parameter [26].

The microscopic approach, within the same type of dipole-exchange formalism as described in section 3.3.2 for films, has also been modified to apply to individual rectangular nanowires (stripes) [47] with width W considerably smaller than those described in the previous paragraph. These authors compared their predicted results with BLS experimental data [48] obtained in the Voigt geometry. The experimental samples were for arrays of stripes that had a sufficiently large lateral separation for interactions to be negligible. Excellent agreement was obtained. The lateral quantization effects within each stripe are now much more pronounced than described in the previous paragraph, and the spread of wave vectors is greater. This theoretical formalism can be usefully generalized to situations where the static magnetization is spatially inhomogeneous, e.g. to apply to BLS studies of bilayer and trilayer magnetic stripes with dipolar coupling through nonmagnetic spacers [49, 50].

4.5 Magnetic disks and rings

In this section we study zero-dimensional (0D) magnetic elements that are ring-like in their geometric structure. In their simplest form, they have a circular cross section as shown schematically in figure 4.7(a). We will denote the inner and outer radii by R_1 and R_2, respectively, and the thickness in the perpendicular direction by L. The special case of a magnetic disk corresponds to taking the limit $R_1 \to 0$. Often the rings or disks are fabricated as being relatively 'flat', meaning that $L \ll R_2$ which is the opposite limit to that for the long tubes and wires considered earlier.

Magnetic rings and disks have attracted a great deal of attention, both theoretical and experimental, in the last two decades. In part, this is because their circular

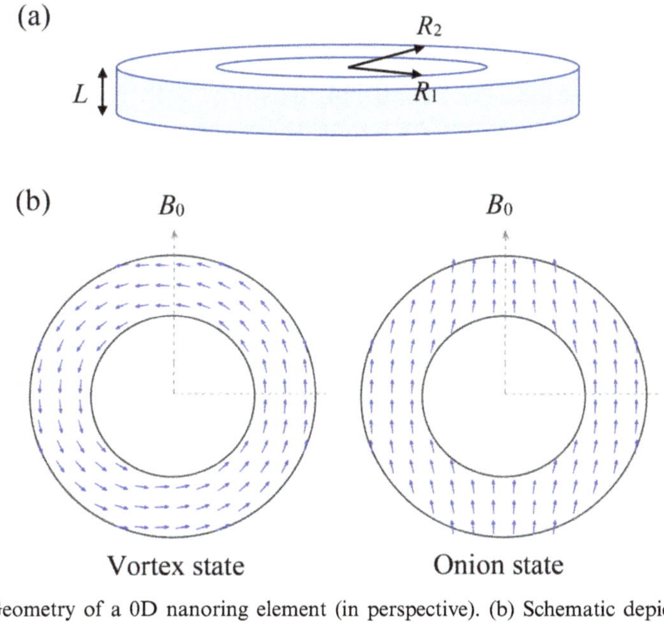

Figure 4.7. (a) Geometry of a 0D nanoring element (in perspective). (b) Schematic depiction (viewed from above) of the two stable magnetization states (vortex and onion) typical of nanorings where the thickness is small compared with the outer radius. The applied magnetic field is B_0 and the arrows indicate the direction of the local magnetization.

symmetry is advantageous for technological applications due to their fundamental properties and potential for high-frequency switching and storage devices. A distinctive feature of the circular nanorings is the existence of two states of magnetic ordering (see e.g. [51, 52] for reviews), depending on the interplay of the external magnetic field with the exchange and dipole–dipole interactions, including the shape anisotropy. These states are the vortex state with flux closure and the bi-domain onion state, which are each shown schematically in figure 4.7(b). Other magnetic states become possible when the nanorings have a larger thickness in the direction perpendicular to the plane of the rings, i.e. as in finite-length nanotubes where L is larger than R_2. In this case the out-of-plane equilibrium spin orientations may become large, leading to other states being possible [42]. For the flat nanorings with only the two stable magnetic states, the earlier studies mentioned above showed that the value of the applied magnetic field (in the plane of the ring) needed to switch states from the low-field vortex state to the higher-field onion state could be sensitively controlled by adjusting the physical dimensions of the ring, such as its thickness and/or its wall width (the difference between the outer and inner radii). The occurrence of a vortex state tends to be less favoured energetically in circular nanodisks, due to the magnetic energy associated with the core region, compared to the nanoring case [51, 52].

In the majority of the experimental works cited for nanorings, the emphasis has been on the static magnetic properties and there have been relatively fewer studies for the magnetization dynamics, including the SW properties of the nanorings. As an exception, we mention a BLS study by Gubbiotti *et al* [53] of the SWs in permalloy nanorings. These authors observed a distinctly different magnetic-field dependence of the SW frequencies in the vortex and onion regions and interpreted their results using a macroscopic continuum dipole-exchange theory. Very recently, the micro-focused BLS technique has been used [54] to study both the frequencies and the intensities of the SWs in a permalloy nanoring for the vortex and onion regions. In these measurements the area of laser beam spot on the sample surface was made sufficiently small (e.g. of order 250 nm across) that the spatial variations in intensity across the ring structure could be sampled.

Calculations of the SW mode frequencies and their spatial intensity variation in permalloy nanorings have been presented using the microscopic dipole-exchange method in [55]. Some results are shown in figure 4.8, where the frequencies for the several lowest SW branches are plotted versus applied magnetic field B_0, to illustrate the different behaviour in the low-field (vortex state) region compared with the high-field (onion state) region. Because the ring is finite in size in all dimensions, the spatially quantized modes are not characterized in terms of any wave-vector components. The predicted behaviour here is very similar to that in the BLS experiments [53], which were for larger ring sizes.

4.6 Magnetic superlattices

We now consider SWs in multilayer magnetic systems, where the dipolar interactions play a role in providing the coupling between magnetic layers, as well as for

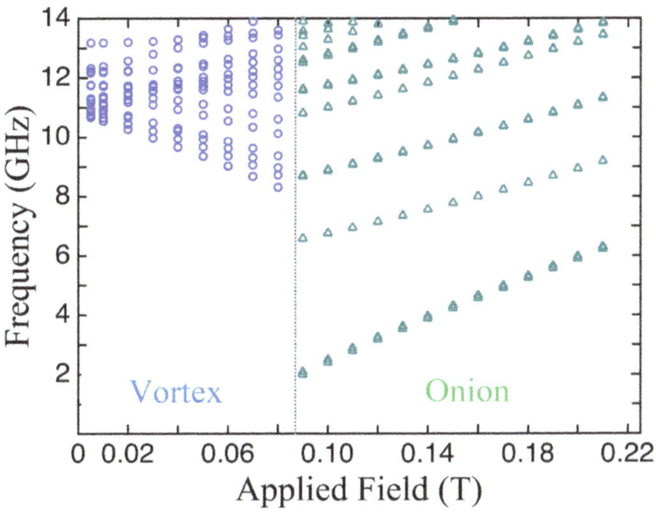

Figure 4.8. SW frequencies calculated using the microscopic dipole-exchange theory and plotted versus the applied magnetic field for a permalloy nanoring with $R_1 = 60$ nm, $R_2 = 100$ nm and $L = 20$ nm. The dotted vertical line represents the transition field between the low-field vortex states (see blue circles) and the higher-field onion states (see green triangles). (Reproduced with permission from [55].)

the behaviour within the layers. Here we focus on superlattices that are formed with a periodic or quasiperiodic growth rule from individual layers. As before, the SW behaviour depends on whether we are dealing with the dipole-dominated (magnetostatic) regime or the dipole-exchange regime. The exchange-dominated regime for superlattices was already covered in section 3.6. Some review accounts giving theoretical and experimental details of this topic are, for example, in [8, 20, 56, 57] for mainly periodic superlattices and [58] for a focus on quasiperiodic superlattices.

The simplest physical system in either the dipolar or dipole-exchange regimes corresponds to superlattices where ferromagnetic layers alternate with nonmagnetic spacer layers, and the magnetic interlayer coupling is provided only via the long-range dipolar fields across the spacers. The theory for the superlattice modes may be worked out separately for the two cases where the magnetization is either parallel [59, 60] or perpendicular [61] to the planar interfaces. Here we outline the magnetostatic calculation for a semi-infinite superlattice with in-plane magnetization in the Voigt configuration as sketched in figure 4.9. Then, with the magnetostatic scalar potential written as $\psi_1(x)\exp(i\mathbf{q}_{\parallel} \cdot \mathbf{r}_{\parallel})$ as in section 4.1.1, we have in the ferromagnetic layer of the lth cell, i.e. when $-(l-1)D > x > -(l-1)D - d_B$, the dependence that

$$\psi_l(x) = a_l \exp\{iq_x[x + (l-1)D]\} + b_l \exp\{-iq_x[x + (l-1)D]\}. \tag{4.51}$$

This expression has a form analogous to that in section 1.5.1 for photonic periodic multilayers, where a_l and b_l are constants. The theory proceeds using a transfer-matrix formalism in a very similar manner to previously, in which a_l and b_l in the lth cell define a two-component column matrix. Here q_x has solutions that are discussed

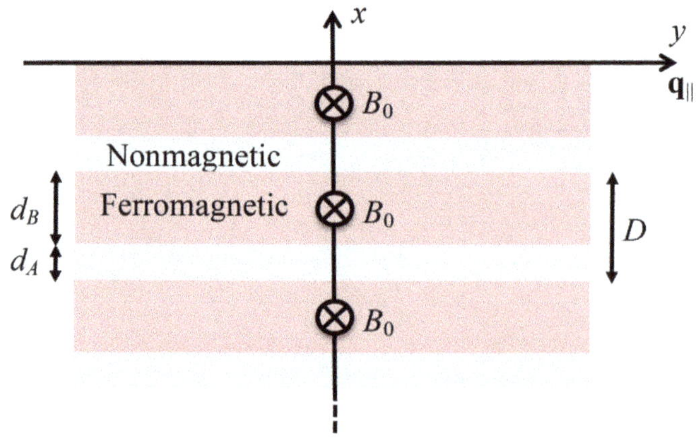

Figure 4.9. Schematic model used for a semi-infinite superlattice in the magnetostatic and dipole-exchange regimes in which there are alternating ferromagnetic and nonmagnetic spacer layers. The Voigt geometry is assumed, for which the in-plane wave vector is perpendicular to the in-plane applied field direction.

in equation (4.8) and the following text. An analogous expression to equation (4.51) can be written down for the potential $\psi_l(x)$ in the nonmagnetic spacer region of the lth cell (replacing iq_x by $q_{\|}$ and introducing two new constants before the exponential terms). We also require the magnetostatic boundary conditions at each interface, and these are the same as discussed in section 4.1.1.

Next, by explicit evaluation of the transfer matrix and the use of equation (1.68), we eventually find that the bulk superlattice modes correspond to the solutions, for real values of the Bloch wave vector Q, of

$$\cos(QD) = \cosh(q_{\|}d_A)\cosh(q_{\|}d_B) + g\sinh(q_{\|}d_A)\sinh(q_{\|}d_B), \tag{4.52}$$

where

$$g = (1 + \chi_a) - \frac{\chi_b^2}{2(1 + \chi_a)}. \tag{4.53}$$

Here equation (4.52) is the analogue of the previous Rytov equations (1.70) and (2.46) for optical modes and phonons, respectively. The fact that (4.52) involves the hyperbolic sinh and cosh functions, instead of sin and cos functions as in the previous examples, is a consequence of the superlattice bulk mode being related to a linear combination of the DE *surface* mode for individual films, instead of a combination of bulk modes [62], as beforehand. The susceptibility terms χ_a and χ_b for a ferromagnet are defined in equation (4.2).

The possible existence of a surface mode of the superlattice for the special case of a semi-infinite superlattice can also be deduced from the above formalism leading to the remarkably simple conclusion that a surface magnetostatic mode exists only when $d_B > d_A$ (i.e. the magnetic films are thicker than the spacer layers), in which case its frequency is given by $\omega_S = \omega_0 + \frac{1}{2}\omega_M$. This frequency is the same as the frequency for the DE mode in a thick film and is independent of the values of d_B and d_A. On the other hand, if $d_B < d_A$ there is no superlattice surface mode, because the localization condition for a spatially decaying mode cannot be satisfied. This predicted behaviour was verified in BLS experiments by Grimsditch *et al* [63] using Ni/Mo superlattices with different layer thicknesses. They observed an extra peak (the surface mode) in the BLS spectra when $d_B > d_A$. This feature occurred on one side only of the spectrum, which demonstrated the property of nonreciprocal propagation, as also predicted by the theory. Subsequently, a series of experiments by Hillebrands *et al* [64] using Fe/Pd superlattices illustrated the same effects in more detail.

When the similar magnetostatic theory is applied to a ferromagnetic/nonmagnetic superlattice in the case where the magnetization M_0 is perpendicular to the interfaces [61], it leads to multiple bands of bulk modes and also, depending on the parameters, multiple surface modes. Nonreciprocal propagation does not occur in this case, which in part is a consequence of the fact that the superlattice modes are now formed from linear combinations of the *bulk* modes of the individual magnetic film.

In the dipole-exchange regime for ferromagnetic/nonmagnetic superlattices, most of the earlier theoretical work focused on the macroscopic approaches. The first calculations were made by Vayhinger and Kronmuller [65], essentially generalizing the perturbative tensorial GF method used for films (see section 4.3.1) by Kalinikos and Slavin [28]. Subsequently, a more detailed macroscopic theory was developed by Hillebrands and others (see e.g. [66]). It was based on the macroscopic torque equation-of-motion method for the magnetization described in section 4.3.1. One of the challenges in the dipole-exchange theories is the proper inclusion of the additional (exchange- and anisotropy-dependent) boundary conditions at the interfaces. Details are given in the review articles on magnetic superlattices mentioned earlier. An example of a macroscopic dipole-exchange calculation for a finite Co/Pd superlattice with nine periods is given in figure 4.10. Further discussion and examples for dipole-exchange superlattices, following both macroscopic and microscopic approaches for the theory, can be found (e.g.) in [7, 8].

Finally, we mention that many examples of experimental and theoretical studies for SWs in quasiperiodic superlattices are described in [58]. Generally, the results for SWs are more complicated than in the phonon case (see section 2.5) because in the magnetic cases there can be competing interactions that lead to a modified ground state for the equilibrium orientation of the magnetizations, as well as the competition between the short-range exchange and long-range dipolar interactions.

4.7 Magnonic crystals

In the last fifteen years magnonic crystals (MCs) have represented a burgeoning field of study due to their basic physics and their potential for device applications. They typically consist of lateral arrays of, for example, nanowires or nanodots arranged in either 1D or 2D with various geometries. Earlier progress was reviewed in a series of articles that appeared in *Journal of Physics D: Applied Physics* in 2010, comprising a general survey [67] and then specialized articles on YIG magnonics [68], Brillouin light scattering studies [69], micromagnetic modelling [70] and logic device applications [71]. More recently there have been multi-author books on magnonics edited

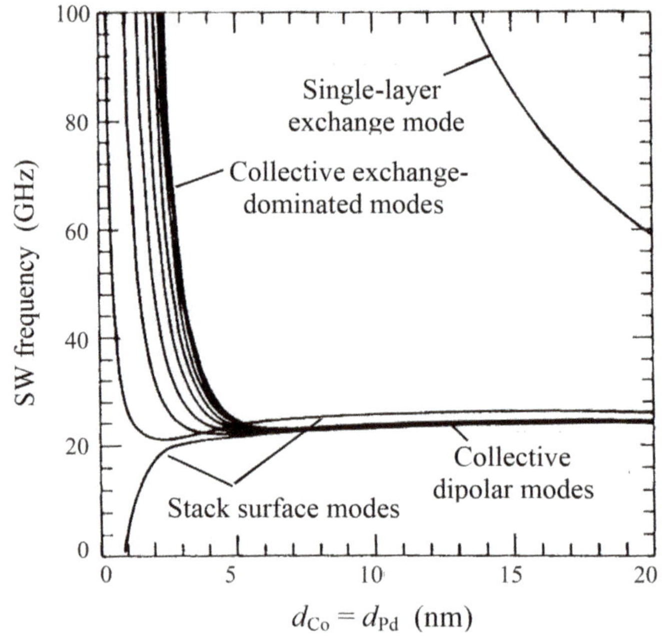

Figure 4.10. Predicted frequencies of dipole-exchange SWs in a finite Co/Pd superlattice with nine periods calculated as a function of the thickness of the magnetic Co layers, taking $d_{Co} = d_{Pd}$. Other fitting parameters in the theory were deduced from BLS measurements. (Reproduced with permission from [66]. Copyright 1999 American Physical Society.)

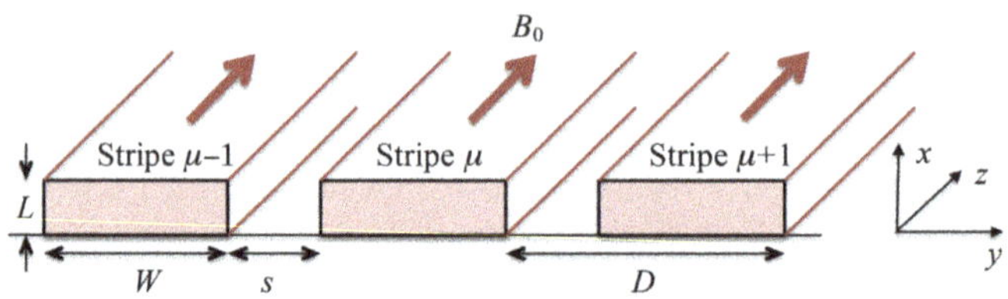

Figure 4.11. Simple geometry for a magnonic crystal (MC) consisting of a 1D periodic array of ferromagnetic stripes of width W and thickness L laterally separated by a nonmagnetic spacer of width s. The periodic distance is $D = W + s$.

by Gubbiotti [8] and by Demokritov and Slavin [72]. Most applications are for the dipole-exchange regime of SW behaviour, and so that will be our focus here.

In order to highlight some of the key features of the theoretical methods to incorporate the lateral periodicity of MCs we take the example of a basic 1D model of an infinitely extended array of nanowire stripes arranged side-by-side with a nonmagnetic intervening material (see figure 4.11). For simplicity the applied field B_0 and hence the static magnetization are taken to be along the longitudinal z-axis of each stripe. The methods that we outline here will generalize to other geometric arrangements of nanoelements.

On the macroscopic side, the theories for MCs (as described, for example, in the review given in [69]) have consisted of using micromagnetic simulations, such as adaptations of the OOMMF package mentioned in section 4.1, or solving analytically a set of dipole-exchange equations analogous to those in section 4.3.1 for ferromagnetic films. In the latter case the connection between the fluctuating dipolar field **h** and the fluctuating magnetization **m** can be conveniently expressed in terms of a tensorial GF relationship, instead of the previous susceptibility equation (4.1). Following the approach of Kostylev and co-workers [73, 74] the required result can be written in a spatial integral form as

$$\mathbf{h}_{q,\,Q}(x,\,y) = \int_A \hat{G}_{q,\,Q}(x - x',\,y - y')\mathbf{m}_{q,\,Q}(x',\,z')\mathrm{d}x'\mathrm{d}z'. \tag{4.54}$$

Here $\mathbf{h}_{q,Q}$ and $\mathbf{m}_{q,Q}$ are Fourier components of the above fields with respect to the longitudinal wave number q in the z-direction and the Bloch wave number Q in the y-direction (see figure 4.11). The integration is over the cross-sectional area A of any one stripe in the infinite array and $\hat{G}_{q,Q}$ is a tensorial GF that has elements given explicitly by

$$\begin{aligned}\hat{G}^{\alpha,\,\beta}_{q,\,Q}(x - x',\,y - y') = &\sum_{\mu=-\infty}^{\infty} \exp[\mathrm{i}Q(y - y' - \mu D)]\\ &\times \frac{\partial^2}{\partial\alpha\partial\beta}K_0(q[(x - x')^2 + (y - y')^2]^{1/2}),\end{aligned} \tag{4.55}$$

where α and β are Cartesian components, and K_0 is a modified Bessel function of the second kind. In practice, approximate forms of the GF can be used in order to deduce the magnonic bands representing the behaviour of the SW frequencies as a function of Q [73, 74].

In contrast, in the microscopic dipole-exchange theory for MCs [75], the starting point is a Hamiltonian having exactly the same form as equation (4.39), but now both interwire and intrawire dipole–dipole interactions must be included, as well as the short-range exchange terms. As in the ferromagnetic film case, a transformation can be made to boson operators and then it is the bilinear term $\mathcal{H}^{(2)}$ that is of interest in describing the SW modes. By analogy with the representation used for the dipolar fields and the GFs in the preceding paragraph, a Fourier transform is made for the dipolar coupling term $D^{\alpha,\beta}_{i,j}$, which was defined in equation (4.40). Specifically, the transform is with respect to the longitudinal wave number q in the z-direction and the Bloch wave number Q in the y-direction. This leads eventually to an expression for $\mathcal{H}^{(2)}$ analogous to that found in equation (4.45):

$$\begin{aligned}\mathcal{H}^{(2)} = \sum_{q,Q}\sum_{n,m}\Big[&A^{(2)}_{n,\,m}(q,\,Q)a^{\dagger}_n(q,\,Q)a_m(q,\,Q)\\ &+ \{B^{(2)}_{n,\,m}(q,\,Q)a^{\dagger}_n(q,\,Q)a^{\dagger}_m(-q,\,-Q) + \mathrm{H.c.}\}\Big],\end{aligned} \tag{4.56}$$

where $A^{(2)}$ and $B^{(2)}$ can be expressed in terms of the dipole–dipole and exchange parameters, as well as the wave vectors. This equation can then be solved using matrix techniques by following the same formal steps as in the film case [75].

We next provide a few examples of experimental results for MCs and their comparison with theory. Some BLS data are shown in figure 4.12, where four magnonic bands were observed in the experiment with the lower two exhibiting significant dispersion with respect to the Bloch wave number Q. The results of macroscopic and microscopic theories are also presented, and it is seen that they both lead to satisfactory agreement with experiment. For an example of a MC composed of alternating stripes of two different ferromagnetic materials (Fe and permalloy in this case) we may refer back to figure 2.20 where there was discussion of the results given in chapter 2 in the context of the phononic behaviour of this structure. Specifically the lower two panels show the BLS spectra of intensity plotted versus the SW frequency for two different values of Q. The corresponding magnonic bands deduced from BLS are next displayed in figure 4.13. The blue lines are from micromagnetic simulations and give reasonable agreement with the lowest two bands observed. There have also been dipole-exchange SW studies, mostly from a theoretical perspective, of MCs where the stripes are arranged according to a quasiperiodic Fibonacci or Thue–Morse sequence [76, 77].

In more recent experimental and theoretical studies, the attention has shifted to nanowire MCs in the dipole-exchange regime where the wire elements are composite in structure [79, 80], rather than the single wires as depicted in figure 4.11. This means that each wire element might actually consist of two or more magnetic layers in the vertical (or x) direction, allowing for several possible magnetic states with the switching between magnetic states controlled by varying the applied magnetic field. Such investigations open up prospects for device applications using 3D magnonics and for magnetic-field-controlled reconfigurable MCs with multimodal frequency transmission characteristics [8].

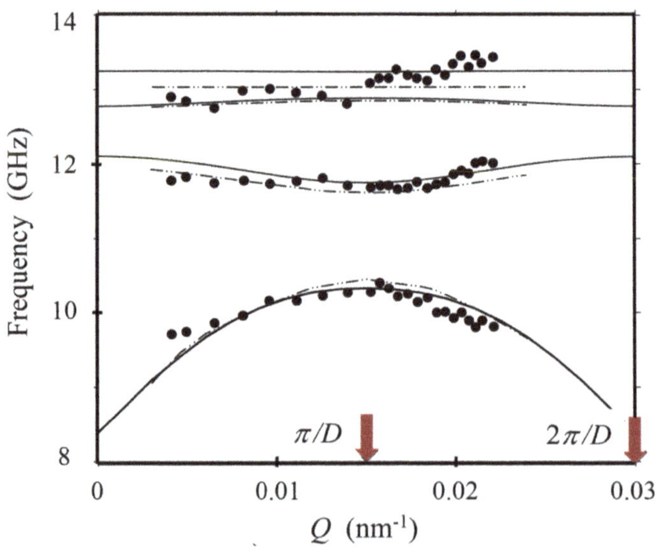

Figure 4.12. Comparison of the macroscopic theory (dashed lines) [69] and the microscopic theory (solid lines) [75] with BLS experimental data (black circles), showing the SW frequencies plotted versus Bloch wave vector Q for an array of permalloy stripes with W = 175 nm, L = 20 nm and s = 35 nm for the spacer (giving a periodic length D = 210 nm). Reprinted with permission from [75].

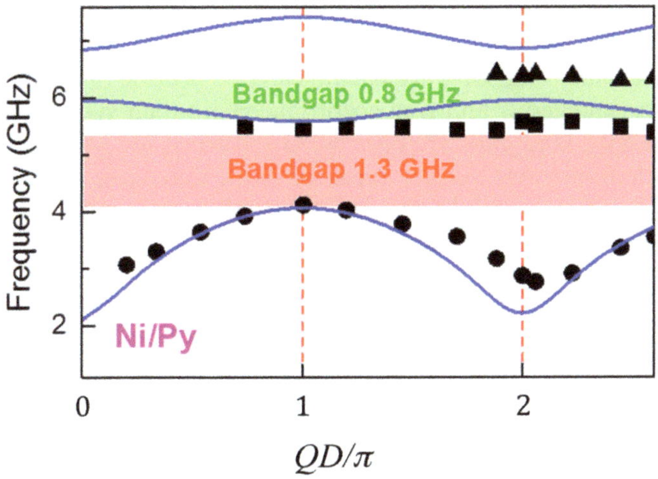

Figure 4.13. Plot of frequency versus reduced wave vector QD/π for the SW bands as deduced from BLS data (see figure 2.20) for a dual phononic/magnonic crystal composed of alternating Fe and $Ni_{80}Fe_{20}$ (permalloy or Py) nanostripes. (Reproduced with permission from [78]. Copyright 2012 American Institute of Physics.)

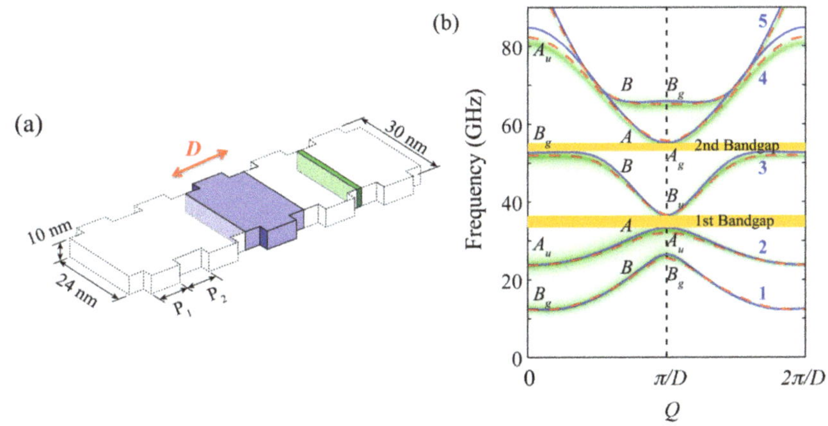

Figure 4.14. Magnonic crystal waveguide consisting of a permalloy stripe with periodic modulations in the width. (a) Geometry and dimensions. The shaded blue block represents the computational unit cell used in the finite-element and microscopic calculations. The green bar indicates where the excitation field is applied in the OOMMF simulations. (b) The magnonic dispersion curves (1 to 5) plotted versus Bloch wave vector Q. The blue solid, red dashed and green bold lines are data obtained from macroscopic, microscopic and OOMMF calculations, respectively. (Reproduced with permission from [81]. Copyright 2014 American Institute of Physics.)

Finally, we mention that another type of MC of practical interest with a very different geometry can be derived from a periodically modulated waveguide. For example, a single ferromagnetic stripe (with width W that is large compared to the thickness L) could be considered, in which periodic lengthwise modulations of W are introduced. Such a case is illustrated in figure 4.14(a), and this has been employed in figure 4.14(b) to make comparisons between three different theoretical approaches, i.e. the macroscopic and microscopic methods mentioned earlier as well as micromagnetic simulations using software packages. This is done in a situation where the magnonic bands (labelled *A* and *B* in accordance with their symmetry [81]) partially overlap and the SW band gaps are narrow. Good consistency between the different approaches is demonstrated. Other types of MCs that have been studied include structures that are periodic in the two transverse directions, and these are described in the general references given earlier in this section.

References

[1] Gurevich A G and Melkov G A 1996 *Magnetization Oscillations and Waves* (Boca Raton: CRC Press)

[2] Stancil D D and Prabhakar A 2009 *Spin Waves: Theory and Application* (Heidelberg: Springer)

[3] Wolfram T and Dewames R E 1972 *Prog. Surf. Sci.* **2** 233

[4] Mills D L 1984 *Surface Excitations* ed V M Agranovich and R Loudon (Amsterdam: North-Holland) p 379

[5] Cottam M G (ed) 1994 *Linear and Nonlinear Spin Waves in Magnetic Films and Superlattices* (Singapore: World Scientific)

[6] Hillebrands B and Ounadjela K (ed) 2001 *Spin Dynamics in Confined Magnetic Structures* (Berlin: Springer)

[7] Barman A and Haldar A 2014 *Solid State Phys.* **65** 1

[8] Gubbiotti G (ed) 2019 *Three-Dimensional Magnonics* (Singapore: Jenny Stanford)

[9] Damon R W and Eshbach J R 1961 *J. Phys. Chem. Solids* **19** 308

[10] Jackson J D 1998 *Classical Electrodynamics* 3rd edn (New York: Wiley)

[11] Griffiths D J 2012 *Introduction to Electrodynamics* 4th edn (Boston: Addison-Wesley)

[12] Sandercock J R and Wettling W 1979 *J. Appl. Phys.* **50** 7784

[13] Cottam M G 1983 *J. Phys. C: Solid State Phys.* **16** 1573

[14] Grünberg P 1980 *J. Appl. Phys.* **51** 4338

[15] Grünberg P 1981 *J. Appl. Phys.* **52** 6824

[16] Camley R E and Maradudin A A 1982 *Solid State Commun.* **41** 585

[17] Grünberg P, Cottam M G, Vach W, Mayr C M and Camley R E 1982 *J. Appl. Phys.* **53** 2078

[18] Cochran J F, Rudd J, Muir W B, Heinrich B and Celinski Z 1990 *Phys. Rev.* B **42** 508

[19] Liu X M, Nguyen H T, Ding J, Cottam M G and Adeyeye A O 2014 *Phys. Rev.* B **90** 064428

[20] Cottam M G and Tilley D R 2005 *Introduction to Surface and Superlattice Excitations* 2nd edn (Bristol: IOP Publishing)

[21] Camley R E 1980 *Phys. Rev. Lett.* **45** 283

[22] Stamps R L and Camley R E 1984 *J. Appl. Phys.* **56** 3497

[23] Sharon T M and Maradudin A A 1977 *J. Phys. Chem. Solids* **38** 977

[24] Das T K and Cottam M G 2007 *J. Magn. Magn. Mater.* **310** 2183

[25] Kalinikos B A and Slavin A N 1986 *J. Phys. C: Solid State Phys.* **19** 7033

[26] K Yu G and Slavin A N 2005 *Phys. Rev.* B **72** 014463

[27] Stamps R L and Camley R E 1987 *Phys. Rev.* B **35** 1919

[28] Kalinikos B A 1994 *Linear and Nonlinear Spin Waves in Magnetic Films and Superlattices* ed M G Cottam (Singapore: World Scientific) p 89

[29] Benson H and Mills D L 1969 *Phys. Rev.* **178** 839

[30] Costa Filho R N, Cottam M G and Farias G A 2000 *Phys. Rev.* B **62** 6545
[31] Stamps R L and Hillebrands B 1991 *Phys. Rev.* B **44** 12417
[32] Pereira J M and Cottam M G 1999 *J. Appl. Phys.* **85** 4949
[33] Holstein T and Primakoff H 1940 *Phys. Rev.* **58** 1098
[34] Griffiths D J and Schroeter 2018 *Introduction to Quantum Mechanics* 3rd edn (Cambridge: Cambridge University Press)
[35] Sakurai J J and Napolitano J 2020 *Modern Quantum Mechanics* 3rd edn (Cambridge: Cambridge University Press)
[36] Tacchi S, Albini L, Gubbiotti G, Madami M and Carlotti G 2002 *Surf. Sci.* **507-510** 324
[37] Nguyen H T and Cottam M G 2011 *J. Phys.: Condens. Matter* **23** 126004
[38] Mikeska H-J and Kolezhuk A K 2004 *Quantum Magnetism* (Lecture Notes in Physics vol 645) (Berlin: Springer) ch 1 pp 1–83
[39] Parkinson J B and Farnell D J J 2010 *An Introduction to Quantum Spin Systems* (Lecture Notes in Physics vol 816) (Berlin: Springer)
[40] Arias R and Mills D L 2001 *Phys. Rev.* B **63** 134439
[41] Wang Z, Kuok M H, Ng S C, Lockwood D J, Cottam M G, Nielsch K, Wehrspohn R B and Gösele U 2002 *Phys. Rev. Lett.* **89** 027201
[42] Wang Z *et al* 2005 *Phys. Rev. Lett.* **94** 137208
[43] Nguyen T M, Cottam M G, Liu H Y, Wang Z K, Ng S C, Kuok M H, Lockwood D J, Nielsch K and Gösele U 2006 *Phys. Rev.* B **73** 140402(R)
[44] Jorzick J, Demokritov S O, Mathieu C, Hillebrands B, Bartenlian B, Chappert C, Decanini D, Rousseaux F and Slavin A N 1999 *Phys. Rev.* B **60** 15194
[45] Demokritov S O, Hillebrands B and Slavin A N 2001 *Phys. Rep.* **348** 441
[46] Demokritov S O and Hillebrands B 2002 *Spin Dynamics in Confined Magnetic Structures* ed B Hillebrands and K Ounadjela vol 1 (Berlin: Springer) p 65
[47] Nguyen H T, Nguyen T M and Cottam M G 2007 *Phys. Rev.* B **76** 134413
[48] Gubbiotti G, Tacchi S, Carlotti G, Vavassori P, Singh N, Goolaup S, Adeyeye A O, Stashkevich A and Kostylev M 2005 *Phys. Rev.* B **72** 224413
[49] Gubbiotti G, Nguyen H T, Hiramatsu R, Tacchi S, Madami M, Cottam M G and Ono T 2014 *J. Phys. D: Appl. Phys.* **47** 365001
[50] Gubbiotti G, Nguyen H T, Hiramatsu R, Tacchi S, Cottam M G and Ono T 2015 *J. Magn. Magn. Mater.* **384** 45
[51] Kläui M, Vaz C A F, Lopez-Diaz L and Bland J A C 2003 *J. Phys.: Condens. Matter* **15** R985
[52] Vaz C A F *et al* 2007 *J. Phys.: Condens. Matter* **19** 255207
[53] Gubbiotti G, Madami M, Tacchi S, Carlotti G, Tanigawa H, Ono T, Giovannini L, Montoncello F and Nizzoli F 2006 *Phys. Rev. Lett.* **97** 247203
[54] Tian C, Chauduri U, Singh N and Adeyeye A O 2020 *Nanotechnology* **31** 145714
[55] Hussain B and Cottam M G 2021 *J. Phys. D: Appl. Phys.* **54** 165002
[56] Mills D L 2000 Collective excitations in superlattice structures *Light Scattering in Solids* vol 5 ed M Cardona and G Güntherodt (Berlin: Springer) pp 13–48
[57] Hillebrands B 2000 Brillouin light scattering from layered magnetic structures *Light Scattering in Solids* vol 7 ed M Cardona and G Güntherodt (Berlin: Springer) pp 174–289
[58] Albuquerque E L and Cottam M G 2003 *Phys. Rep.* **376** 225
[59] Camley R E, Rahman T S and Mills D L 1983 *Phys. Rev.* B **27** 261
[60] Grünberg P and Mika K 1983 *Phys. Rev.* B **27** 2955

[61] Camley R E and Cottam M G 1987 *Phys. Rev.* B **35** 189
[62] Barnás J 1994 Spin waves in multilayers *Linear and Nonlinear Spin Waves in Magnetic Films and Superlattices* ed M G Cottam (Singapore: World Scientific) p 157
[63] Grimsditch M, Khan M R, Kueny A and Schuller I K 1983 *Phys. Rev. Lett.* **51** 498
[64] Hillebrands B, Boufelfel A, Falco C M, Baumgart P, Güntherodt G, Zirngiebl E and Thompson J D 1988 *J. Appl. Phys.* **63** 3880
[65] Vayhinger K and Kronmüller H 1986 *J. Magn. Magn. Mater.* **62** 159
[66] Hillebrands B, Harzer J V, Güntherodt G, England C D and Falco C M 1990 *Phys. Rev.* B **42** 6839
[67] Kruglyak V V, Demokritov S O and Grundler D 2010 *J. Phys. D: Appl. Phys.* **43** 264001
[68] Serga A A, Chumak A V and Hillebrands B 2010 *J. Phys. D: Appl. Phys.* **43** 264002
[69] Gubbiotti G, Tacchi S, Madami M, Carlotti G, Adeyeye A O and Kostylev M 2010 *J. Phys. D: Appl. Phys.* **43** 264003
[70] Kim S-K 2010 *J. Phys. D: Appl. Phys.* **43** 264004
[71] Khitun A, Bao M and Wang K L 2010 *J. Phys. D: Appl. Phys.* **43** 264005
[72] Demokritov S O and Slavin A N (ed) 2013 *Magnonics: From Fundamentals to Applications* (Berlin: Springer)
[73] Kostylev M P, Stashkevich A A and Sergeeva N A 2004 *Phys. Rev.* B **69** 064408
[74] Kostylev M P, Schrader P, Stamps R L, Gubbiotti G, Carlotti G, Adeyeye A O, Goolaup S and Singh N 2008 *Appl. Phys. Lett.* **92** 132504
[75] Nguyen H T and Cottam M G 2011 *J. Phys. D: Appl. Phys.* **44** 315001
[76] Hussain B and Cottam M G 2018 *J. Phys. D: Appl. Phys.* **51** 355003
[77] Hussain B, Cottam M G and Ge B 2019 *J. Phys. D: Appl. Phys.* **52** 095001
[78] Zhang V L, Ma F S, Pan H H, Lin C S, Lim H S, Ng S C, Kuok M H, Jain S and Adeyeye A O 2012 *Appl. Phys. Lett.* **100** 163118
[79] Gubbiotti G, Zhou X, Haghshenasfard Z, Cottam M G and Adeyeye A O 2018 *Phys. Rev.* B **97** 134428
[80] Gubbiotti G, Zhou X, Haghshenasfard Z, Cottam M G, Adeyeye A O and Kostylev M 2019 *Sci. Rep.* **9** 4617
[81] Di K, Lim H S, Zhang V L, Ng S C, Kuok M H, Nguyen H T and Cottam M G 2014 *J. Appl. Phys.* **115** 053904

IOP Publishing

Dynamical Properties in Nanostructured and Low-Dimensional Materials (Second Edition)

Michael G Cottam

Chapter 5

Electronic and plasmonic excitations

In this chapter we examine the properties of a number of electronic-related excitations in situations of low dimensionality. We start by considering single-electron states localized at surfaces, initially in terms of some simplified 1D models with arrays of potential wells. This is followed by an analysis of the electronic states in 2D graphene, where we focus on complete atomic monolayers and finite-width nanoribbons with edge effects, giving rise to localized edge modes. Then we turn attention to the *collective* electronic excitations, or plasmons, as part of an analysis of the dispersion properties associated with the dielectric functions in various different types of solids. Some of these topics were already introduced briefly in chapter 1, but here the concepts are developed further. The derived dielectric properties will be utilized further in chapter 6 for the discussion of polaritons, where the crystal excitations are modified due to the coupling to a light wave (or photons).

5.1 Electronic surface states

The possibility of electronic surface states localized at a free planar surface was first investigated in the 1930s in calculations by Tamm [1] and shortly thereafter by Goodwin [2] and Shockley [3]. They all employed a 1D model in terms of the coordinate axis perpendicular to the surface(s). Tamm studied solutions of Schrödinger's equation for an electron in a semi-infinite lattice (or array) of square potential wells. In other words, it was a generalization of the periodic Kronig–Penney model that is typically described in textbooks on solid-state physics (see e.g. [4, 5]) to the case of a semi-infinite array, allowing for the termination of a sample at a surface. In contrast, both Shockley and Goodwin consider finite 1D arrays, and they chose different forms for the atomic potentials as will be discussed. Comprehensive reviews of electric surface states and the relevant band-structure theory can be found, for example, in [6, 7].

doi:10.1088/978-0-7503-3903-2ch5

5.1.1 Kronig–Penney model

We start here with an account of the Kronig–Penney (KP) model as applied to solving Schrödinger's equation for the one-electron states in a lattice of atomic potentials. The calculation assumes a periodic lattice in 1D with period d equal to the spacing between atoms and with the spatial coordinate denoted by x. The situation is as shown schematically in figure 5.1, except that for the present we will consider infinite arrays without any terminating surface(s). Hence, Bloch's theorem applies along the x-direction, giving us the property that $\psi(x + d) = \exp(\mathrm{i}Qd)\psi(x)$ for the electronic wave function, where Q is the Bloch wave-number (see section 1.2). We now seek solutions of Schrödinger's equation for $\psi(x)$ in any cell of length d, as in figure 5.1. Realistic choices for the potentials associated with each cell will be discussed later in this section, but in the KP model described next a square-shaped potential is typically adopted, as in figure 5.1(a), for which an exact analytic solution may be obtained (see e.g. [4, 5]).

Here we will follow a slightly different approach, which retains all the essential physics of the KP model, by considering an array of delta-function potentials, as sketched in figure 5.1(b). For the 1D case of equation (1.37) for the Hamiltonian, we simply replace $\nabla^2 \to \mathrm{d}^2/\mathrm{d}x^2$ and $V(\mathbf{r}) \to V(x)$ where

$$V(x) = W_0 \sum_{n=-\infty}^{\infty} \delta(x - nd). \tag{5.1}$$

The quantity W_0 is a strength factor associated with the delta-function spikes at $x = nd$, and the previous Schrödinger's equation (1.38) becomes

$$\frac{\mathrm{d}^2\psi}{\mathrm{d}x^2} + \frac{2m}{\hbar^2}\left\{E - W_0 \sum_{n=-\infty}^{\infty} \delta(x - nd)\right\}\psi = 0. \tag{5.2}$$

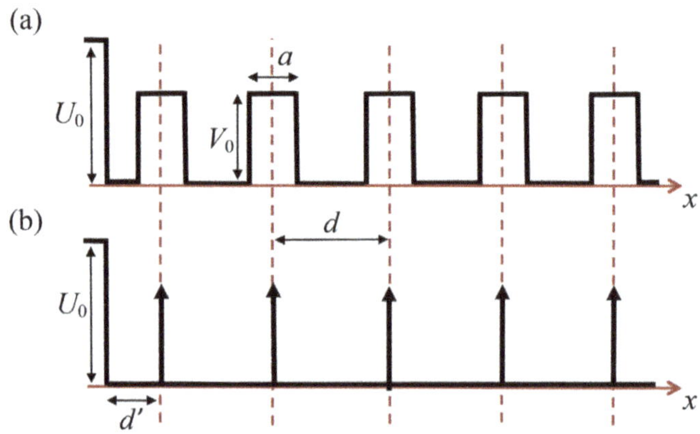

Figure 5.1. Schematic form of some 1D potentials used for discussion of the Kronig–Penney model: arrays of (a) square-shaped barriers and (b) delta-function barriers.

If we focus on any one periodic cell (taken as the interval of x from $-\eta$ to $d - \eta$, where η is a positive infinitesimal), we may write down for all x, except for $x = 0$ where the delta function is located, the expression

$$\psi(x) = A\cos(\lambda x) + B\sin(\lambda x) \tag{5.3}$$

with A and B denoting constants. Also E is the energy eigenvalue for an electron in the system and $\lambda^2 = 2mE/\hbar^2$. One boundary condition is that the wave function is continuous around $x = 0$, so

$$\psi(\eta) = \psi(-\eta) = \exp(-Qd)\psi(d - \eta), \tag{5.4}$$

where the last step follows from applying Bloch's theorem to adjacent cells. The second boundary condition is for the slope discontinuity of the wave function around $x = 0$; it is obtained by integrating Schrödinger's equation (5.2) over the infinitesimal interval from $-\eta$ to η as

$$\left(\frac{\mathrm{d}\psi}{\mathrm{d}x}\right)_{x=\eta} - \left(\frac{\mathrm{d}\psi}{\mathrm{d}x}\right)_{x=-\eta} = \frac{2mW_0}{\hbar^2}\int_{-\eta}^{\eta}\delta(x)\mathrm{d}x. \tag{5.5}$$

After using Bloch's theorem again and completing the integration, this yields

$$\left(\frac{\mathrm{d}\psi}{\mathrm{d}x}\right)_{x=\eta} - \exp(-Qd)\left(\frac{\mathrm{d}\psi}{\mathrm{d}x}\right)_{x=d-\eta} = \frac{2mW_0}{\hbar^2}. \tag{5.6}$$

When equation (5.3) for $\psi(x)$ is substituted into the boundary condition equations (5.4) and (5.6), we obtain two separate expressions connecting the constants A and B. After some straightforward manipulation the consistency condition leads to

$$\cos(Qd) = \cos(\lambda d) + W\frac{\sin(\lambda d)}{\lambda d}, \tag{5.7}$$

where $W = W_0\, dm/\hbar^2$ is a dimensionless constant. This is our main result in this subsection, giving us an implicit dispersion relation for E versus Q. A numerical plot to illustrate the features is shown in figure 5.2 taking $W = 1$. The overall behaviour is reminiscent of that found earlier in figures 1.10 and 2.13 for other 1D periodic systems (photonic crystals and acoustic superlattices, respectively), where there are bands and band gaps. In the present case, the gaps appear whenever $W > 0$ and they generally become more pronounced when W is increased. In particular, we note that there is a band gap around zero energy, which is as expected because a confined electron in quantum mechanics will have a zero-point energy (of the same order as the quantity E_0 used as a scaling factor in figure 5.2).

It is useful also to quote the more general KP dispersion relation in the specific case of square-shaped potential wells, as depicted in figure 5.1(a):

$$\cos(Qd) = \cosh(\kappa a)\cos[\lambda(d-a)] + \left(\frac{\kappa^2 - \lambda^2}{2\kappa\lambda}\right)\sinh(\kappa a)\sin[\lambda(d-a)], \tag{5.8}$$

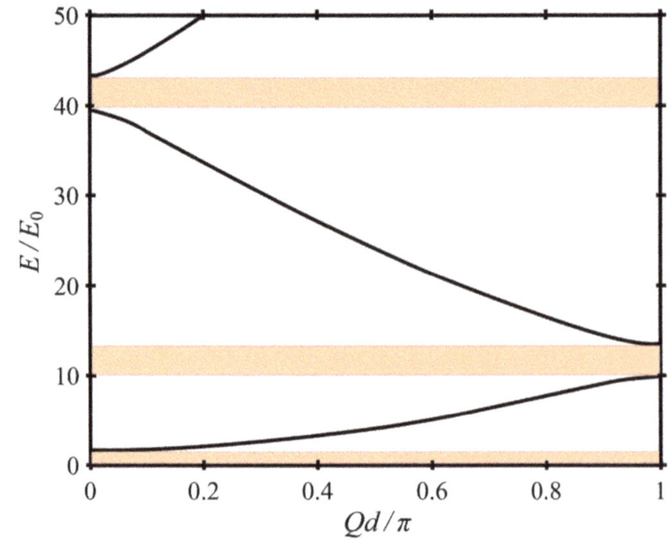

Figure 5.2. Example of dispersion curves (solid lines) showing the energy eigenvalues E plotted versus Bloch wave-number Q for the Kronig–Penney model using equation (5.7) with $W = 1$. Here the energy is scaled relative to the quantity $E_0 = \hbar^2/2md^2$. The shaded areas denote the band gaps.

where λ is defined as before and we denote $\kappa^2 = 2m(V_0 - E)/\hbar^2$. This result is more cumbersome to derive because it involves applying boundary conditions at an additional interface. We note that it simplifies to the result just derived in equation (5.7) when the limit is taken of $a \to 0$ such that $aV_0 = W_0$ (a constant).

Tamm surface states

As mentioned earlier, the KP model was extended in the seminal work by Tamm [1] to a semi-infinite geometry to study localized surface electronic states. Essentially, he considered the situation depicted in figure 5.1(b) with a 1D array of delta-function potentials terminating at one end (the surface). It was assumed that the surface was offset by a distance d' from the end delta function and that the exterior potential has a value U_0 that was larger than inside the solid. The KP calculation given earlier in this subsection for the periodic case can be extended to the semi-infinite array: the main difference is that Bloch's theorem with the wave-number Q is no longer applicable, but it can be replaced by an expression of the form $\psi(x + d) = \exp(-\Lambda d)\psi(x)$ where Re $(\Lambda) > 0$ for localization.

Taking $d' = \frac{1}{2}d$ and U_0 to be sufficiently large, Tamm showed that localized surface states are predicted in this model. As might be expected, the surface states occur at energies within the band-gap regions. Similar results for surface states can be obtained using the form of the potential in figure 5.1(a). A clear exposition of these calculations can be found in [7], where a generalization of the semi-infinite geometry to the finite-thickness case (with two surfaces) is also outlined.

An interesting experimental verification of Tamm states was given much later by Ohno *et al* [8] in an artificially fabricated layered nanostructure. They employed a superlattice of alternating $Al_{0.2}Ga_{0.8}As/GaAs$ layers with a terminating layer of AlAs for the perturbed 'surface', which is well-described by figure 5.1(a). From photoluminescence measurements they were able to demonstrate bulk and surface modes in a controlled fashion.

5.1.2 Other 1D calculations for surface states

Other 1D calculations (see e.g. [2, 3]) have employed more realistic forms for the atomic potentials than in the case of the KP theory; examples can be seen drawn schematically in figure 5.3. The following outline of the basic theory is broadly based on Shockley's work [3], but we refer also to the description given in [9].

Shockley surface states

We consider a model with N repeats of a potential well with cells of width d, as represented in figure 5.3(b). Inside any interior cell (with its centre chosen to be at $x = 0$), the general solution of Schrödinger's equation may be written as a linear combination of two parts $u(x)$ and $g(x)$ that are, respectively, odd and even functions of x:

$$\psi(x) = a_1 g(x) + \mathrm{i}a_2 u(x). \tag{5.9}$$

Here a_1 and a_2 are complex coefficients, and the insertion of the factor 'i' is just for later mathematical convenience. If the 1D Bloch's theorem is assumed to hold, then the expression for the wave function in the neighbouring cell centred at $x = d$ is

$$\psi(x) = \exp(\mathrm{i}Qd)[a_1 g(x - d) + \mathrm{i}a_2 u(x - d)]. \tag{5.10}$$

The boundary conditions on the wave function at the $x = d/2$ interface are that ψ and $\mathrm{d}\psi/\mathrm{d}x$ must be continuous there (assuming the potential remains finite), giving

$$\begin{aligned} a_1 g + \mathrm{i}a_2 u &= \exp(\mathrm{i}Qd)(a_1 g - \mathrm{i}a_2 u), \\ a_1 g' + \mathrm{i}a_2 u' &= \exp(\mathrm{i}Qd)(-a_1 g' + \mathrm{i}a_2 u'). \end{aligned} \tag{5.11}$$

Here u and g, and their derivatives u' and g' with respect to x, are all evaluated at $x = d/2$ and their symmetry properties have been utilized. The solvability condition for this pair of equations in a_1 and a_2 leads to the simple condition that

$$\tan^2(Qd/2) = -\left(\frac{g'u}{gu'}\right). \tag{5.12}$$

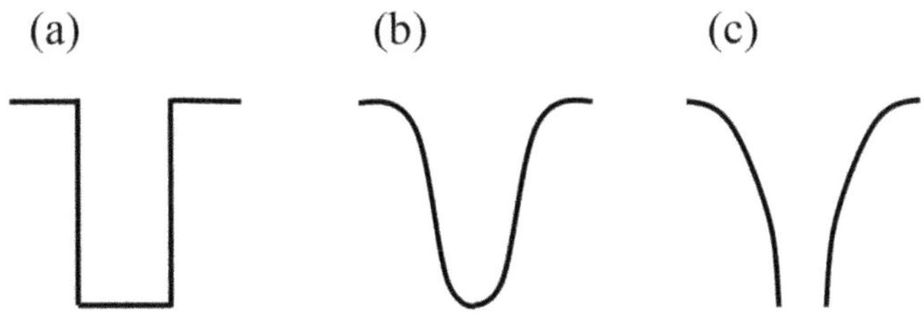

Figure 5.3. Schematic form of the 1D potentials used in calculations of electronic surface states: (a) square well; (b) Shockley-type well; (c) Goodwin-type well.

If we consider first the limit of $N \to \infty$, which is when Bloch's theorem with the use of a real wave vector Q is applicable, equations (5.10) and (5.12) provide us with information about the bulk energy bands in this 1D example. So, for a given $u(x)$ and $g(x)$, if the right-hand side of equation (5.12) is positive then there is real solution for Q. The electron energy given by $E = \hbar^2 Q^2/2m$ must lie within a pass band of electronic states (denoting m as the electronic mass or effective mass). A band gap (or stop band) corresponds to the situation when the right-hand side of equation (5.12) is negative.

By analogy with similar calculations made in chapters 2 to 4 for other excitations in 1D, the possibility of complex solutions for Q arises when N is finite. This situation can be investigated by matching the Bloch form of the wave functions inside the material to the wave function in the outside regions, where the potential can be assumed to have a constant value V_0. Hence the exterior wave functions will have the form of $\exp(\kappa x)$ on the left as $x \to -\infty$ and $\exp(-\kappa x)$ on the right as $x \to \infty$, where κ is a real and positive decay constant given by

$$\kappa = \sqrt{2m(V_0 - E)/\hbar^2}\,. \tag{5.13}$$

The outcome of treating the finite-N case in this manner is that the continuum of energy states in the pass bands now consists of discrete (or 'quantized') levels and also surface branches might occur in the stop bands, depending on details of the modelling. An numerical example, as obtained by Shockley [3], is shown in figure 5.4 taking $N = 8$ and atomic-type potentials. When the cell size d is very large, the energy-level diagram consists simply of the three levels (each one being eight-fold degenerate) for the isolated atoms. As d is reduced to a value of order d_1, as indicated, the degeneracy is lifted and the 8 discrete levels in each band are evident. Then for much smaller d ($\sim d_2$) it is seen that overlap effects begin to occur and that surface branches may occur in the band-gap regions.

A symmetry-related distinction that is sometimes made between Tamm and Shockley surface states is that the mode localization in the former case depends on there being a perturbation of the potential associated with the end cell(s) whereas in the latter case it is attributable to the termination of the array. A discussion of this characteristic property, along with the conditions for the coexistence of Tamm and Shockley states, is presented in [10]. An optical technique was employed by Malkova *et al* [11] in an experimental observation of Shockley-like surface states in a semi-infinite photonic superlattice. By varying either the external boundary condition (through control of the surface impurities) or by increasing the power of the optical probe beam to a nonlinear regime, a transition between Shockley and Tamm states could be achieved.

Other similar analytic 1D calculations are presented in the books by Lüth [6], Zangwill [12], and Lannoo and Friedel [13]. These references also provide details of generalizations for surface states in the formalism of electron-band theory applied to 3D structures, including schemes of surface reconstruction (where appropriate). Electronic states that are localized with interfaces between different materials are also analysed.

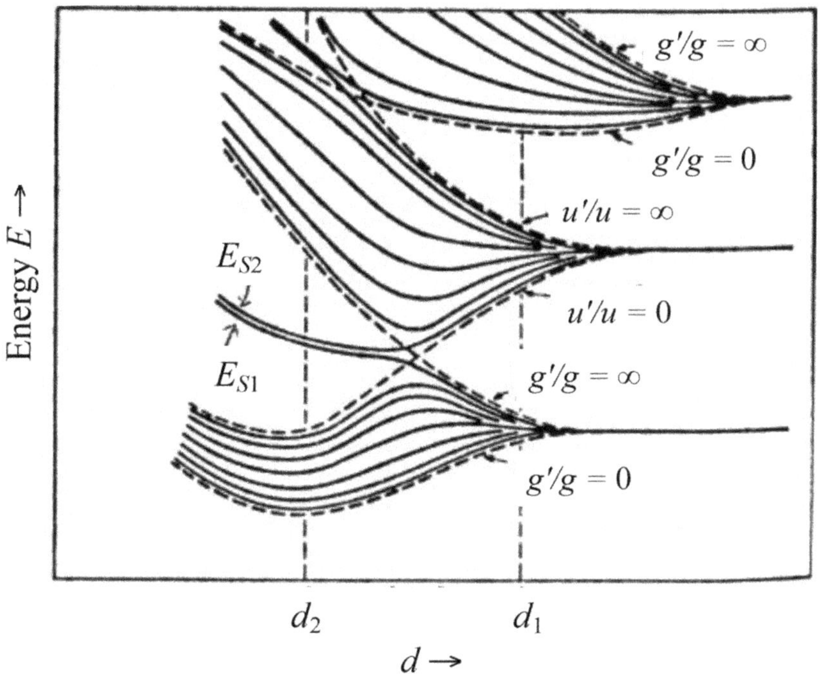

Figure 5.4. Plot of electronic energy E versus cell size d for a 1D model with $N = 8$ atomic potential wells. The dashed lines represent the boundaries of the bands corresponding to $N \rightarrow \infty$. (Reproduced with permission from [3]. Copyright 1939 American Physical Society.)

Extrinsic surface states

A distinction can often be usefully made between *intrinsic* and *extrinsic* electronic surface states. The former are those described so far in this section. In other words, the intrinsic surface states are a property of microscopically clean surfaces (even if surface reconstruction of the atoms and bonds may have occurred, as in pure Si surfaces [13]). By contrast with this case, extrinsic surface states are associated with adsorbed or absorbed atoms or molecules at the surface. The latter have also been widely studied both theoretically and experimentally [13]. An interesting case is provided by the (111) surface of Si, which can undergo a variety of surface reconstructions in both the pure state, when cleaved from a bulk Si crystal, or when the surface is patterned by a chosen coverage of (e.g.) Al, Ga, or In. As a specific example, we show in figure 5.5 a comparison of theory with experiment for a clean Si(111) surface with a 2×1 reconstruction [14]. This type of designation means that the lengths of the basic vectors at the surface ($\mathbf{a}_1$ and $\mathbf{a}_2$ in the notation of section 1.1.1) are scaled by factors of 2 and 1, respectively, in the reconstructed 2D surface layer. There are two cases shown for the energy of the surface states, depending on the directions chosen for the wave vectors in the 2D Brillouin zone (see the inset for the standard notation of symmetry points in the 2D Brillouin zone). In figure 5.5(a) some experimental data deduced from angle-resolved photoemission spectroscopy (ARPES) are indicated, giving reasonable agreement with theory. Some more recent

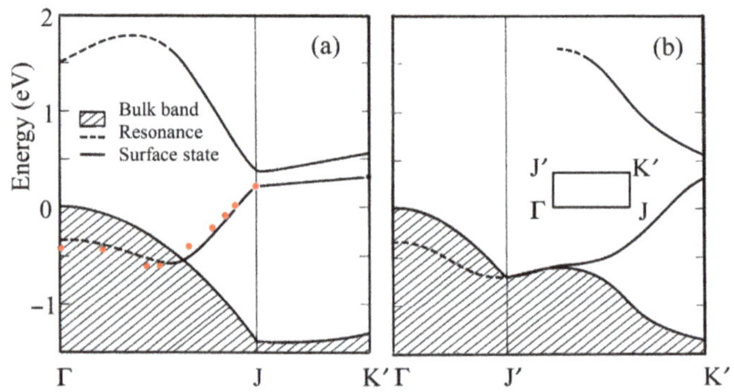

Figure 5.5. The dispersion with wave vector of surface electronic states for a Si(111) sample with a 2×1 reconstruction. (a) Comparison of theory for the energy of surface states (full line) and associated resonances (broken line) with angle-resolved photoemission measurements (data points in red) for the Γ–J–K′ directions. (b) Theory for Γ–J′–K′ directions. The labelling of symmetry points in the 2D surface Brillouin zone is shown as an inset in (b). (Reproduced with permission from [14]. Copyright 1982 American Physical Society.)

references relating to surface electronic states in semiconductor structures including some wires and dots, and mostly in the context of photonics applications, can be found in [15].

5.2 Graphene sheets and ribbons

Graphene is a 2D monolayer form of carbon that has been intensively studied, particularly in the last two decades, for its remarkable electronic band structure that includes zero-gap features in its energy spectrum at specific nonzero wave vectors. It shows amazing versatility in its electronic properties depending on whether it is produced in a pristine form or whether edges, impurities, adatoms, vacancies (or other forms of defects) are introduced. Likewise, the tailoring of magnetic properties (e.g. at edges) has attracted an enormous amount of attention. Although graphene was not successfully isolated and fabricated experimentally until 2004 in work [16] that led to the award of the 2010 Nobel Prize in Physics to Geim and Novoselov, there were some theoretical investigations long beforehand due to Wallace (see [17]). Some excellent reviews of graphene properties have been given by Castro-Neto *et al* [18] and by Power and Ferreira [19], for example.

The lattice structure of graphene was discussed already in chapter 1 (see figure 1.2) in terms of a 2D honeycomb space lattice with two interpenetrating sublattices of carbon atoms, labelled as A and B. Choices for the basic lattice vectors of a unit cell with two carbon atoms per cell were given there, as well as the corresponding basic vectors of the reciprocal lattice. The σ bond between adjacent carbon atoms that are separated by a distance $a_0 = 0.142$ nm results from the sp^2 hybridization. Additionally there is also a weaker π bonding between electrons with delocalized wave functions perpendicular to the planar structure, and it is these π electrons' properties that are conventionally studied for the low-lying electronic excitations. The edge states of graphene have been of particular interest following seminal results reported in [20, 21]. Apart from the edges being of intrinsic novelty they can also be functionalized by other chemical groups.

Graphene nanoribbons (GNRs) can be envisioned as long structures having a pair of edges along the same chiral direction in the bipartite lattice at a finite width apart. The simplest two cases, which are known as zigzag and armchair GNRs, are depicted in figures 5.6(a) and (b), respectively. Their differences can be highlighted by considering the rows of atoms formed by and parallel to the edges. The rows can be labelled by integer n (= 1, 2, 3, …N), as in the zigzag case shown here for $N = 8$. In the zigzag case each row contains only A or only B atoms. By contrast, each row parallel to an edge in the armchair case contains equal numbers of A and B atoms. Also it can be seen that the spacing of the rows follows a different scheme in the two cases.

A calculation of the bulk and edge modes in these GNRs will now be outlined, showing that the latter modes have quite distinctive properties. For symmetry reasons, it is the zigzag case that turns out to be of greater interest for the edge modes [20, 21], and so we focus on that structure. The starting point is the Hamiltonian, which we take as having the electronic tight-binding form in second quantization as

$$\mathcal{H} = -\sum_{i,j} t_{ij}(a_i^\dagger a_j + \text{H.c.}). \tag{5.14}$$

Here the fermion operators denoted by $a_i^\dagger$ and a_i create and annihilate, respectively, an electron at any site i, corresponding to an A sublattice site if i is located on an odd n row and a B sublattice site if i is on an even n row. Due to their different statistics, the fermion operators satisfy *anticommutation* properties among themselves [22], instead of the commutation properties used for boson operators in the previous chapters. This means that they have the properties that

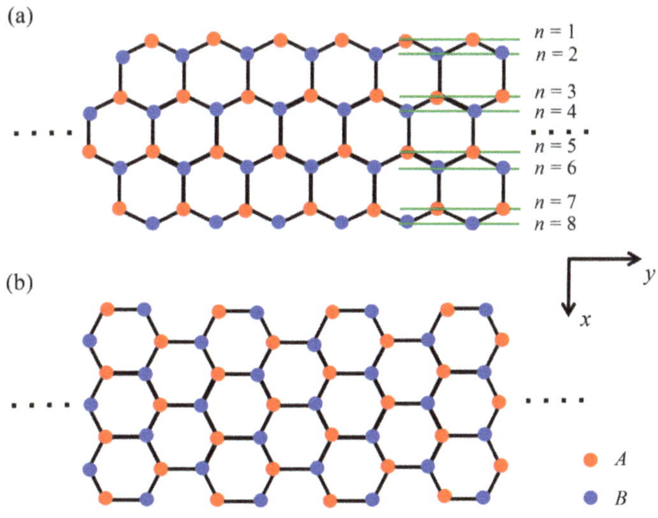

Figure 5.6. Assumed geometry of graphene nanoribbons (GNRs) with (a) zigzag edges and (b) armchair edges. Each carbon atom lies either on sublattice A or B as indicated. The GNRs are of finite width in the x-direction and they are assumed to be of infinite length in the y-direction.

$$\left\{a_i, a_j^\dagger\right\} \equiv a_i a_j^\dagger + a_j^\dagger a_i = \delta_{i,j}, \qquad \left\{a_i, a_j\right\} = \left\{a_i^\dagger, a_j^\dagger\right\} = 0. \tag{5.15}$$

Also t_{ij} is the hopping energy for an electron to move to any one of the nearest neighbours, which are on opposite sublattices (in fact, on adjacent rows). This hopping energy in graphene has a value denoted by t (and is well known [18] to be approximately 2.8 eV), except possibly at the zigzag edge where we assume that it may take a modified value denoted by t_e. We may now make a Fourier transform in the usual manner from site labels to a representation with a 1D wave vector q_y along the y-direction and a row number n. Equation (5.14) can then be rewritten as

$$\mathcal{H} = -\sum_{q_y, n, n'} \left[\tau_{nn'}(q_y) a_{q_y, n}^\dagger a_{q_y, n'} + \tau_{nn'}(-q_y) a_{q_y, n} a_{q_y, n'}^\dagger\right], \tag{5.16}$$

where the transformed hopping factor $\tau_{nn'}(q_y)$ for the zigzag structure has the form

$$\tau_{nn'}(q_y) = t\left[\beta(q_y)\delta_{n',n\pm1} + \delta_{n',n\mp1}\right]. \tag{5.17}$$

The upper (lower) signs refer n odd (even), and we have defined

$$\beta(q_y) = 2\cos\left(\frac{\sqrt{3}}{2} q_y a_0\right). \tag{5.18}$$

The excitations can now be investigated by using *either* the operator equation of motion (1.39) for the $a_{q_y, n}^\dagger$ or $a_{q_y, n}$ fermion operators, *or* the Green's function (GF) equation of motion, as in equation (A.24). Both methods involve working out the commutators of the creation or annihilation operators with the above Hamiltonian [23, 24]. However, the latter method (which is outlined in a general way in section A.2 of the appendix) is preferable here, because it eventually provides us with fuller information about the localized edge modes in terms of their position-dependent amplitudes. From equations (5.16) and (A.24) we may deduce that

$$\omega\, G(a_{q_y, n}^\dagger; a_{q_y, n'} \mid \omega) = \frac{1}{2\pi}\delta_{n,n'} + \sum_m \tau_{nm}(q_y)\, G(a_{q_y, m}^\dagger; a_{q_y, n'} \mid \omega). \tag{5.19}$$

This expression can be recognized as equivalent to a set of finite-difference equations in terms of integers n and n'. For any given (fixed) value of n' the finite-difference equations for the variable n can be reorganized such that the odd-n terms (referring to sublattice A) are eliminated in terms of the even-n terms (referring to sublattice B). The resulting equations can then be recast in a matrix form as $\mathbf{A}\,\mathbf{G}_{n'} = \mathbf{b}_{n'}$, where $\mathbf{G}_{n'}$ is the column matrix formed from the GFs $G(a_{q_y, n}^\dagger; a_{q_y, n'} \mid \omega)$ with n even, while $\mathbf{b}_{n'}$ is another column matrix with elements defined by

$$(\mathbf{b}_{n'})_n = \frac{\omega}{2\pi t^2 \beta(q_y)}\delta_{n,n'}. \tag{5.20}$$

Also, $\mathbf{A}$ is a tridiagonal matrix (TDM) that can be separated into two parts as $\mathbf{A}_0 + \mathbf{D}$ as described in section A.1 of the appendix.

To simplify matters, if we take the case of a wide GNR (by studying the semi-infinite limit that corresponds to $N \to \infty$), it is found that $\mathbf{A}_0$ takes precisely the form of equation (A.3) with the definitions

$$d = [(\omega/t)^2 - \beta^2(q_y) - 1]/\beta(q_y), \tag{5.21}$$

and the edge perturbation matrix $\mathbf{D}$ has only one nonzero element given by

$$D_{1,1} = D_e \equiv \left(1 - \frac{t_e^2}{t^2}\right)\beta(q_y). \tag{5.22}$$

We note that the edge parameter D_e vanishes if $t_e = t$ for the hopping terms.

It is now straightforward to solve for the GFs, and hence to deduce the frequencies of the excitations, by following the approach outlined in section A.1 (see also [24]). In particular, it is helpful to introduce the complex variable x, which satisfies $x + x^{-1} = d$ and is formally related to the frequency ω through equations (5.21) and (A.5). Then equation (A.4) is utilized to deduce the inverse of the matrix $\mathbf{A}$ and hence the GF matrix. To simplify the discussion we will quote here the results only for the special case where one of the GF labels refers to the zigzag edge. In other words, we choose $n' = 1$ to find

$$\begin{aligned}
G(a_{q_y,1}^{\dagger}; a_{q_y,1} \mid \omega) &= \frac{1}{2\pi\omega} + \frac{t_e^2\beta(q_y)}{2\pi t^2\omega(1 + xD_e)}x \quad (n = 1),\\
G(a_{q_y,n}^{\dagger}; a_{q_y,1} \mid \omega) &= \frac{t_e x^{1/2}[\beta(q_y) + x^{-1}]}{2\pi t\omega(1 + xD_e)}x^{n/2} \quad (\text{odd } n > 1),\\
G(a_{q_y,n}^{\dagger}; a_{q_y,1} \mid \omega) &= \frac{-t_e}{2\pi t^2(1 + xD_e)}x^{n/2} \quad (\text{even } n).
\end{aligned} \tag{5.23}$$

There are several points to make regarding the physical interpretation of these rather formal expressions. First, each GF involves overall factors of x to an integer power. As explained in section A.1 for the TDM method, the bulk-like (or propagating) modes are associated with $|x| = 1$. In fact, if we now put $x = \exp(i3q_xa_0/2)$ as a phase factor and use equations (5.21) and (A.5) we may deduce that $\omega = \omega_B$ where

$$\omega_B(q_x, q_y) = \pm t\sqrt{\beta^2(q_y) + 2\beta(q_y)\cos(3q_xa_0/2) + 1}. \tag{5.24}$$

This is identical to the standard form of the dispersion relation for the electronic band in an infinite graphene sheet [18], where q_x is a real wave-vector component in the x-direction. In contrast, any poles of the above GFs with $|x| < 1$ will correspond to localized modes, i.e. the edge modes in this graphene example. One way in which this can occur is through the factor $(1 + xD_e)$ that appears in all cases in equation (5.23). We may conclude that there is an edge mode for $x = -1/D_e$, but only provided the localization condition that $|D_e| > 1$ can be satisfied. Solving for the dispersion relation then gives $\omega = \omega_e$ where

$$\omega_e(q_y) = \pm t\sqrt{(t_e^2/t^2)\beta^2(q_y) + [t_e^2/(t_e^2 - t^2)]}. \tag{5.25}$$

The localization condition just mentioned can be satisfied for this type of edge mode in two ways: *either* we have $0 < (t_e/t)^2 < 0.5$ (when $D_e > 1$) *or* we have $(t_e/t)^2 > 1.5$ (when $D_e < -1$). These cases represent, respectively, acoustic and optic edge modes, as illustrated by the numerical examples in figure 5.7. We note, in particular, that there are no edge modes of this type if $t_e = t$, so they arise specifically as a consequence of perturbations in the edge hopping.

There is, however, another more subtle edge-mode effect that arises only from the odd n terms in equation (5.23) for the semi-infinite GNR. It can be seen that these GFs also have poles (vanishing denominators) at $\omega = 0$, and this so even when $t_e = t$. A simple rearrangement of equations (5.21) and (A.5) leads to an expression for ω^2 in terms of x as

$$\omega^2 = t^2[\beta(q_y) + x][\beta(q_y) + x^{-1}]. \tag{5.26}$$

This shows that when $\omega = 0$ we have either $x = -\beta(q_y)$ or $x = -1/\beta(q_y)$. In fact, it is the former condition that correspond to an edge mode, as can be seen by rewriting the second line of equation (5.23) in the form

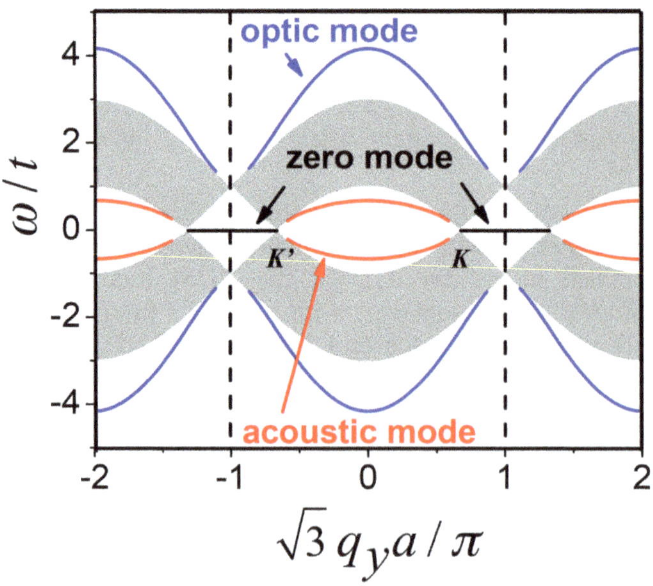

Figure 5.7. Dispersion curves for the mode frequencies versus longitudinal wave-number q_y for a semi-infinite graphene nanoribbon (GNR) with a zigzag edge. The two regions of bulk modes are shown shaded, with K and K' indicating the Dirac points. The vertical dashed lines correspond to the boundaries of the first Brillouin zone. Examples of optic edge modes when $t_e/t = 1.6$ and acoustic edge modes when $t_e/t = 0.4$ are shown, as well as the zero mode discussed in the text. (Reproduced with permission from [24]. Copyright 2014 American Institute of Physics.)

$$G(a^{\dagger}_{q_y, n}; a_{q_y,1} \mid \omega) = \frac{t_e}{2\pi t^2(1 + xD_e)} \sqrt{\frac{\beta(q_y) + x^{-1}}{\beta(q_y) + x}}\, x^{(n+1)/2} \text{ (odd } n > 1). \quad (5.27)$$

We therefore conclude that this type of edge mode exists only on the *odd* rows of the semi-infinite GNR. Also, for the localization condition to be satisfied, we must have $|\beta(q_y)| < 1$, which is the case if $q_y a_0$ lies in the range $2\pi/3\sqrt{3} < |q_y a_0| < \pi/\sqrt{3}$ in the first Brillouin zone. This excitation is often referred to as the *zero edge mode* and it was first predicted by Fujita *et al* [20]. It is seen as the flat line in figure 5.7 that extends from the Dirac points (labelled K and K') to the Brillouin zone boundary. The above analysis can be extended to show that the conclusions for the frequency and existence of the zero mode still apply when $t_e \neq t$, but the spectral intensity (which can be deduced from the GFs) is modified. The occurrence of these Dirac points, corresponding to 2D wave vectors where the electronic spectrum is gapless, is one of the properties that makes graphene of such special interest (see [18]). The wave vectors for K and K' in our notation are $(q_x, q_y) = (2\pi/3a_0, \pm 2\pi/3\sqrt{3}\,a_0)$. For an excitation situated close to the K point we can easily deduce from equation (5.24) that the approximate dispersion relation takes the form

$$\omega_B(q_x, q_y) \simeq \pm t\sqrt{\delta_x^2 + 3\delta_y^2} \qquad (|\delta_x| \ll 1), (|\delta_y| \ll 1), \quad (5.28)$$

where $\delta_x = \pi - \frac{3}{2}q_x a_0$ and $\delta_y = \frac{\pi}{3} - \frac{\sqrt{3}}{2}q_y a_0$ are the rescaled dimensionless wave numbers used in making the expansions.

A similar analysis can be followed to investigate the excitations in a semi-infinite GNR with an armchair edge, as illustrated in figure 5.6(b). It is found that the bulk modes can again be described by equation (5.24), *provided* that the roles of wave-vector components q_x and q_y are formally interchanged (so that q_y still denotes the longitudinal wave vector). In this case the zero-gap point in the bulk spectrum occurs at $q_y = 0$. However, the most important difference applies to the edge modes. It is found that there is no analogue of the zero mode for armchair edges. Further discussion is provided in [21, 25], including examples of ribbon edges with different orientations and GNRs with finite width.

The experimental evidence for edge modes in GNRs is indirect, coming from their influence on the electronic density of states which can be probed by electron spin resonance (ESR) from any paramagnetic impurity centres at the zigzag edges [26]. Lines of impurities or defects, along either the zigzag or armchair direction, are also of interest in GNRs. We mention that topological line defects are discussed, for example in [27], while self-assembly of lines of metal atoms on graphene is explored in [28, 29].

5.3 Bulk dielectric functions

In section 1.1.2 we introduced the concept of the (linear) dielectric function written as $\epsilon(\mathbf{q}, \omega)$, or more usually as $\epsilon(\omega)$ if spatial dispersion can be neglected. Also, a simple expression for $\epsilon(\omega)$ was quoted in equation (1.21) for the case of an electron-gas plasma and this was employed to illustrate the concept of a plasmon-polariton.

We now present an extended discussion of the dielectric function for a range of different types of materials and excitations, including some cases of spatial dispersion. Further accounts of this topic can be found in [9, 30–32], for example.

Plasmonic materials

We begin with a re-examination of the dielectric behaviour of an electron gas as in a metal or a doped semiconductor. A full quantum-mechanical treatment can be made in terms of the Lindhard function [33], but here we will follow a more straightforward physical approach where the electron gas is described hydrodynamically as a continuous fluid of electrons [31] with a particle density $n(\mathbf{r}, t)$, velocity $\mathbf{v}(\mathbf{r}, t)$ and therefore charge current density $\mathbf{j}(\mathbf{r}, t) = -n(\mathbf{r}, t)e\mathbf{v}(\mathbf{r}, t)$. The continuity equation of hydrodynamics requires that

$$\frac{\partial n}{\partial t} + \nabla \cdot (n\mathbf{v}) = 0, \tag{5.29}$$

and the velocity obeys the acceleration equation in the form

$$m\frac{D\mathbf{v}}{Dt} = -e(\mathbf{E} + \mathbf{v} \times \mathbf{B}) - \frac{m\mathbf{v}}{\tau} - \left(\frac{m\beta^2}{n}\right)\nabla n. \tag{5.30}$$

Here we are employing the total derivative operator $D/Dt = \partial/\partial t + \mathbf{v} \cdot \nabla$ and we have included electric- and magnetic-field terms for the force on an electron. Damping is added through the term involving the collision time τ and the final term allows for the compressibility of the electron gas, where it is often assumed that $\beta^2 = 3v_F^2/5$ with v_F denoting the Fermi velocity [31].

The next step is to linearize the above expressions by taking the electric field as the driving term, so $\mathbf{E}$ and $\mathbf{v}$ are first-order quantities and we write the fluctuating density as $n = n_0 + n_1$, where n_0 is a constant. Also we assume that the time-dependent quantities vary as $\exp(-\mathrm{i}\omega t)$, leading to the linearized equations in the form $\mathbf{j} = -n_0 e\mathbf{v}$ and

$$-\mathrm{i}\omega n_1 + n_0\nabla \cdot \mathbf{v} = 0, \tag{5.31}$$

$$\mathrm{i}\omega m\mathbf{v} = e(\mathbf{E} + \mathbf{v} \times \mathbf{B}_0) + m\mathbf{v}/\tau + (m\beta^2/n_0)\nabla n_1. \tag{5.32}$$

The term with $\mathbf{B}_0$ in equation (5.32) allows for the effect of a static magnetic field in addition to the electric field. Then, eliminating n_1 yields

$$\mathrm{i}\omega m\mathbf{v} = e(\mathbf{E} + \mathbf{v} \times \mathbf{B}_0) + m\mathbf{v}/\tau - (\mathrm{i}m\beta^2/\omega)\nabla\nabla \cdot \mathbf{v}, \tag{5.33}$$

from which we can now deduce the dielectric function as follows. Taking $\mathbf{E}$ to have a plane-wave form proportional to $\exp(\mathrm{i}\mathbf{q} \cdot \mathbf{r})$ in a bulk material, and putting the external magnetic field $\mathbf{B}_0 = 0$ for the present, we find

$$\mathbf{j} = \left(\frac{\mathrm{i}e^2 n_0}{m}\right)\frac{\mathbf{E}}{\omega^2 + \mathrm{i}\omega/\tau - \beta^2 q^2}. \tag{5.34}$$

From Maxwell's equation for curl **H**, when expressed in its two equivalent forms, we have the relationships

$$\mathrm{i}\mathbf{q} \times \mathbf{H} = \mathbf{j} - \mathrm{i}\omega\epsilon_0\epsilon_\infty\, \mathbf{E} = -\mathrm{i}\omega\epsilon_0\epsilon(\mathbf{q}, \omega)\, \mathbf{E}. \tag{5.35}$$

Here the constant ϵ_∞, which may differ from unity, has been included to allow for the effects of excitations with higher frequencies than the plasma mode. The modification is typically unimportant for metals, but may be useful in the case of doped semiconductors. From equations (5.34) and (5.35) we arrive at

$$\epsilon(\mathbf{q}, \omega) = \epsilon_\infty\left(1 - \frac{\omega_p^2}{\omega^2 + \mathrm{i}\omega/\tau - \beta^2 q^2}\right) \tag{5.36}$$

in terms of the plasma frequency ω_p defined earlier. This equation represents the dielectric function for an electron-gas plasma with effects of both spatial dispersion (the q dependence) and damping included. It is a generalization of the simpler result quoted in equation (1.21), which is now plotted in figure 5.8(a).

The extension of the theory to retain the effects of a static applied magnetic field, by using the full form of equation (5.33), is straightforward but algebraically lengthy. In fact, the dielectric function is found to generalize from a scalar to a gyrotropic matrix (or tensor) quantity. If the applied magnetic field is along the z-direction and spatial dispersion is ignored, the expression is

$$\bar{\epsilon} = \begin{pmatrix} \epsilon_a & \mathrm{i}\epsilon_b & 0 \\ -\mathrm{i}\epsilon_b & \epsilon_a & 0 \\ 0 & 0 & \epsilon_z \end{pmatrix}, \tag{5.37}$$

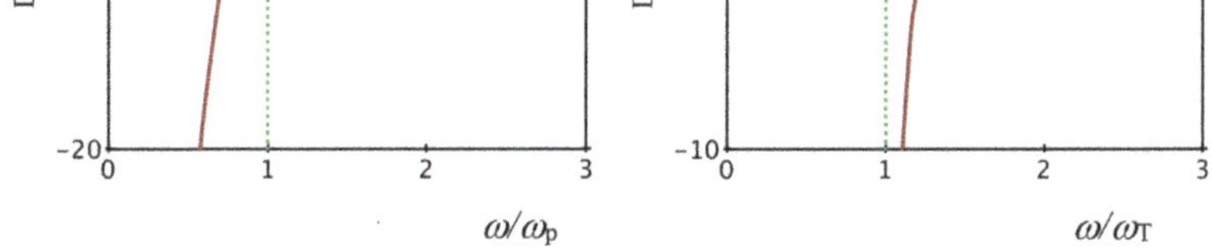

Figure 5.8. Contrasting forms of the scalar dielectric function $\epsilon(\omega)$ plotted versus frequency for (a) an electron-gas plasma and (b) an ionic crystal, taking spatial dispersion and damping to be negligible in both cases and choosing $\epsilon_\infty = 10$. In (a) we used equation (1.21) to plot $\epsilon(\omega)$ versus ω/ω_p, whereas in (b) we used equation (5.44) with $\omega_L/\omega_T = 1.2$ to plot $\epsilon(\omega)$ versus ω/ω_T.

where the terms are

$$\epsilon_a(\omega) = \epsilon_\infty\left(1 - \frac{\omega_p^2(\omega + \mathrm{i}/\tau)}{\omega[(\omega + \mathrm{i}/\tau)^2 - \omega_c^2]}\right), \tag{5.38}$$

$$\epsilon_b(\omega) = \epsilon_\infty\frac{\omega_p^2\omega_c}{\omega[(\omega + \mathrm{i}/\tau)^2 - \omega_c^2]}, \tag{5.39}$$

with ϵ_∞ unchanged from the scalar value in equation (5.36). The new angular frequency ω_c is the *cyclotron resonance frequency* (see e.g. [5]) for a charged particle moving in a magnetic field of magnitude B_0. We have

$$\omega_c = eB_0/m. \tag{5.40}$$

Ionic materials

Next we consider the form of the dielectric function in ionic crystals by following the lattice dynamics approach used by Born and Huang [32]. We use a simplified model in terms of an infinite diatomic lattice (e.g. as in section 2.1.2 for the 1D case). As well as the two sublattices having different masses for the atoms, they are now also associated with opposite charges. It is straightforward to see that it is the long-wavelength ($q \approx 0$) optic phonon that couples to the electromagnetic radiation. This is because a strong electric dipole moment is created when the two sublattices vibrate in antiphase. Hence if $\mathbf{u}$ denotes the relative vector displacement for the two sublattices, there will be two contributions to the polarization $\mathbf{P}$. One contribution will be proportional to $\mathbf{u}$ and the other will involve the electronic susceptibility χ and the electric-field vector $\mathbf{E}$, resulting in

$$\mathbf{P} = \epsilon_0(\alpha\mathbf{u} + \chi\mathbf{E}), \tag{5.41}$$

where α is a constant. Assuming a time dependence such as $\exp(-\mathrm{i}\omega t)$, the equation of motion for the displacement gives

$$(-\omega^2 - \mathrm{i}\omega\Gamma + \omega_T^2)\mathbf{u} = \beta\mathbf{E}_{\mathrm{loc}} = \gamma\mathbf{E}. \tag{5.42}$$

Here the left-hand side contains the usual oscillator terms (with $\Gamma \neq 0$ if there is damping) and ω_T denotes the frequency of the transverse optic (TO) phonon. The first term on the right-hand side strictly involves the *local* electric field $\mathbf{E}_{\mathrm{loc}}$ rather than the macroscopic electric field $\mathbf{E}$, which is obtained by averaging $\mathbf{E}_{\mathrm{loc}}$ over several unit cells of the crystal. However, it has been argued [30, 32] that these two fields are linearly related, and so the final step in equation (5.42) can be made where γ is a new proportionality factor.

Using equations (5.41) and (5.42) the polarization can be expressed as

$$\mathbf{P} = \epsilon_0\left(\frac{\alpha\gamma}{\omega_T^2 - \mathrm{i}\omega\Gamma - \omega^2} + \chi\right)\mathbf{E}. \tag{5.43}$$

It remains only to use equation (1.17) to conclude that the dielectric function, when spatial dispersion is ignored, is given by

$$\epsilon(\omega) = \epsilon_\infty\left(1 + \frac{\omega_L^2 - \omega_T^2}{\omega_T^2 - \mathrm{i}\omega\Gamma - \omega^2}\right). \tag{5.44}$$

To achieve this conventional form of the expression, we have defined $\epsilon_\infty = 1 + \chi$ and

$$\omega_L^2 = \omega_T^2 + \frac{\alpha\gamma}{\epsilon_\infty}, \tag{5.45}$$

where equation (5.45) defines the LO phonon frequency. In the static limit ($\omega \to 0$) of equation (5.44) we obtain

$$\epsilon(0) = \epsilon_\infty\left(\frac{\omega_L}{\omega_T}\right)^2, \tag{5.46}$$

which is sometimes known as the Lyddane–Sachs–Teller (LST) relation. A plot of the frequency dependence predicted by equation (5.44) for an ionic crystal with negligibly small damping is given in figure 5.8(b), where it can be compared with the behaviour in the case of an electron-gas plasma. Specifically, it has been verified that equation (5.44) provides an accurate description of the real and imaginary parts of $\epsilon(\omega)$ for undoped GaAs [34].

In some materials, such as semiconductors at an appropriate doping level, the plasma frequency and TO-phonon frequency might occur in a broadly similar range. In such cases it is useful to employ a combined form of equations (5.36) and (5.44) that has both plasma and TO-phonon resonances (see [35] for details).

Excitonic materials

Before concluding this section, we briefly consider excitonic materials. These are of interest because their dielectric functions typically exhibit strong spatial dispersion (or dependence on wave vector), contrasting with the previous examples covered. In semiconductors *excitons* can occur as electron–hole pairs in bound states, which arise as a consequence of the electrostatic attraction. Such a bound state may move through a crystal transporting the energy of excitation but not transporting any net charge. Most studies of excitons have focused on the following two limiting cases, e.g. see the book edited by Rashba and Sturge [36]. When the electron–hole attraction is strong (e.g. as in ionic crystals) we have the case of a Frenkel exciton where the electron and hole are close together, effectively on or near the same atom. In contrast, in most semiconductors the Coulomb interaction is strongly screened since the electron and hole are further apart (typically many lattice spacings) and they are only weakly bound to form a Wannier–Mott exciton. It is the latter case that we will explore further here.

The Wannier–Mott exciton occurs in a direct-gap, polar semiconductor such as ZnSe or GaAs, so it of relevance for many photonics and photoluminescence studies. It is formed from an electron (of effective mass m_e) near the bottom of the conduction band and a hole (of effective mass m_h) near the top of the valence band. In a simplified model the excitation energy $\hbar\omega(q)$ of the exciton can be written as

$$\hbar\omega(q) = E_g - \frac{m^* e^4}{2\hbar^2 \epsilon_s n^2} + \frac{\hbar^2 q^2}{2M}. \tag{5.47}$$

Here E_g is the band-gap energy and the second term on the right is a hydrogenic-type energy term, in which $m^* = m_e m_h/(m_e + m_h)$ denotes the reduced effective mass, ϵ_s denotes the static dielectric constant of the semiconductor, and $n = 1, 2, \ldots$ is a quantum number corresponding to the states of the system. The third term is an exciton kinetic energy contribution, where $M = m_e + m_h$ is the total exciton mass and $\hbar q$ is the momentum of the centre of mass. Equation (5.47) can be used to estimate the size of the exciton as represented by its effective Bohr radius, which will be large compared with the lattice spacing when ϵ_s is large and m^* is small. The exciton couples strongly to light, giving a contribution to the dielectric function which is q-dependent and has the form

$$\epsilon(q, \omega) = \epsilon_\infty + \frac{S}{\omega_e^2 + Dq^2 - \omega^2 - \mathrm{i}\omega\Gamma}, \tag{5.48}$$

where ω_e is a shorthand for the quantity $\omega(0)$ coming from the first two terms on the right-hand side of the dispersion equation (5.47), constant $D = \hbar/2M$, and S is an oscillator strength for the exciton. The main difference in form between the above result and the earlier result for phonons is the strong q-dependent term in the denominator, and apart from that the method of derivation via Maxwell's equations is rather similar.

The results for the dielectric functions obtained in this section, along with those for the gyromagnetic susceptibility matrix already found for ferromagnets and antiferromagnets in chapter 4, will be employed in chapter 6 to discuss polaritons in the electromagnetic regime where the excitations couple to photons.

5.4 Plasmonics in two dimensions

Here we examine some limiting cases of plasmonics in 2D by taking cases where the charges, which may be electrons or holes, are confined to a sheet of negligible thickness. For many years, a physical realization of such cases would typically be achieved at semiconductor heterojunctions, such as at the interface between GaAs and $Al_xGa_{1-x}As$ layers for certain Al concentrations $0 < x < 1$, or as charges trapped at the surface of liquid helium due to the image potential formed there. Nowadays, a graphene layer (with a monolayer thickness of order only 0.33 nm) at the interface between two other materials provides a more idealized realization for 2D plasmonics. We start here by examining the effects of a single charge sheet, and subsequently we generalize the results to the case of a superlattice formed by a periodic array of charge sheets.

5.4.1 Single charge sheet

First, we consider a single 2D electron gas (or 2DEG) localized at the $z = 0$ planar interface between two media, taking without loss of generality a wave propagating in the x-direction with wave vector q_x. The calculation for the case of a 2D hole gas

(2DHG) is similar. Two isotropic dielectric media, which we label as A and B, are assumed to fill the half-spaces above ($z > 0$) and below ($z < 0$) the charge sheet, respectively. These media are characterized here as having positive dielectric constants ϵ_A and ϵ_B.

The calculation proceeds by considering the electric and magnetic fields in regions A and B subject to boundary conditions at $z = 0$ for the charge sheet. We take the case where the electric field has the form

$$\mathbf{E}_J(\mathbf{r}, t) = (E_{Jx}, 0, E_{Jz}) \exp(\mathrm{i}q_x x) \exp(\mathrm{i}q_{Jz} z) \exp(-\mathrm{i}\omega t) \quad (J = A, B), \tag{5.49}$$

where there is a common q_x and ω, but we need to solve for the components q_{Jz} in the perpendicular direction. In general, they will satisfy

$$q_x^2 + q_{Jz}^2 = \frac{\epsilon_J \omega^2}{c^2} \quad (J = A, B). \tag{5.50}$$

If q_x is sufficiently large that $q_x^2 \gg \epsilon_J \omega^2/c^2$, corresponding to the so-called *electrostatic approximation*, then we have $q_{Az}^2 \simeq q_{Bz}^2 \simeq -q_x^2$. The appropriate choice of signs assumed here for localization as $z \to \pm\infty$ in this approximation is $q_{Az} = -q_{Bz} = \mathrm{i}q_x$ if $q_x > 0$. From Maxwell's equations the corresponding magnetic field must be along the y-direction, i.e. perpendicular to the x-direction of propagation. This corresponds to a transverse-magnetic (TM) wave, or p-polarization. We note in passing that the other possible choice for the fields, which is the transverse-electric (TE) wave or s-polarization, turns out to be uninteresting for achieving mode localization.

The electromagnetic boundary conditions at $z = 0$ require continuity of the tangential electric fields, so $E_{Ax} = E_{Bx}$, while the tangential y-component of the magnetic field are related to the interface charge current and satisfy

$$H_{By} - H_{Ay} = j_x. \tag{5.51}$$

By analogy with the case of the 3D plasma already discussed in section 5.3 (specifically comparing with equation (5.33) when collisions and spatial dispersion are ignored in the hydrodynamic limit), we have the classical linear-response result that $j_x = \sigma(\omega) E_{Ax}$ where

$$\sigma(\omega) = \frac{\mathrm{i}\nu e^2}{m^* \omega}. \tag{5.52}$$

Here ν is the number of particles per unit area in the charge sheet and m^* denotes an effective mass. From equations (5.50)–(5.52) and the assumed electric and magnetic-field polarizations, it is straightforward to show in general that the frequency ω of the localized plasmon satisfies

$$\frac{\epsilon_A}{\sqrt{q_x^2 - \epsilon_A(\omega/c)^2}} + \frac{\epsilon_B}{\sqrt{q_x^2 - \epsilon_B(\omega/c)^2}} = \frac{\Omega_0 c}{\omega^2}, \tag{5.53}$$

where $\Omega_0 = \nu e^2/\epsilon_0 mc$ is a characteristic frequency. Now, applying the electrostatic approximation as earlier, the above expression reduces to give

$$\omega = \sqrt{\frac{\Omega_0 c\, q_x}{\epsilon_A + \epsilon_B}}, \tag{5.54}$$

This last result, which was first derived by Stern [37], shows a gapless behaviour at $q_x = 0$ with a proportionality to $\sqrt{q_x}$. This behaviour is quite different from the 3D case, as can be seen on comparing with equation (1.22) at zero wave vector. The inclusion of retardation effects, leading to plasmon-polaritons, will be explored in chapter 6.

The analysis for a graphene sheet at an interface proceeds in a broadly analogous fashion, with the main difference being that the expression for the frequency-dependent conductivity $\sigma(\omega)$ in graphene is typically more complicated than quoted in equation (5.52). This is because of the band structure of graphene, which leads to there being intraband and interband contributions to σ. In fact, writing $\sigma(\omega) = \sigma_{\text{intra}}(\omega) + \sigma_{\text{inter}}(\omega)$ it is found (see e.g. [38, 39] and references therein) that

$$\begin{aligned} \sigma_{\text{intra}} &= \frac{2ie^2 k_B T}{\pi \hbar^2 \omega} \ln\left[2\cosh\left(\frac{\mu_c}{k_B T}\right)\right], \\ \sigma_{\text{inter}} &= \frac{e^2}{4\hbar}\left\{\frac{1}{2} + \frac{1}{\pi}\tan^{-1}\left(\frac{\hbar\omega - 2\mu_c}{2k_B T}\right) - \ln\left[\frac{(\hbar\omega + 2\mu_c)^2}{(\hbar\omega - 2\mu_c)^2 - (2k_B T)^2}\right]\right\}. \end{aligned} \tag{5.55}$$

Here the chemical potential μ_c can be controlled by an gate voltage applied between the edges of the graphene sheets. Note that the dependence on temperature T has been introduced in the above expressions. We stress that the interband contribution plays the leading role around the absorption threshold, which occurs at $\omega \approx 2\mu_c$, while the intraband contribution is important at relatively low frequencies compared with μ_c, so there is an aspect of competition between the two terms.

The dispersion relation for the 2D plasmon in graphene is more complicated than that in equation (5.54). It is found, however, that the frequency ω is still proportional to $\sqrt{q_x}$ as was the case for the 2DEG, whereas $\omega \propto \nu^{1/4}$ for the dependence on areal charge density, which is different from the 2DEG (see the review article by Huang *et al* [40]).

5.4.2 Superlattices of charge sheets

The analysis just carried out for a single charge sheet (either 2DEG or graphene) can be generalized to a periodic array of 2D charge sheets, each of negligible thickness, separated by one or more types of dielectric media. The calculations can readily be accomplished by following the transfer-matrix approach as described in section 1.5. The simplest model is to take an infinite array of identical charge sheets with separation D and embedded in the same material (denoted by A) with positive dielectric constant ϵ_A. This is depicted in figure 5.9, where we show the labelling of the interfaces (and hence the superlattice cells).

The z-axis is chosen to be perpendicular to the interfaces and, as in the previous subsection, we choose the in-plane propagation to be along the x-direction with wave vector q_x. Again we are choosing to consider TM electromagnetic waves, or p-polarization, with **E** in the xz-plane and **H** in the y-direction. As in the previous

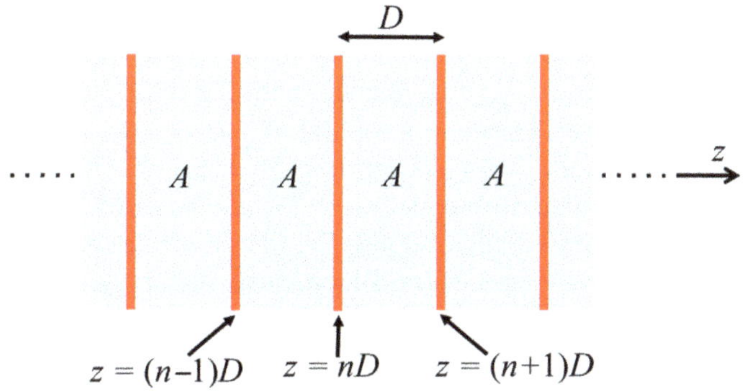

Figure 5.9. Model for an infinitely extended charge-sheet superlattice with periodicity length D. The charge sheets (shown in red) are embedded in a material A.

subsection we employ the electrostatic approximation, so the allowed values for the analogue of q_{Az} in equation (5.49) are $\pm iq_x$. Therefore, within any superlattice cell we need to form solutions for the electric and magnetic-field components that involve a linear combination of terms such as $\exp(q_x z)$ and $\exp(-q_x z)$ that express the z-dependence. For example, the E_x field in the nth cell, taken as the one between the charge sheets at $z = (n - 1)D$ and $z = nD$, can be written as

$$E_x = \exp(iq_x x)\left[a_n \exp[q_x(z - nD)] + b_n \exp[-q_x(z - nD])\right] \exp(-i\omega t), \quad (5.56)$$

where a_n and b_n are constants. By defining a column vector with the components a_n and b_n we have the basis of a 2×2 transfer-matrix method, following the analysis in section 1.5. We use Bloch's theorem, by analogy with equation (1.64), and also we need to apply the boundary conditions at the interfaces where the charge sheets are located. These boundary conditions are just the same as in the previous subsection, requiring continuity of the tangential electric fields and a discontinuity in the tangential magnetic field as in equation (5.51). The final result for the frequency of the superlattice bulk plasmons when the layers are 2DEG or 2DHG is

$$\omega(q_x, Q) = \Omega\left(\frac{q_x D \sinh(q_x D)}{2[\cosh(q_x D) - \cos(QD)]}\right)^{1/2}, \quad (5.57)$$

where Ω is a characteristic frequency given by

$$\Omega = \sqrt{\frac{\nu e^2}{\epsilon_0 \epsilon_A m^* D}}. \quad (5.58)$$

It is controlled by the areal density ν of charges and the periodic distance D for the superlattice. Equations (5.57) and (5.58) would need to be modified slightly when the charge layers are graphene, due to the different behaviour of the conductivity.

The above result for the plasmon modes in the electrostatic (or unretarded) approximated were first obtained by Fetter [41] and Das Sarma and Quinn [42]. The inclusion of retardation effects will be considered in chapter 6, along with further discussion of graphene sheets, in the context of plasmon-polaritons. Some limiting cases of equation (5.57) can be examined analytically. If $q_x D \gg 1$ it follows that ω is approximately proportional to $\sqrt{q_x}$ in agreement with the result in equation (5.54) for a single embedded charge sheet. On the other hand, if $q_x D \ll 1$ we conclude that, except when $QD = 0$, frequency $\omega \propto q_x$ with a proportionality factor that depends strongly on Q. The general behaviour predicted by equation (5.57) is illustrated in figure 5.10, where the shaded area shows the bulk band and also the lines indicate the frequency versus q_x dependence for several fixed values of the Bloch wave vector Q.

Several developments are possible, starting from this model. One extension is to the case of a semi-infinite superlattice, corresponding again to the structure in figure 5.9 but filling only the half-space $z \leqslant 0$. The surface superlattice modes were investigated for this case in [43], where two parameters are found to characterize the behaviour. These are the dielectric constant ϵ_s of the external medium filling $z > 0$ and the (possibly modified) charge density ν_s for the charge sheet at the external interface $z = 0$. The surface mode frequency ω then satisfies the equation

$$\begin{aligned}&\mu_s(\mu_s - 1)\sinh(q_x D)(q_x D\omega)^2 + \Omega^2(r_s^2 - 1)\sinh(q_x D)\\&+\Omega\left[r_s(1 - 2\mu_s)\sinh(q_x D) + \cosh(q_x D)\right] = 0,\end{aligned} \tag{5.59}$$

where we denote $r_s = \epsilon_s/\epsilon_A$ and $\mu_s = \nu_s/\nu$. Equation (5.59) is applicable provided ω is real and the following condition for localization near the $z = 0$ interface is satisfied:

$$\left|\left[r_s - (\mu_s q_x D\omega/\Omega)\right]\sinh(q_x D) + \cosh(q_x D)\right| < 1. \tag{5.60}$$

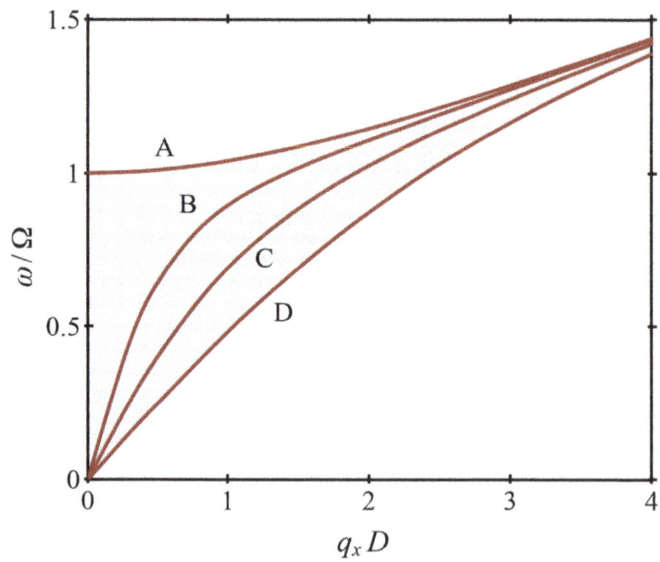

Figure 5.10. Numerical results for the reduced frequency ω/Ω plotted as a function of $q_x D$ for an infinite charge-sheet superlattice with periodicity length D using equation (5.57). The plasmon band is shown by the shaded area, and curves A, B, C and D correspond to QD values of 0, 0.2π, 0.4π and π, respectively.

Depending on the values of r_s and μ_s, a surface plasmon mode may occur either above or below the bulk plasmon superlattice region. Some numerical examples for these surface modes are given in figure 5.11.

In the case when graphene layers are used to form the charge sheets at interfaces, the theory is very similar to that just described. The main difference is that both intraband and interband contributions to the conductivity in graphene must be included, as in equation (5.55). Details can be found in the article by Zhan *et al* [44].

Other extensions of the basic charge-sheet model have included cases where the unit cell is more complicated than in figure 5.9. For example, the charge sheets might be separated by alternating layers *A* and *B* of two different materials and/or the same material but different thicknesses (see [43]) or the sheets might consist of alternating electron and hole layers [45]. Another application of interest is to n–i–p–i semiconductor superlattices [46], which have four components to the unit cell: n-type layer, i-layer (insulating), p-type layer and another i-layer (with the electron and hole sheets at the interfaces of the n and p regions, respectively). Various cases of quasiperiodic superlattices, with different growth rules, have also been studied [47].

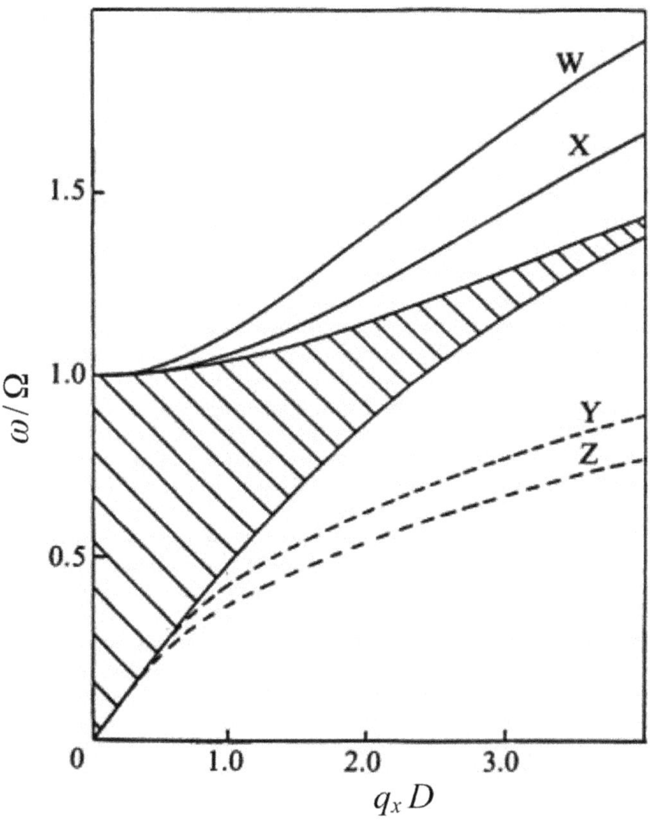

Figure 5.11. Numerical results for ω/Ω plotted as a function of q_xD for a semi-infinite charge-sheet superlattice with periodicity length D using equation (5.59). The bulk band is shown by the shaded area, and the curves correspond to surface plasmons: W, $r_s = 0.08$ and $\mu_s = 1$; X, $r_s = 0.08$ and $\mu_s = 0.75$; Y, $r_s = 4$ and $\mu_s = 1$; Z, $r_s = 4$ and $\mu_s = 0.75$. (Reproduced with permission from [43].)

5.5 Bulk-slab model for superlattice plasmons

The superlattice calculations in the previous section are mainly directed to what may be called *intrasubband plasmons*, since it is assumed that the electrons or holes form 2D mobile sheets. For the 2DEG and 2DHG cases, however, the charge carriers at a semiconductor interface are trapped by a potential well that has a small spatial width in the z-direction perpendicular to the interfaces, so the idealization only holds approximately. In practice, there is some degree of delocalization of the surface states, and when this is taken into account there may be multiple bound states for the carriers in this potential, leading to more than one subband. The plasmons that occur after allowance for the modifying influence of transitions between these subbands were termed *intersubband plasmons* (see e.g. [48]). In the case of graphene layers there are both interband and intraband effects, as already noted.

The bulk-slab model of a superlattice was introduced in section 1.5, where we were mainly concerned with the optical properties (photonics), and the dielectric functions of the constituent layers were taken as positive constants. For the applications here to plasmons it is now assumed that one or more of the constituent layers contains a volume distribution of mobile charges, so the dielectric function takes the form given in equation (5.36) in general, or the simpler form given in equation (1.21) if damping and spatial dispersion can be ignored.

We start with the simple geometry of a two-component *ABABAB*... superlattice, as in section 1.5, and we take the non-retarded limit as in the previous section of this chapter. This step allows us to simplify the previous bulk superlattice dispersion equation (1.70) by taking q_{Az} and q_{Bz} as $\pm iq_x$. Also we will assume p-polarization, since this is known to provide the more interesting case experimentally for plasmons, whereupon the required implicit dispersion relation becomes

$$\cos(QD) = \cosh(q_x d_A)\cosh(q_x d_B) + \frac{1}{2}\left(\frac{\epsilon_A}{\epsilon_B} + \frac{\epsilon_B}{\epsilon_A}\right)\sinh(q_x d_A)\sinh(q_x d_B). \quad (5.61)$$

A convenient alternative form of this expression [49] is

$$\epsilon_A/\epsilon_B = -\beta \pm (\beta^2 - 1)^{1/2} \quad (5.62)$$

with

$$\beta = \frac{\cosh(q_x d_A)\cosh(q_x d_B) - \cos(QD)}{\sinh(q_x d_A)\sinh(q_x d_B)}. \quad (5.63)$$

Then, when ϵ_A and ϵ_B both have the bulk plasma form of equation (1.21) with plasma frequencies ω_{pA} and ω_{pB}, respectively, we conclude that there are two solutions for ω (denoted by ω^+ and ω^-). These represent two bulk superlattice bands and their frequencies are given explicitly by

$$\omega^\pm = \omega_{pB}\left(1 - \frac{s(1 - r^2)}{s + \beta \mp (\beta^2 - 1)^{1/2}}\right)^{1/2}, \quad (5.64)$$

where we have defined the ratios $r = \omega_{pA}/\omega_{pB}$ and $s = \epsilon_{\infty A}/\epsilon_{\infty B}$.

Much of the experimental work on plasmons in superlattices (see e.g. [9, 50] for reviews) was based on Raman scattering (RS). The semiconductor system GaAs/ $Al_xGa_{1-x}As$ has been the focus of many studies because multilayers with good-quality epitaxial interfaces can be grown for a range of values of the Al concentration x. Typically the multi-quantum-well (MQW) structure is such that the quantum wells (for the electrons) correspond to the GaAs layers, so these will have a volume concentration of electrons while the $Al_xGa_{1-x}As$ layers do not. We therefore have a realization of the above theory in which we take medium A as $Al_xGa_{1-x}As$ with $\omega_{pA} = 0$ and medium B as GaAs with $\omega_{pB} \neq 0$. Hence $r = 0$, and usually it is a good approximation to take $s = 1$.

The first RS measurements to detect plasmons in GaAs/$Al_xGa_{1-x}As$ were reported by Olego *et al* [51]. They employed a geometry for the experiments in which the incident light was backscattered at 90° from a planar surface. This is the case of $\theta + \phi = 90°$ in the inset (lower left) to figure 5.12, where RS spectra are also

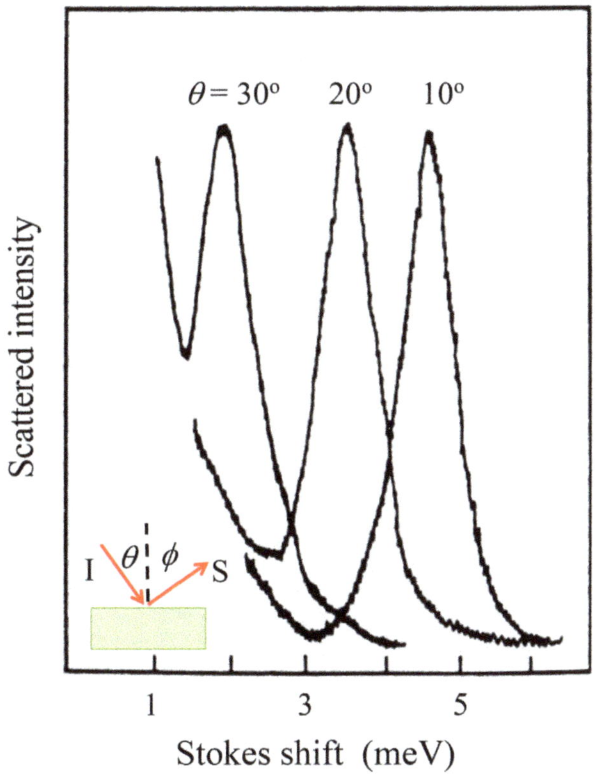

Figure 5.12. Raman scattering spectra from plasmons in a superlattice of alternating layers of A as $Al_xGa_{1-x}As$ (medium A) and GaAs (medium B) for three values of the incident angle θ. The layer thicknesses are $d_A \simeq 63$ nm and $d_B \simeq 26$ nm (so $D \simeq 89$ nm). The scattering geometry is illustrated in the inset, where I and S refer to the incident and scattered light, respectively, and $\theta + \phi = 90°$ in the experiments. (Adapted with permission from [51]. Copyright 1982 American Physical Society.)

shown for three different values of the incident angle θ. The magnitude of in-plane wave vector $q_{\|}$ imparted to the plasmon is easily seen to be

$$q_{\|} = (2\pi/\lambda)|\sin\theta - \cos\theta| \tag{5.65}$$

where λ denotes the vacuum wavelength of the incident light. The three spectra in figure 5.12 clearly illustrate a strong variation in the superlattice plasmon frequency as θ (and hence $q_{\|}$) are changed. A theoretical analysis was made by Babiker *et al* [52] and some results are shown in figure 5.13, where the plasmon frequencies are calculated as described here using the bulk-slab model. The theory lines are characterized in terms of the parameter $\delta \equiv d_B/D$, which is just the fractional thickness of the charge layer. The full lines (V, W and Y for different δ) correspond to the 'acoustic', or lower-band ω^-, solutions in equation (5.64), whereas the dashed lines (W, Y and Z) correspond to the 'optic', or upper-band ω^+ solutions. In the limit where $\delta \to 0$ (case V) the results of the bulk-slab model yield the same as the 2DEG charge sheets model for the ω^- frequency while $\omega^+ \to \infty$ for the other branch. When $\delta = 0.29$ (case W), corresponding to the experimental data shown in figure 5.12, the ω^- theory curve shifts downwards and is seen to be in good agreement with the RS

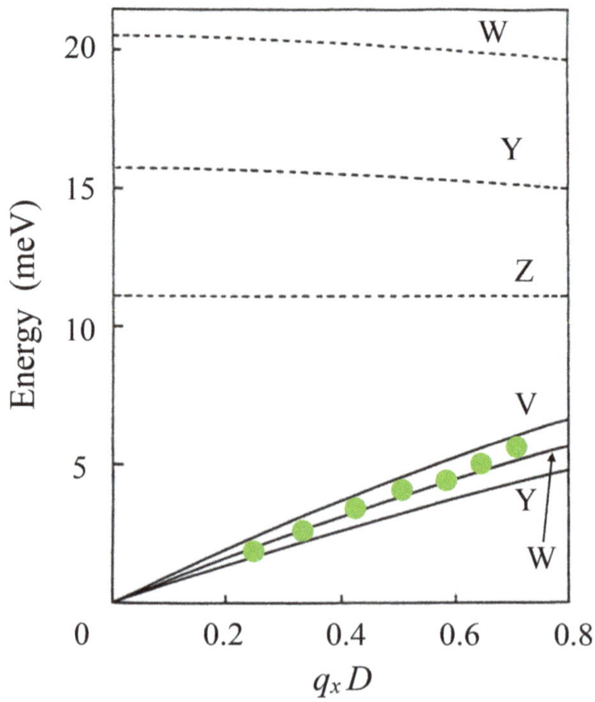

Figure 5.13. Theoretical plasmon dispersion curves for GaAs/Al$_x$Ga$_{1-x}$ As superlattices, showing upper (ω^+, dashed lines) and lower (ω^-, full lines) branches for several values of $\delta \equiv d_B/D$: V, 0.0; W, 0.29; Y, 0.6; Z, 1.0. The corresponding experimental ω^- results (crosses) for $\delta = 0.29$ are included [51]. (Adapted with permission from [52]. Copyright 1986 Elsevier.)

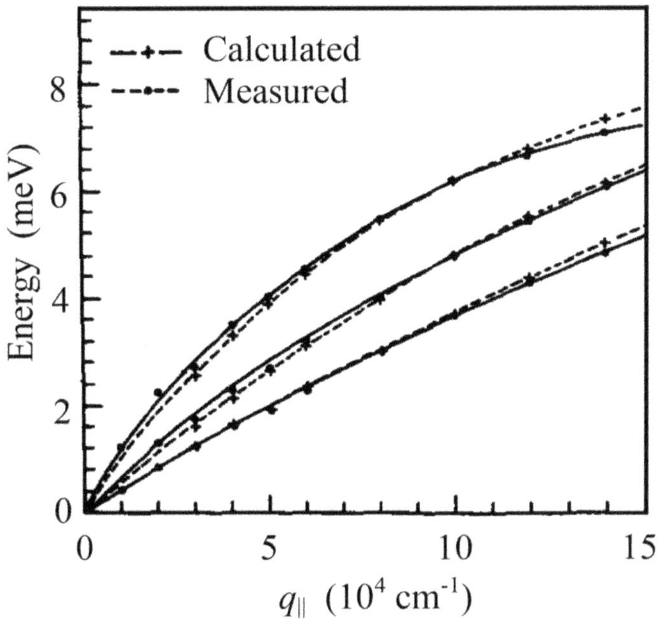

Figure 5.14. Experiment and theory for the lowest three plasmon bands in a GaAs/$Al_{0.3}Ga_{0.7}As$ Fibonacci superlattice of multi-quantum wells (MQWs). Calculated and measured (by Raman scattering) values of energy $\hbar\omega(q_{\|})$ at discrete values of the in-plane wave vector $q_{\|}$. The solid lines are a guide to the eye. (Reproduced with permission from [53]. Copyright 1992 Elsevier.)

data. Calculations were also made for the light scattering integrated intensities by means of electric-field GF methods.

Finally, we mention that plasmons in quasiperiodic superlattices have been extensively studied, both experimentally (e.g. by RS) and theoretically (using the bulk-slab model and its charge sheets limit). A review can be found in [47]. In figure 5.14 a comparison is made between calculated and measured plasmon frequencies in a GaAs/$Al_{0.3}Ga_{0.7}As$ Fibonacci superlattice [53]. In this case the building blocks A and B, assembled in sequence as described in section 2.4.1 for phonons and in figure 2.15(a) generally, were *compound* structures. Specifically block A consisted of 27 nm of GaAS and 80 nm of $Al_{0.3}Ga_{0.7}As$, while block B was 27 nm of GaAS and 43 nm of $Al_{0.3}Ga_{0.7}As$. In the Fibonacci superlattice the $Al_{0.3}Ga_{0.7}As$ barriers were doped so that the multi-quantum wells (MQWs) were formed by trapping of the charge carriers in the 27 nm thick GaAs layers, which are unequally spaced due to the growth rule. The theory was carried out using a modified tight-binding approximation for MQWs. It can be seen from the figure that the agreement between experiment and theory for the energy $\hbar\omega(q_{\|})$ of the three lowest plasmon bands is excellent.

References

[1] Tamm I E 1932 *Sov. Fiz. Zhurn.* **1** 733
[2] Goodwin E T 1939 *Proc. Camb. Phil. Soc.* **35** 232

[3] Shockley W 1939 *Phys. Rev.* **56** 317
[4] Snoke D W 2020 *Solid State Physics: Essential Concepts* 2nd edn (Cambridge: Cambridge University Press)
[5] Kittel C 2004 *Introduction to Solid State Physics* 8th edn (New York: Wiley)
[6] Lüth H 1995 *Surfaces and Interfaces of Solid Materials* (Berlin: Springer)
[7] Davison S G and Stęślica M 1996 *Basic Theory of Surface States* (Oxford: Oxford University Press)
[8] Ohno H, Mendez E E, Brum J A, Hong J M, Agulló-Rueda F, Chang L L and Esaki L 1990 *Phys. Rev. Lett.* **64** 2555
[9] Cottam M G and Tilley D R 2005 *Introduction to Surface and Superlattice Excitations* 2nd edn (Bristol: IOP Publishing)
[10] Kłos J and Puszkarski H 2003 *Phys. Rev.* B **68** 045316
[11] Malkova N, Hromada I, Wang X, Bryant G and Chen Z 2009 *Opt. Lett.* **34** 1633
[12] Zangwill A 1988 *Physics at Surfaces* (Cambridge: Cambridge University Press)
[13] Lannoo M and Friedel P 1991 *Atomic and Electronic Structure of Surfaces* (Berlin: Springer)
[14] Northrup J E and Cohen M L 1982 *Phys. Rev. Lett.* **49** 1349
[15] Pavesi L and Lockwood D J (ed) 2004 *Silicon Photonics* (Berlin: Springer)
[16] Novosolev K S, Geim A K, Morozov S V, Jiang D, Zhang Y, Dubonos S V, Gregorieva I V and Firsov A A 2004 *Science* **306** 666
[17] Wallace P R 1947 *Phys. Rev.* **71** 622
[18] Castro Neto A H, Guinea F, Peres N M R, Novosolov K S and Geim A K 2009 *Rev. Mod. Phys.* **81** 109
[19] Power S R and Ferreira M S 2013 *Crystals* **3** 49
[20] Fujita M, Wakabayashi K, Nakada K and Kusakabe K 1996 *J. Phys. Soc. Japan* **65** 1920
[21] Nakada K, Fujita M, Dresselhaus G and Dresselhaus M S 1996 *Phys. Rev.* B **54** 17954
[22] Landau L D and Lifshitz E M 1980 *Statistical Physics* (Oxford: Pergamon)
[23] Cunha A M C, Ahmed M Z, Cottam M G and Costa Filho R N 2014 *J. Appl. Phys.* **116** 013704
[24] Akbari-Sharbaf A and Cottam M G 2014 *J. Appl. Phys.* **116** 194309
[25] Costa-Filho R N, Farias G and Peeters F 2007 *Phys. Rev.* B **76** 193409
[26] Akbari-Sharbaf A, Cottam M G and Fanchini G 2013 *J. Appl. Phys.* **114** 024309
[27] Lahiri J, Lin Y, Bozkurt P, Oleynik I I and Batzill M 2010 *Nat. Nanotechnol.* **5** 326
[28] Zhou J, Zhang S, Wang Q, Kawazoe Y and Jena P 2015 *Nanoscale* **7** 2352
[29] Ouyang T, Akbari-Sharbaf A, Park J, Bauld R, Cottam M G and Fanchini G 2015 *RSC Adv.* **5** 98814
[30] Ashcroft N W and Mermin N D 1976 *Solid State Physics* (New York: Saunders College Publishing)
[31] Boardman A D (ed) 1982 *Electromagnetic Surface Modes* (Chichester: Wiley)
[32] Born M and Huang K 1985 *Dynamical Theory of Crystal Lattices* (Oxford: Clarendon)
[33] Lindhard J 1954 *Kgl. Danske Mat.-Fys. Medd.* **28** 8
[34] Kim O K and Spitzer W G 1979 *J. Appl. Phys.* **50** 4362
[35] Palik E D, Kaplan R, Gammon R W, Kaplan H, Wallis R F and Quinn J J 1976 *Phys. Rev.* B **13** 2497
[36] Rashba E I and Sturge M D (ed) 1982 *Excitons* (Amsterdam: North-Holland)
[37] Stern F 1967 *Phys. Rev. Lett.* **18** 546
[38] Madani A and Entezar S R 2013 *Physica* B **431** 1

[39] Vasconcelos M S and Cottam M G 2019 *J. Phys. D: Appl. Phys.* **52** 285104
[40] Huang S, Song C, Zhang G and Yan H 2017 *Nanophotonics* **6** 1191
[41] Fetter A L 1974 *Ann. Phys., NY* **88** 1
[42] Das Sarma S and Quinn J J 1982 *Phys. Rev.* B **25** 7603
[43] Constantinou N C and Cottam M G 1986 *J. Phys. C: Solid State Phys.* **19** 739
[44] Zhan T, Shi X, Dai Y, Liu X and Zi J 2013 *J. Phys.: Condens. Matter* **25** 215301
[45] Qin G, Giuliani G F and Quinn J J 1983 *Phys. Rev.* B **28** 6144
[46] Farias G A, Auto M M and Albuquerque E L 1988 *Phys. Rev.* B **38** 12540
[47] Albuquerque E L and Cottam M G 2003 *Phys. Rep.* **376** 225
[48] Eliasson G, Hawrylak P and Quinn J J 1987 *Phys. Rev.* B **35** 5569
[49] Camley R E and Mills D L 1984 *Phys. Rev.* B **29** 1695
[50] Pinczuk A 1984 *J. Physique* **45** C5–477
[51] Olego D, Pinczuk A, Gossard A C and Wiegmann W 1982 *Phys. Rev.* B **25** 7867
[52] Babiker M, Constantinou N C and Cottam M G 1986 *Solid State Commun.* **59** 751
[53] Merlin R, Valladares J P, Pinczuk A, Gossard A C and English J H 1992 *Solid State Commun.* **84** 87

Michael G Cottam

Chapter 6

Polaritons

The basic concept of polaritons was introduced in chapter 1 in terms of the coupling of an electromagnetic wave (a photon) to a crystal excitation in an unbounded homogeneous dielectric medium. For example, simple results for a bulk plasmon-polariton were presented in equation (1.22) and figure 1.3 for unbounded media. In this chapter we discuss various different kinds of polaritons, both nonmagnetic and magnetic, at surfaces and interfaces and in nanostructured materials. The full form of Maxwell's equations will be employed, allowing the effects of electromagnetic retardation to be included for the excitations. The role of the dielectric function of the constituent materials (or the magnetic susceptibility function for magnetic materials) will be important, and we will make extensive use of the results derived for those quantities in chapters 4 and 5.

Some excellent accounts of the earlier work on polaritons can be found in the volumes edited by Burstein and de Martini [1], Agranovich and Mills [2] and Boardman [3], while the book by Albuquerque and Cottam [4] focuses on multilayer structures that are either periodic or quasiperiodic. Related experimental methods, including the method of attenuated total reflection (ATR) and other optical techniques, are reviewed in [5, 6]. A novel development for polaritons in recent years has come from using graphene sheets, either as an added layer on the surface of a film or with multiple embedded layers in a material. This has been particularly relevant for plasmon-polaritons, and has led to the burgeoning field of graphene plasmonics described later.

6.1 Phonon-polaritons

We start with the well-known result found in most solid-state physics textbooks (see [7]) for the bulk phonon-polaritons in an ionic crystal. This can be obtained

doi:10.1088/978-0-7503-3903-2ch6

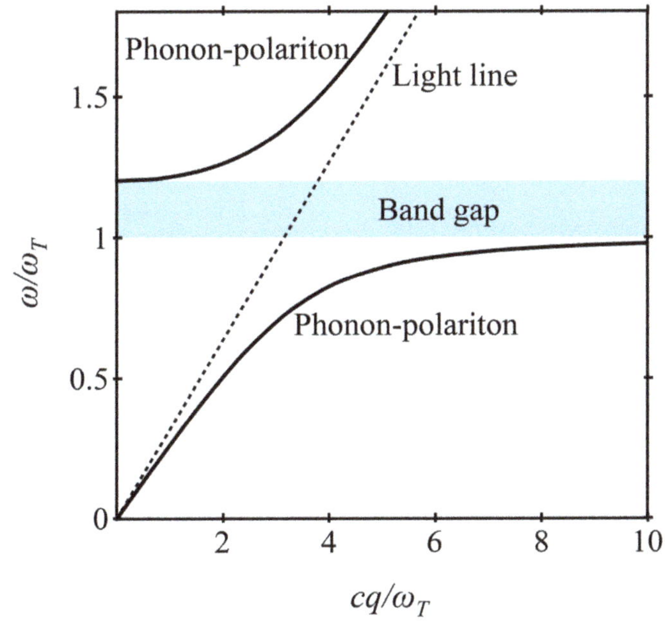

Figure 6.1. Dispersion relation for the two branches of a linear phonon-polariton (solid lines) in a 3D ionic material, as given by equations (1.22) and (6.2). The frequency, in terms of dimensionless ω/ω_T is plotted versus dimensionless wave number, in terms of cq/ω_T taking $\omega_L/\omega_T = 1.2$ and $\epsilon_\infty = 10$.

straightforwardly by putting the damping equal to zero in equation (5.44) for the dielectric function, which then simplifies to

$$\epsilon(\omega) = \epsilon_\infty \left(\frac{\omega_L^2 - \omega^2}{\omega_T^2 - \omega^2} \right). \tag{6.1}$$

Then we substitute $\epsilon(\omega)$ into the general expression in equation (1.22) to obtain the dispersion relation as

$$\frac{q^2c^2}{\epsilon_\infty \omega^2} = \frac{\omega_L^2 - \omega^2}{\omega_T^2 - \omega^2}. \tag{6.2}$$

On rearranging, the quadratic equation in ω^2 yields two solutions, corresponding to a dispersion relation with two branches, as shown in figure 6.1. The light line corresponds to $\omega = cq/\sqrt{\epsilon_\infty}$. This line represents one of the asymptotic solutions as $q \to \infty$, the other being $\omega = \omega_T$. There is a stop band (or band gap) for the range $\omega_T < \omega < \omega_L$; this is where we would have $q^2 < 0$ and hence where we might expect to find localized modes when surfaces are introduced.

6.1.1 One-interface geometry

We suppose now that there is a single planar interface at $z = 0$ between two semi-infinite dielectric media described by isotropic dielectric functions $\epsilon_A(\omega)$ and $\epsilon_B(\omega)$ as

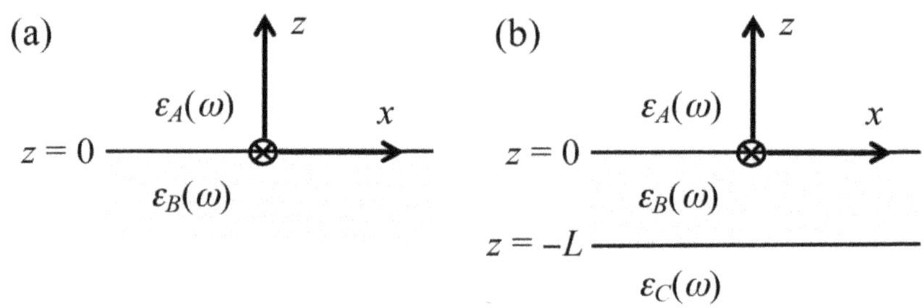

Figure 6.2. Assumed geometry and notation for polariton calculations in nonmagnetic materials. (a) Case of a single interface (at $z = 0$) between two dielectric media A and B. (b) Case of two interfaces (at $z = 0$ and $z = -L$) involving dielectric media A, B and C.

illustrated in figure 6.2(a). The mode propagation parallel to the interface will be taken along the x-axis with wave vector denoted by q_x. It can be verified that there is no surface mode corresponding to the electric field in the y-direction, and therefore we restrict attention here to the case where the electric field $\mathbf{E}_J(\mathbf{r}, t)$ in each of the media $J = A, B$ is in the xz-plane, written as in equation (5.49) for a p-polarized or TM wave. Also we must have $\text{Im}(q_{Az}) > 0$ and $\text{Im}(q_{Bz}) < 0$ for localization, where both quantities satisfy equation (5.50). If the ϵ_J are real, it follows that $q_x^2 > \epsilon_J \omega^2/c^2$ for localization. In fact, we may write $q_{A_z} = \text{i}\kappa_A$ and $q_{B_z} = -\text{i}\kappa_B$ where

$$\kappa_J = \sqrt{q_x^2 - \epsilon_J(\omega^2/c^2)} > 0 \quad (J = A, B). \tag{6.3}$$

One of Maxwell's equations to be satisfied at the $z = 0$ interface is $\nabla \cdot \mathbf{D} = 0$, since there is no charge sheet in the present calculation. This implies for the electric-field components that

$$q_x E_{Jx} + q_{J_z} E_{zJ} = 0 \quad (J = A, B). \tag{6.4}$$

The other boundary conditions require the continuity of the tangential $\mathbf{E}$ field and the normal $\mathbf{D}$ field, giving

$$E_{Ax} = E_{Bx}, \qquad \epsilon_A E_{Az} = \epsilon_B E_{Bz}. \tag{6.5}$$

The four equations in (6.4) and (6.5) constitute a solvability condition for the four electric-field components, leading to the condition that

$$q_{A_z}/\epsilon_A = q_{B_z}/\epsilon_B. \tag{6.6}$$

Then substitution into equation (5.50) in terms of q_x^2 gives the surface-polariton dispersion relation in the form

$$q_x^2 = \frac{\omega^2}{c^2}\left(\frac{\epsilon_A \epsilon_B}{\epsilon_A + \epsilon_B}\right). \tag{6.7}$$

We have already established that q_{Az} and q_{Bz} must have opposite signs, so it follows from equation (6.6) that this must also be true for ϵ_A and ϵ_B at the surface-mode frequency. We therefore conclude that two necessary conditions are $\epsilon_A \epsilon_B < 0$

and $\epsilon_A + \epsilon_B < 0$, where the latter condition is deduced from equation (6.7). These requirements can be satisfied in various ways, and in many experimental situations it may be the case that one of the media (say A) has a positive and frequency-independent dielectric constant ϵ_A, while the other medium B (referred to as the *surface-active medium*) has the negative and frequency-dependent dielectric function.

So far, our discussion of surface polaritons has been quite general, but now we particularize to the phonon-polariton case. Specifically we choose medium A to be vacuum or air, so $\epsilon_A = 1$, and medium B to be an ionic crystal with real dielectric function given by equation (6.1). It follows from our discussion in the previous paragraphs that a surface phonon-polariton can exist in the frequency range corresponding to $-\infty < \epsilon(\omega) < -1$, which means explicitly a range of ω from ω_T to ω_S, where we define

$$\omega_S = \sqrt{\frac{\epsilon_\infty \omega_L^2 + \omega_T^2}{\epsilon_\infty + 1}}. \tag{6.8}$$

This characteristic frequency ω_S lies in the band gap between ω_T and ω_L. The predicted dispersion of the surface phonon-polariton is illustrated in figure 6.3 for the same choice of active-medium parameters as made previously for the bulk case in figure 6.1. We note that the surface polariton exists only to the right of the *vacuum* light line, from which it emerges at $\omega = \omega_T$. The explicit algebraic form of the ω versus q_x relationship for the surface mode is quoted, for example, in [4].

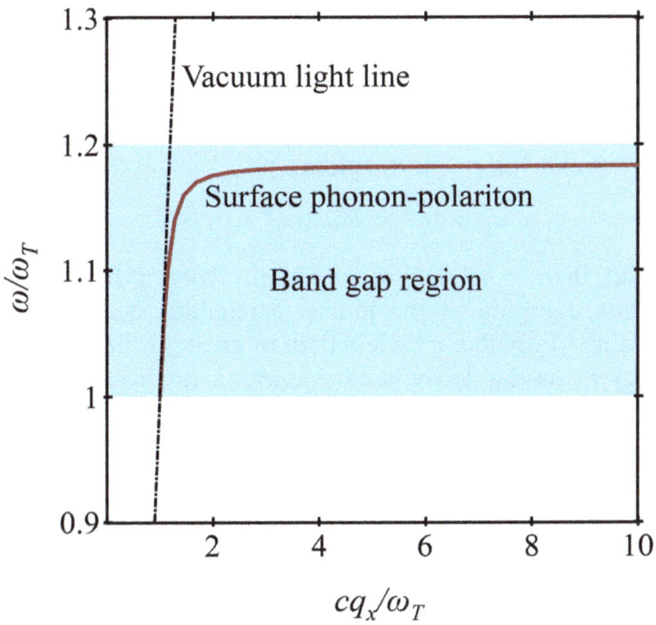

Figure 6.3. Dispersion relation for a surface phonon-polariton (solid line) localized near the planar interface between vacuum and an isotropic ionic material, assumed to be described by the same parameters as in figure 6.1. The vacuum light line corresponding to $\omega = cq_x$ is also shown.

We note that the results derived here become much more complicated when one or both of the media are anisotropic. Discussions of some of the cases that arise under these circumstances can be found, for example, in [2, 4].

6.1.2 Two-interface geometries

Next we turn to the case of two-interface modes by considering the geometry depicted in figure 6.2(b). Medium B is now a slab of finite thickness L, and it is bounded on either side by semi-infinite media labelled as A and C. The analysis can be carried out essentially as described for polaritons in the previous single-interface geometry, except that we now write down solutions for the electric-field vector in all three regions and the analogous electromagnetic boundary conditions are applied at the $z = 0$ and $z = -L$ interfaces. Many different cases can arise, depending on factors such as the choice for polarizations of the modes, which of the media are surface active (i.e. have an explicitly ω-dependent dielectric function), whether the media are isotropic or otherwise, whether damping can be ignored, and so forth. Results for most of these cases can be found in the very extensive literature now available. Some reviews with many further references are given in [2, 4, 8], while some particular examples of interest are highlighted below.

When all three media are isotropic and described by different real, scalar dielectric functions $\epsilon_J(\omega)$ with $J = A, B, C$, the type of calculation described above leads to the implicit dispersion relation for the surface polaritons in the form

$$\left(\frac{\epsilon_A(\omega)q_{Bz} + \epsilon_B(\omega)q_{Az}}{\epsilon_A(\omega)q_{Bz} - \epsilon_B(\omega)q_{Az}}\right)\exp(-2iq_{Bz}L) = \left(\frac{\epsilon_C(\omega)q_{Bz} + \epsilon_B(\omega)q_{Cz}}{\epsilon_C(\omega)q_{Bz} - \epsilon_B(\omega)q_{Cz}}\right). \tag{6.9}$$

The complex quantities q_{Jz} are defined analogously to equation (5.50), but the localization criteria are modified appropriately for the three-layer structure. We require $\mathrm{Im}(q_{Az}) > 0$ and $\mathrm{Im}(q_{Cz}) < 0$, leading to

$$q_x^2 > (\omega^2/c^2)\max(\epsilon_A, \epsilon_C). \tag{6.10}$$

On the other hand, there is no such requirement involving $\mathrm{Im}(q_{Bz})$. Assuming $\epsilon_B(\omega)$ to be real (i.e. the damping in the film is negligible), q_{Bz} can be either real or imaginary according to whether q_x^2 is less than or greater than $(\omega^2/c^2)\epsilon_B$, respectively. If q_{Bz} is imaginary we have a decay-like behaviour away from both interfaces, which can be interpreted as the surface polariton at one interface being perturbed by the surface polariton at the other interface. The two surface modes overlap in their amplitude terms and become shifted in frequency giving upper- and lower-frequency surface branches. In the other situation, where q_{Bz} is real, there is a wave-like behaviour for the z-dependence of the electric-field components within the film. This corresponds to a *guided-wave polariton*, where the light wave is effectively trapped in the film by total internal reflections at the interfaces.

The general dispersion relation given by equation (6.9) is complicated, but there is a useful simplification for a symmetric geometry in which the bounding media A and C are the same and are dispersionless, i.e. $\epsilon_A(\omega) = \epsilon_C(\omega) \equiv \epsilon_A$. The localization condition requires that $q_{Az} = -q_{Cz} = i\kappa_A$, with κ_A defined as in equation (6.3), whereupon equation (6.9) can be rearranged and factorized to give two solutions. For the surface-polariton case (when the imaginary $q_{zB} \equiv i\kappa_B$) the two solutions correspond to

$$\begin{aligned} \epsilon_B(\omega) &= -\,(\epsilon_A\kappa_B/\kappa_A)\tanh(\kappa_B L/2), \quad \text{or} \\ \epsilon_B(\omega) &= -\,(\epsilon_A\kappa_B/\kappa_A)\coth(\kappa_B L/2). \end{aligned} \tag{6.11}$$

The hyperbolic functions in the above equations both tend to unity as the film thickness $L \to \infty$, and so we recover the dispersion relations for single-interface modes by analogy with equation (6.6). For the p-polarized guided-wave polaritons where q_{Bz} is real in this symmetric geometry the two results become

$$\begin{aligned} \epsilon_B(\omega) &= -\,(\epsilon_A q_{Bz}/\kappa_A)\tan(q_{Bz} L/2), \quad \text{or} \\ \epsilon_B(\omega) &= -\,(\epsilon_A q_{Bz}/\kappa_A)\cot(q_{Bz} L/2). \end{aligned} \tag{6.12}$$

Numerical results deduced for the surface phonon-polariton frequencies at a finite film thickness L are shown in figure 6.4 using equation (6.1) for $\epsilon_B(\omega)$ and parameters appropriate for LiF films bounded by different materials [9]. Various cases that emphasize the distinction between the upper and lower modes are included, taking examples for both symmetric and asymmetric geometries through choices of different substrate materials. The two modes of an unsupported film (with media A and C being vacuum) correspond to the curves labelled a and b which show quite

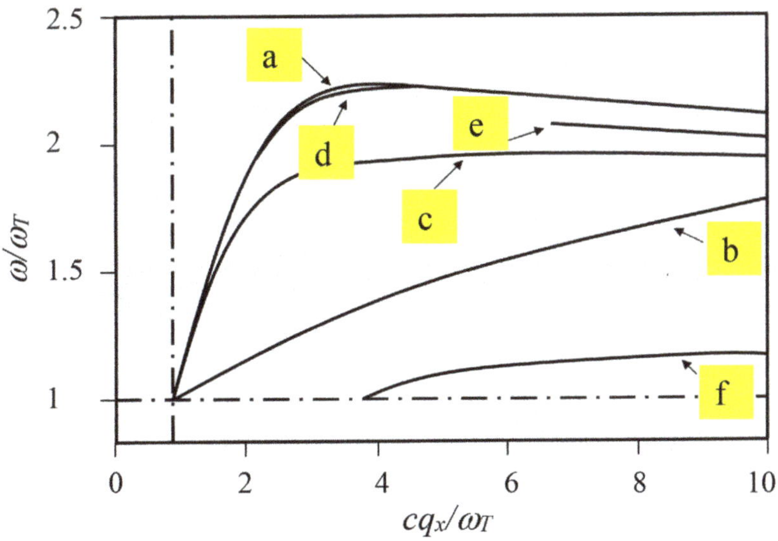

Figure 6.4. Upper and lower surface phonon-polaritons for a LiF film bounded by various media. The curves are as follows: a, upper branch for an unsupported film with $\omega_T L/c = 0.1$; b, lower branch for an unsupported film with $\omega_T L/c = 0.1$; c, upper branch for $L \to \infty$ on any substrate; d, upper branch for a film with $\omega_T L/c = 0.05$ on a K substrate (with $\omega_p = 3.8$ eV); e, upper branch for a film with $\omega_T L/c = 0.1$ on a Si substrate (with $\epsilon_C = 11.7$); f, lower branch for a film with $\omega_T L/c = 0.1$ on a Si substrate. (Reproduced with permission from [9]. Copyright 1981 Wiley.)

contrasting behaviours. Curve c applies for an infinitely thick film, irrespective of the substrate material. Curve d illustrated the minimal difference made (compared to case a) when the lower half of the film is replaced by a metal substrate. Curves e and f refer to an asymmetric geometry where the film is deposited on a Si substrate.

It has been assumed so far that damping effects are negligible in all the media, and so the dielectric functions are real. The extensions of the theory to include complex dielectric functions in the two-interface geometry were made first by Fukui *et al* [10] and Sarid [11] in the context of plasmon-polaritons. In particular, they identified a mode referred to as the *long-range surface plasmon* (LRSP), which will be mentioned in section 6.2 on plasmon-polaritons. Analogous investigations of the role of damping in the phonon-polariton case are described in [12].

The observation of surface polaritons required the development of specialized techniques. This is because, in a planar geometry as discussed so far, the q_x and ω values satisfy the inequality $q_x^2 > \epsilon_B\omega^2/c^2$ in a medium B. On the other hand, a conventionally propagating light beam incident in the medium satisfies $q_x^2 + q_z^2 = \epsilon_B\omega^2/c^2$ with q_z real. This means that the surface polariton cannot be excited directly by the light beam and is therefore referred to as a *non-radiative* mode. The technique of attenuated total reflection (ATR) was developed to overcome this limitation, essentially by removing the restriction on q_z to be real. The usual configuration, as introduced by Otto [5], is sketched in figure 6.5. It involves a coupling prism with real positive dielectric constant ϵ_P on top of a spacer medium (often air or vacuum) to form a gap, with the surface-active medium (the sample being studied) as the third medium. The prism is chosen to have a relatively large refractive index, so that $\epsilon_P > \epsilon_A$. Then, the direction of the incident light is chosen such that angle θ exceeds the critical angle θ_c for total internal reflection at the prism/gap interface, where

$$\theta_c = \sin^{-1}(\sqrt{\epsilon_A/\epsilon_P}). \tag{6.13}$$

Consequently, there will be an evanescent wave in the gap (decaying with respect to the z-direction as previously defined). This can excite a surface polariton in the active-layer B (the sample), which will be observed as a dip in the measured intensity reflection coefficient when ω and $q_x = \sqrt{\epsilon_P}(\omega/c)\sin\theta$ both lie on the surface-

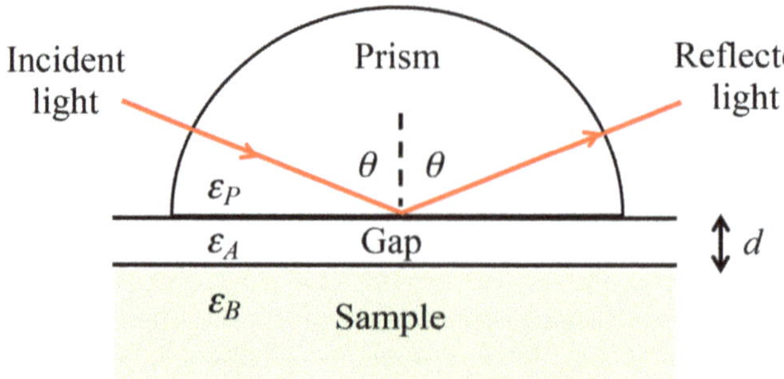

Figure 6.5. Schematic geometry for an attenuated total reflection (ATR) experiment in the Otto configuration. The surface-active medium is shown shaded.

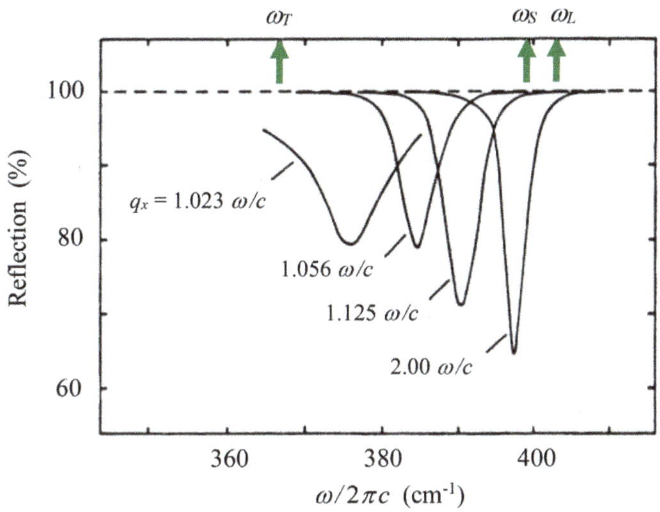

Figure 6.6. ATR spectra for the intensity reflection coefficient plotted against frequency for surface phonon-polaritons in GaP. Each curve is labelled with its q_x value as the angle θ is varied. (Reproduced with permission from [13]. Copyright 1972 American Physical Society.)

polariton dispersion curve. It is straightforward, based on the analysis earlier in this subsection, to calculate an expression for the dependence of the reflection coefficient on frequency (see e.g. [8]). An example showing several ATR spectra at different q_x values is given in figure 6.6. The dips correspond to surface phonon-polaritons excited in GaP [13]. In accordance with the theory presented earlier, the frequencies where the reflection is a minimum lie between ω_T and ω_S, where the latter quantity is defined in equation (6.8).

Variations of the standard Otto ATR method include the Raether–Kretschmann ATR method, in which the surface-active medium is grown by evaporation directly on to the prism (i.e. there is no gap), and the grating method, in which coupling to the non-radiative mode is achieved through having a diffraction grating laid down on the surface of the active medium. Thorough review accounts of these methods, together with examples of applications, can be found in [6, 14].

As mentioned earlier, Raman scattering is a suitable technique in general for observing the phonon-polaritons in the geometries under discussion. With surface phonon-polaritons, however, the conditions for the scattering cross section to be sufficiently large for observations are more stringent. The usual backscattering configuration is unfavourable, but it was demonstrated by Evans *et al* [15] that the near-forward RS by transmission through a film leads to a reinforced scattered signal.

6.1.3 Multiple-interface geometries

For the case of a periodic superlattice we have already derived a general result for the frequencies of the optical modes in section 1.5.1. There we considered a two-component *ABABAB*... structure with alternating layers as depicted in figure 1.9. Although we had in mind the photonics properties of the structure, for which ϵ_A and ϵ_B in this bulk-slab model are usually positive constants, the formal results are quite general and include cases where either (or both) dielectric functions have a frequency dependence as in equation (6.1). Therefore in the present context, equations (1.70)–(1.72) for the bulk modes of the superlattice having either the s- or p-polarization are still applicable.

The analytic expressions become quite formidable, even when just one of the media (say B) has a frequency dispersion with $\epsilon_B(\omega)$ given by equation (6.1) while ϵ_A is a constant independent of ω. Consequently, approximations are frequently used, depending on the superlattice periodic length D and the wavelengths of the superlattice modes that are relevant in any experiment.

One useful simplification is known as the *effective-medium approximation*, and broadly it applies when D is large compared with the interatomic spacings in the media A and B but small compared with any of the relevant optical wavelengths. In practice, for our assumed geometry and notation, the latter requirement means that $QD \ll 1$, $q_x D \ll 1$, $q_{Az} D \ll 1$ and $q_{Bz} D \ll 1$. We may then expand the trigonometric terms in the Rytov equation (1.70) for small values of their arguments, neglecting terms that are cubic and higher order in the small quantities. The result for the dispersion relation, as found by Raj and Tilley [16], depends on the polarization and takes the form

$$\frac{Q^2 + q_x^2}{\bar{\epsilon}_{xx}} = \frac{\omega^2}{c^2} \tag{6.14}$$

for s-polarization, and

$$\frac{Q^2}{\bar{\epsilon}_{xx}} + \frac{q_x^2}{\bar{\epsilon}_{zz}} = \frac{\omega^2}{c^2} \tag{6.15}$$

for p-polarization. Here $\bar{\epsilon}_{xx}$ and $\bar{\epsilon}_{zz}$ are tensor components of an effective dielectric function defined by

$$\begin{aligned} \bar{\epsilon}_{xx} &= \frac{\epsilon_A d_A + \epsilon_B d_B}{D}, \\ \frac{D}{\bar{\epsilon}_{zz}} &= \frac{d_A}{\epsilon_A} + \frac{d_B}{\epsilon_B}. \end{aligned} \tag{6.16}$$

The above result essentially shows that the superlattice can be represented by an effective medium with an anisotropic dielectric function.

As discussed in [16], at the long wavelengths where the above approach is valid the electromagnetic boundary conditions that E_x and D_z are continuous at any interface mean that these field components have essentially the same values (denoted by $\bar{E}_x$ and $\bar{D}_z$, respectively) in either the A or B layers over many periods of the superlattice. The other field components are found from

$$D_{Jx} = \epsilon_0 \epsilon_J E_{Jx}, \quad E_{Jz} = D_{Jz}/\epsilon_0 \epsilon_J, \quad (J = A, B). \tag{6.17}$$

It may then be verified that, on defining $\bar{D}_x$ and $\bar{E}_z$ as the spatial averages of these field components over one superlattice period, we have the effective-medium relationships

$$\bar{D}_x = \epsilon_0 \bar{\epsilon}_{xx} \bar{E}_x, \; \bar{D}_z = \epsilon_0 \bar{\epsilon}_{zz} \bar{E}_z, \tag{6.18}$$

where the tensor components of the effective dielectric function are those found in equation (6.16). Other formulations of the effective-medium approximation for layered superlattices are given in [17, 18]. Also the method has been extended to surface modes of superlattices (see e.g. [16, 19]. As an example of some experimental data on superlattice phonon-polaritons, along with its successful interpretation using the effective-medium approximation just described, we quote work reported in [20]. The authors made normal-incidence far-infrared reflectivity measurements on a superlattice composed of alternating layers of GaAs with thickness 5.5 nm and $Al_xGa_{1-x}As$ with thickness 17 nm. The experiments utilized dispersive Fourier-transform spectroscopy (DFTS) to provide simultaneous data for the amplitude and phase of the reflectivity. The theory based on the effective-medium approximation gave good agreement with the data.

The periodic layered superlattice is just one example of a composite medium. Other examples occur if we have periodic arrays of spheres or cylindrical wires embedded in a matrix of another material, again leading to effective-medium expressions for components of the dielectric function. A discussion is given by Bergman and Stroud [21].

Apart from the effective-medium approach, the other approximation applied to the general formalism of polaritons in superlattices where the media have frequency-dependent dielectric functions is for the case of short-period superlattices. Specifically, the approximation refers to the properties of *confined modes* in these structures, which occur if one of the quantities q_{Az} and q_{Bz} is real, while the other is not. We take the case where it is q_{Bz} that is real, so from equation (5.50) this means that

$$\epsilon_B > c^2 q_x^2 / \omega^2 > \epsilon_A. \tag{6.19}$$

Then, if the thickness d_B is small (i.e. comparable with the optical wavelengths involved) the confined modes in this case have very nearly the character of standing waves across medium B. If the wave-vector quantization is such that an integer number of half wavelengths fit into the thickness d_B then

$$q_{Bz} = n\pi / d_B, \quad n = 1, 2, \ldots n_B, \tag{6.20}$$

where $n_B = d_B / a$ (with a being the lattice constant) is the number of lattice units in layer B. The mode frequencies are then given to a good approximation by inserting the above quantized q_{Bz} values into the dispersion relation for bulk phonon-polaritons in medium B. In practice, a better fit to the experimental data can be obtained if d_B in equation (6.20) is replaced by $d_B^* = (n_B + \delta)a$ to allow for a small amount of penetration of the phonon-polariton mode into medium A (see e.g. [22]).

6.2 Plasmon-polaritons

We recall that in chapter 1 the example of a bulk electron plasma was used to introduce the concept of a polariton. There we employed equation (1.21) for $\epsilon(\omega)$ of a plasma, in the absence of both damping and spatial dispersion, to obtain the bulk plasmon-polariton dispersion relation quoted in equation (1.22) and depicted in figure 1.3. To summarize, it consists of a single branch with a band gap extending in frequency from 0 to ω_p and its limiting forms for $cq/\omega_p \ll 1$ and $cq/\omega_p \gg 1$ were already given.

By analogy with the phonon-polariton case in the previous section of this chapter we anticipate that the surface plasmon-polariton associated with any interfaces will occur within the above band gap. As in the case where plasmons at surfaces and interfaces were described in chapter 5, two approaches that can be followed for the plasmon-polaritons, depending on the degree of electron localization, are the charge-sheet model (with 2DEG or 2DHG layers) and the bulk-slab model. These are employed in the next two subsections, where we generalize the formalism to include the electromagnetic retardation effects that were neglected in sections 5.4 and 5.5. Then, the results for the much newer field of graphene plasmonics will be presented in section 6.3.

6.2.1 Two-dimensional electron gas model

We start here with the same single-interface model as described in section 5.4.1 which consisted of a 2DEG at the $z = 0$ planar interface between the dielectric media A and B that fill the half-spaces $z > 0$ and $z < 0$, respectively. Equations (5.49)–(5.52) are still applicable when we consider the case of p-polarization (with $\mathbf{E}$ in the xz-plane and $\mathbf{H}$ in the y-direction).

The difference compared with chapter 5 is that we now retain the full form of q_{Jz} (with $J = A, B$), written as $q_{Az} = \mathrm{i}\kappa_A$ and $q_{Bz} = -\mathrm{i}\kappa_B$ with κ_A and κ_B both real and positive and defined as in equation (6.3). By following the same steps as before, we find that equation (5.54) for the dispersion relation in the electrostatic case is replaced by the implicit dispersion relation

$$\frac{\epsilon_A}{\kappa_A} + \frac{\epsilon_B}{\kappa_B} = \frac{\Omega_0 c}{\omega^2}, \tag{6.21}$$

where $\Omega_0 = \nu e^2/\epsilon_0 mc$ as before. It is worth noting that equation (6.21) gives rise to a localized polariton mode even when the bounding media A and B are identical. The general result given above can be rearranged in this special case as

$$\frac{8\epsilon_A\omega^2}{\Omega_0^2} = \left[1 + (4cq_x/\Omega_0)^2\right]^{1/2} - 1. \tag{6.22}$$

This expression describes a single plasmon-polariton branch with an angular frequency ω that increases monotonically from zero as q_x is increased in magnitude. For small q_x, such that $cq_x \ll \Omega_0$, the predicted behaviour is that $\omega \propto q_x$. The dependence becomes $\omega \propto \sqrt{q_x}$ at larger q_x (such that $cq_x \gg \Omega_0$).

Next, for plasmon-polaritons in the charge-sheet model of a superlattice we may straightforwardly generalize the results in section 5.4.2. The model and geometry for the superlattice are the same as in figure 5.9 and the method of calculation is similar, but the results with retardation are more complicated. Instead of the explicit result in equation (5.57), the bulk-mode frequency is given implicitly by the following generalization:

$$\cos(QD) = \cos(q_{Az}D) + \left(\frac{\Omega^2 q_{Az} D}{2\omega^2}\right) \sin(q_{Az}D), \tag{6.23}$$

where Ω is the characteristic frequency related to the areal charge density ν and the periodic length D as specified in equation (5.58). The wave-vector component q_{Az} in medium A is given by

$$q_{Az} = \left[(\epsilon_A \omega^2/c^2) - q_x^2\right]^{1/2}. \tag{6.24}$$

This quantity can be imaginary (as in the limit of negligible retardation) or real depending on the value of the wave vector q_x.

Just as described for the unretarded plasmons in charge-sheet superlattices (see section 5.4.2) the above simple model can be extended in various ways [23–25]. The extensions include studying the surface modes in semi-infinite or finite superlattices, having alternating A and B layers of different media between the charge (electron or hole) sheets, and applications to n–i–p–i semiconductor superlattices. A thorough review account, also covering quasiperiodic structures, is given in [4].

6.2.2 Bulk-slab model

In this model there are no charge sheets (2DEG or 2DHG) at the interfaces, but instead the charges are distributed in the volume of one or more dielectric media. In the single-interface geometry we assume the same geometry as in figure 6.1(a), leading once again to equation (6.7) as giving the formal expression for the surface-polariton dispersion relation, provided the necessary conditions $\epsilon_A \epsilon_B < 0$ and $\epsilon_A + \epsilon_B < 0$ can be satisfied.

As a particular example, we may take medium A as vacuum or air, so $\epsilon_A = 1$, while medium B is an electron plasma as in a metal or doped semiconductor with its real dielectric function given by equation (1.21). We then conclude that the surface plasmon-polariton exists in the frequency interval $0 < \omega < \omega_S$ where now we have

$$\omega_S = \omega_p \sqrt{\frac{\epsilon_\infty}{\epsilon_\infty + 1}}, \tag{6.25}$$

and the dispersion relation of the mode is easily found. We recall that the bulk plasmon-polaritons in medium B satisfy the dispersion relation in equation (1.22), so they occur for $\omega > \omega_p$ as shown in figure 1.3. The surface and bulk polariton frequencies are now plotted in figure 6.7 in terms of cq_x/ω_p, where we assume $\epsilon_\infty = 1$ so $\omega_S/\omega_p = 1/\sqrt{2} \simeq 0.71$.

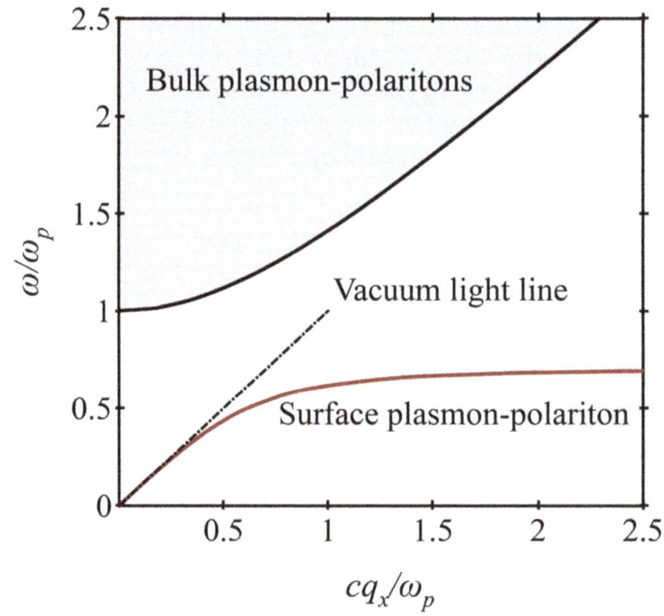

Figure 6.7. Dispersion relation for a surface plasmon-polariton (solid line) localized near the planar interface between vacuum and a plasma material (such as *n*-doped GaAs). The vacuum light line corresponding to $\omega = cq_x$ is shown, as well as the continuum region (shown shaded) of bulk plasmon-polaritons.

Generalizing now to a superlattice, the formal expressions for the dispersion relations are essentially the same as already described for phonon-polaritons in section 6.1.3 on the basis of the bulk-slab model. The only essential difference is that the plasma material form for the dielectric function, either the simplified form in equation (1.22) or the more general expression in equation (5.36), needs to be employed instead of the previous dielectric function for an ionic solid. Likewise the simplifications that were discussed in section 6.1.3 using the effective-medium approximation and/or the treatment of confined modes are applicable also for plasmon-polaritons [4].

Following advances in the techniques for the growth of semiconductor materials in layers with abrupt interfaces (see e.g. [26]), it is nowadays feasible to produce semiconductor superlattices or arrays with a more complicated growth pattern than the periodic ...*ABABAB*... structure. One notable example regarding the study of plasmon-polaritons (either retarded or in the unretarded limit) is the case of periodic n–i–p–i superlattices, which were mentioned briefly in section 5.4. These superlattices involve a repeating pattern with four layers as illustrated in figure 6.8 and defined as: *A*, an n-type semiconductor with a volume distribution of electron carriers; *B*, an insulating (i) medium; *C*, a p-type semiconductor with a volume distribution of hole carriers; and *D*, another i medium. These layers might consist of, for example, layers of n-doped and p-doped GaAs together with intrinsic (pure) GaAs. The doped layers are crystallographically only slightly perturbed by the

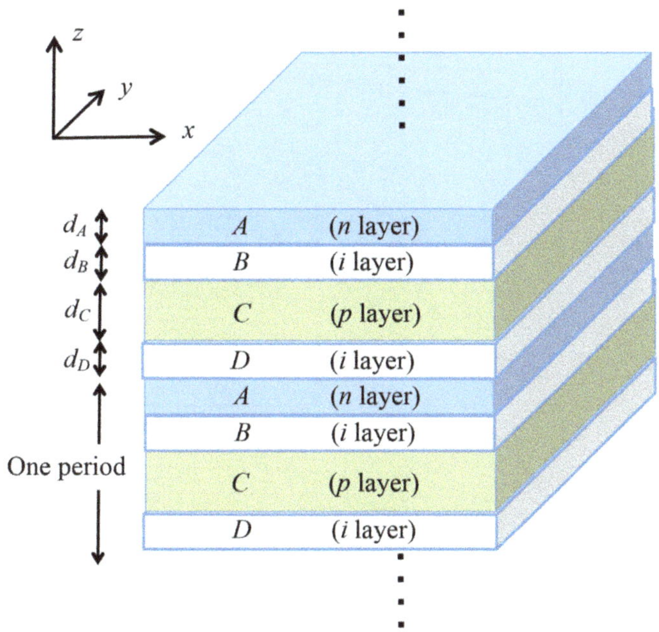

Figure 6.8. Structure of a four-component ...*ABCDABCD*... periodic multilayer taken as a model for a n–i–p–i semiconductor superlattice.

relatively low concentration of dopants, typically 10^{17} to 10^{19} cm^{-3}. The layers might be of any arbitrary thickness, as indicated, and the periodicity length is $D = d_A + d_B + d_C + d_D$. If the thicknesses d_A and d_C of the charge layers are small enough, they might be replaced by 2D charge sheets (2DEG or its hole counterpart 2DHG for the n- and p-type layers, respectively) as an approximation. Due to their more complicated structure and the novel features expected for the plasmon-polaritons, n–i–p–i superlattices have been investigated extensively (see [25, 27]).

The plasmon-polariton modes can be studied by a straightforward extension of the transfer-matrix method described in section 1.5.1. Briefly, a 2 × 2 matrix representation can be defined as before in terms of forward- and backward-travelling waves for the components of the electric field, and the transformation relations across each layer can be formed analogously to equations (1.56)–(1.62). The resulting expression for the transfer matrix T involves a product of more individual matrices than in equation (1.63) because there are now four layers in the repeating periodic arrangement [4, 25, 27].

An example of a calculation for the dependence of the plasmon-polariton frequencies on the in-plane wave vector q_x is given in figure 6.9. It has been assumed [25] that all the constituent layers have the same thickness equal to 20 nm (so $D = 80$ nm), that the A and C layers are Si doped with n and p impurities to the same extent (such that $\omega_p = 76.5$ THz for both layers) and that the B and D layers are SiO_2. The background dielectric constants ϵ_∞ are taken as 11.7 and 12.3 for Si and SiO_2, respectively. It can be seen that there are four bulk bands, as expected for a

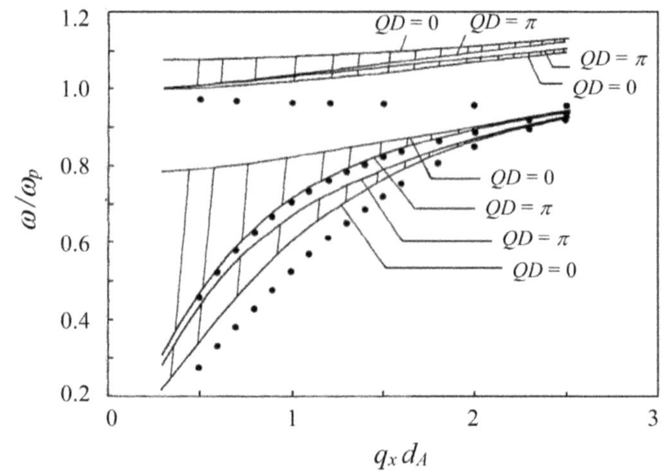

Figure 6.9. Frequencies of plasmon-polaritons plotted versus the dimensionless in-plane wave vector q_xd_A for an n–i–p–i semiconductor superlattice, showing four bulk modes (shaded areas) and three surface-mode branches (dots). See the text for the parameter values. (Reproduced with permission from [25]. Copyright 1988 American Physical Society.)

four-component superlattice. These are shown shaded and have boundaries corresponding to $QD = 0$ and π. For a semi-infinite n–i–p–i superlattice in this case it is found that there are three surface modes (shown by the dotted lines).

6.3 Graphene plasmonics

The topic of graphene plasmonics was touched upon in section 5.4 where we discussed the plasma modes in the non-retarded case that can arise at a charge-layer interfaces between dielectric media. The situation where it is doped graphene at the interface(s) and electromagnetic retardation may be important (as for polaritons) has given rise to an exciting new field. It offers new possibilities, compared with the 2DEG or 2DHG models, because the graphene conductivity is quite different (with competing interband and intraband effects) and the realization of 2D charge layers is more ideal. Some general reviews of graphene plasmonics, covering the optical properties and potential applications, are to be found in review articles by Chang *et al* [28] and Huang *et al* [29] and the book by Liu and Lin [30]. In this section we will focus on the plasmon-polaritons.

Already, in section 5.4.2, we referred to transfer-matrix calculations by Zhan *et al* [31] for plasmons in superlattices formed by periodic arrays of graphene embedded in dielectric media. Various extensions are possible, including the incorporation of retardation effects and the case where the materials between the graphene layers are doped semiconductors, as considered (e.g.) by Vasconcelos and Cottam [32]. These authors considered semi-infinite periodic arrays (superlattices or PCs) of the form *ABABAB*... (e.g. see figure 1.9), where each component *A* and *B* may be a doped semiconductor layers of finite thickness, and additionally there are now identical graphene layers (with electrical conductivity σ) at each *AB* interface within the PC

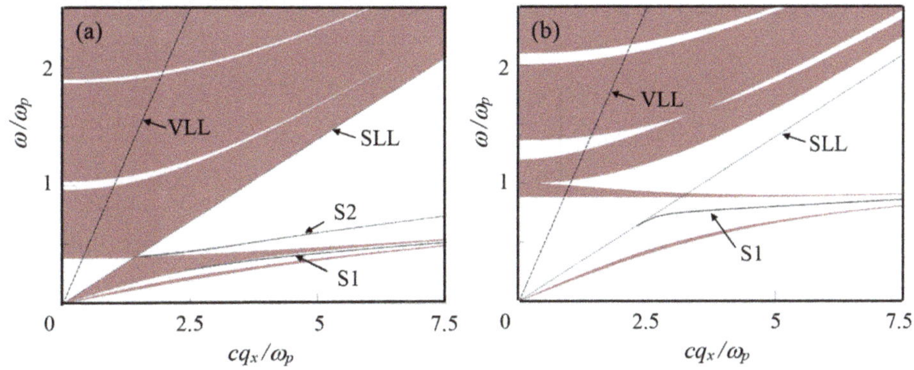

Figure 6.10. Plasmon-polariton bulk and surface dispersion relation for p-polarization in a semi-infinite photonic crystal (PC): (a) all the semiconductor layers are intrinsic GaAs with dielectric constant equal to 12.9; (b) the A layers are intrinsic GaAs and the B layer are n-type doped GaAs. The straight lines VLL and SLL represent the vacuum light line and the semiconductor light line, respectively. Here we have considered the electrical conductivity of the graphene sheet at the external surface to be zero ($\sigma' = 0$). See the text for other parameter values. (Reproduced with permission from [32].)

and another graphene layer (possibly with a different conductivity σ') at the external surface of the PC. By extending previous transfer-matrix methods, they calculated the bulk- and surface-like plasmon-polaritons associated with this structure, finding that the dispersion relation plot of frequency versus in-plane wave number q_x could consist of multiple bulk bands together with surface modes in the band-gap regions. The occurrence of the surface modes, along with their spatial decay characteristics, could be sensitively controlled through the σ and σ' parameters of the doped graphene.

Two examples of dispersion relations for semiconductor PCs are given in figure 6.10. In (a) both of the semiconductor layers A and B are intrinsic GaAs with dielectric constant $\epsilon_\infty = 12.9$, whereas in (b) the A layers are intrinsic GaAs and B layers are n-type GaAs with a charge density 1.8×10^{17} cm^{-3} (which corresponds to a bulk plasma frequency of 9.48 THz). In both cases the frequency and in-plane wave number are expressed in terms of the dimensionless variables ω/ω_p and cq_x/ω_p, respectively, where $\omega_p/2\pi = 9.48$ THz is a convenient scaling factor. The vacuum light line (VLL) and the semiconductor light line (SLL) are given by $\omega = cq_x$ and $\omega = cq_x/\sqrt{\epsilon_\infty}$, respectively. Also, we have taken the conductivity σ for the embedded graphene sheets, given by equation (5.55), to correspond to temperature $T = 300$ K and chemical potential $\mu_c = 0.2$ eV.

6.4 Magnon-polaritons

The magnons or spin waves (SWs) were discussed in chapters 3 and 4 in various regimes of the wave vector, but we have left aside until now the electromagnetic regime in the dipole-dominated case (see figure 3.1). Many aspects of the treatment for nonmagnetic polaritons will be applicable here, but with the role of the dielectric function being replaced by the permeability tensor which has a gyromagnetic form [33]:

$$\bar{\mu} = \begin{pmatrix} \mu_{xx} & \mu_{xy} & 0 \\ -\mu_{xy} & \mu_{xx} & 0 \\ 0 & 0 & \mu_{zz} \end{pmatrix}, \tag{6.26}$$

where $\mu_{xx} = \mu_\infty(1 + \chi_a)$, $\mu_{xy} = i\mu_\infty\chi_b$ and $\mu_{zz} = \mu_\infty$. The magnetic susceptibility terms χ_a and χ_b are defined in equations (4.2) and (4.21) for ferromagnets and uniaxial antiferromagnets, respectively. The inclusion of the scalar factor μ_∞ is to account for the background permeability in a way that is analogous to the introduction of ϵ_∞ in the scalar dielectric function (see section 5.3). In chapter 5 we had, for simplicity, taken $\mu_\infty = 1$.

From Maxwell's equations the magnetic-field vectors **H** and **M** satisfy the wave equation [34, 35]

$$c^2[\nabla(\nabla \cdot \mathbf{H}) - \nabla^2\mathbf{H}] = -\epsilon\frac{\partial^2}{\partial t^2}(\mathbf{H} + \mathbf{M}) \tag{6.27}$$

when retardation effects are included. Assuming now a dependence for the fluctuating field terms as $\exp(i\mathbf{q} \cdot \mathbf{r} - i\omega t)$ for bulk waves in an unbounded magnetic material, where $\mathbf{q}$ is a 3D wave vector, the above equation becomes

$$q^2\mathbf{h} - \mathbf{q}(\mathbf{q} \cdot \mathbf{q}) = (\epsilon\omega^2/c^2)(\mathbf{h} + \mathbf{m}). \tag{6.28}$$

Here **m** and **h** are defined as in equations (3.18) and (3.19) and ϵ is the dielectric constant of the material. If we now take coordinate axes such that the z-axis is along the static magnetization direction for a ferromagnet (or along the uniaxial direction for an antiferromagnet) and we choose $q_y = 0$ without loss of generality, then equation (6.28) can be rewritten in component form as

$$\begin{aligned} h_x + ih_y &= \xi m^+ - \xi' m^-, \\ h_x - ih_y &= -\xi' m^+ + \xi m^-, \end{aligned} \tag{6.29}$$

where

$$\begin{aligned} \xi &= \frac{2(\epsilon\omega^2/c^2) - q_x^2}{2[q^2 - (\epsilon\omega^2/c^2)]}, \\ \xi' &= \frac{q_x^2}{2[q^2 - (\epsilon\omega^2/c^2)]}. \end{aligned} \tag{6.30}$$

When these results are combined with the susceptibility relations derived in section 4.1.1 they lead to the implicit dispersion relation for bulk magnetic polaritons (or magnon-polaritons) as

$$(1 - \xi\chi_a)^2 - \xi^2\chi_b^2 = (\xi')^2(\chi_a^2 - \chi_b^2). \tag{6.31}$$

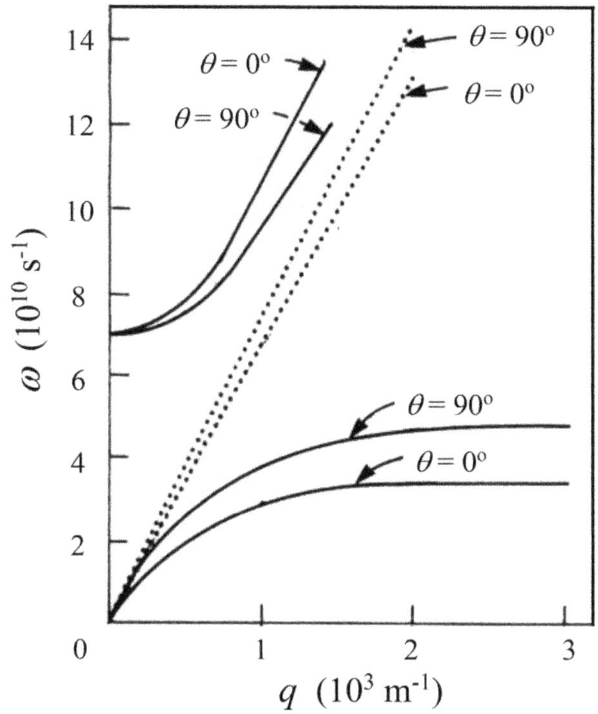

Figure 6.11. Magnetic polariton dispersion curves (solid lines) for a bulk ferromagnet for two different propagation angles $\theta = 0°$ and 90° (see the text). The dashed lines are the light lines. The chosen parameters are $B_0 = \mu_0 M_0 = 0.175$ T. (Adapted with permission from [36]. Copyright 1960 American Institute of Physics.)

This equation is applicable to both ferromagnets and antiferromagnets. In the ferromagnetic case, with χ_a and χ_b given by equation (4.2), the result becomes

$$\omega^2 = \omega_0^2 - 2\xi\omega_0\omega_M + [\xi^2 - (\xi')^2]\omega_M^2, \tag{6.32}$$

which is equivalent to an expression first given by Auld [36]. The two branches to the polariton spectrum, which depend on the angle θ between the propagation vector $\mathbf{q}$ and the z-axis, are illustrated by the numerical examples for $\theta = 0°$ and 90° in figure 6.11. The corresponding dispersion curves in the case of an antiferromagnet are discussed, for example, by Sarmento and Tilley [37].

6.4.1 One- and two-interface geometries

With the results for bulk magnon-polaritons established, we are now in a position to describe the behaviour of these polaritons when interfaces are involved. In this subsection we focus attention on simple geometries with just one or two interfaces. Specifically we will consider the case of a film as depicted in figure 4.1, where the static magnetization along the z-axis is parallel to the interfaces (i.e. in the yz-plane). There are interfaces at $x = 0$ and $x = -L$, and in the semi-infinite limit ($L \to \infty$) only the $x = 0$ interface becomes of interest. The surface magnon-polaritons in this film geometry was first studied by Karsono and Tilley [38] and subsequently their analysis was extended to include both surface and guided magnon-polaritons by Marchand and Caillé [39].

Here we focus, for simplicity, on a ferromagnetic film in the Voigt geometry (the case of angle $\phi = 90°$ in figure 4.1), but the theory has also been carried out for antiferromagnets and for general values of ϕ. We now deal directly with the fluctuating fields **h** and **m** instead of the magnetostatic scalar potential as in section 4.1.1 where retardation was neglected. The analogous expressions to replace equation (4.7) in the three spatial regions are

$$\mathbf{h} = \begin{cases} \mathbf{a}_1 \exp(-\kappa_a x)\exp(\mathrm{i}q_\parallel y - \mathrm{i}\omega t) & \text{for } x > 0, \\ \left[\mathbf{a}_2 \exp(\mathrm{i}q_x x) + \mathbf{a}_3 \exp(-\mathrm{i}q_x x)\right]\exp(\mathrm{i}q_\parallel y - \mathrm{i}\omega t) & \text{for } 0 > x > -L, \\ \mathbf{a}_4 \exp(\kappa_a x)\exp(\mathrm{i}q_\parallel y - \mathrm{i}\omega t) & \text{for } x < -L, \end{cases} \tag{6.33}$$

where the surrounding nonmagnetic media are taken to be air or vacuum, and

$$\kappa_a = \sqrt{q_\parallel^2 - (\omega^2/c^2)}. \tag{6.34}$$

The quantity q_x may be real or imaginary in general and is found from

$$q_x^2 = \epsilon\mu_V(\omega)(\omega^2/c^2) - q_\parallel^2, \tag{6.35}$$

where $\mu_V(\omega)$ is an effective frequency-dependent permeability, sometimes known as the Voigt permeability, given in terms of equation (6.26) for the permeability tensor by

$$\mu_V(\omega) = \mu_{xx} + \frac{\mu_{xy}^2}{\mu_{xx}}. \tag{6.36}$$

The quantity q_x can either be imaginary (whereupon we denote $q_x = -\mathrm{i}\kappa_b$ with $\kappa_b > 0$) or real, depending on the sign of the right-hand side of equation (6.35). These situations correspond, respectively, to surface modes (decaying in the film) and guided modes (wave-like in the film). The calculation for the surface modes is more straightforward and follows as a generalization of the Damon–Eshbach (DE) magnetostatic calculation in section 4.1.1. It is necessary to apply the full electromagnetic boundary conditions for electric and magnetic-field components at the $x = 0$ and $x = -L$ interfaces [38]. The implicit result for the surface magnon-polaritons of the film can eventually be written as (see [39])

$$\left(\kappa_a^2\mu_V^2 + \kappa_b^2 + q_\parallel^2\frac{\mu_{xy}^2}{\mu_{xx}^2}\right)\tanh(\kappa_b L) + 2\kappa_a\kappa_b\mu_V = 0. \tag{6.37}$$

The above result can be regarded as a generalization of equation (4.9) for the magnetostatic DE modes of a magnetic film. It is useful to look at some simplifications in the single-interface case, which is formally obtained if we take the limit of $L \to \infty$. The localization condition for the existence of a surface mode is found (in the absence of damping) to be

$$\kappa_a \mu_V + \kappa_b = \mathrm{i} q_\parallel \frac{\mu_{xy}}{\mu_{xx}}, \tag{6.38}$$

where we recall that μ_{xy} is imaginary so all the above terms are real. Equation (6.38) can be satisfied only in certain intervals of the frequency ω, which will be different depending on whether $q_y = +q_\parallel$ or $-q_\parallel$. This latter property is another manifestation of the nonreciprocal propagation that was mentioned in section 4.1.1 for the magnetostatic (DE) modes. Suppose we take the same sign conventions as before, which are consistent with having $\kappa_b > 0$. It is then found (see [33]) that when $|q_y| = q_\parallel \gg \omega/c$ there is a surface magnon-polariton localized near the $x = 0$ interface when $q_y > 0$ but no surface mode when $q_y < 0$. This is the regime where retardation effects are small and the surface-mode frequency is approximately given by

$$\omega_S \simeq \omega_0 + \frac{\mu_\infty \omega_M}{\mu_\infty + 1}, \tag{6.39}$$

which is just the generalization of equation (4.11) when $L \to \infty$ for the case with $\mu_\infty \neq 1$. On the other hand, when $q_\parallel \sim \omega/c$ the retardation effects are strong; it is found that surface magnon-polaritons may exist for both signs of q_y, and they still show nonreciprocal behaviour.

The corresponding theory has also been developed for the interface modes in antiferromagnetic materials (see e.g. [40]). Generally, the results for the magnon-polaritons are more complicated, but there is a simplification when the applied magnetic field is zero ($B_0 = 0$) because the off-diagonal susceptibility component χ_b vanishes due to cancellations between the two sublattices, as can be seen from equations (4.21) and (4.22). As a consequence, the propagation of the modes becomes reciprocal in this special case, and the dispersion relation of the surface magnon-polaritons is

$$\frac{c^2 q_\parallel^2}{\omega^2} = \frac{\mu_{xx}(\mu_{xx} - \epsilon)}{\mu_{xx}^2 - 1}. \tag{6.40}$$

As mentioned in chapters 3 and 4, the ferromagnetic resonance (FMR) frequencies are in the microwave region, so typical wavelengths are several centimetres. It follows, therefore, that the study of the retarded magnon-polaritons in ferromagnets would require excessively large sample sizes. In contrast, the AFMR frequencies are often in the far infrared, so experimental studies of antiferromagnetic magnon-polaritons are more feasible. The first experimental observation of magnon-polaritons was reported by Jensen *et al* [41] for FeF_2 using high-resolution reflectivity. Subsequently, ATR and improved reflectivity measurements on FeF_2 were reported [42]. A comparison of ATR experiment and theory is shown in figure 6.12. The applied field of magnitude 0.3 T is reversed in direction for the two spectra, and the different features clearly exhibit the nonreciprocity of the magnon-polaritons.

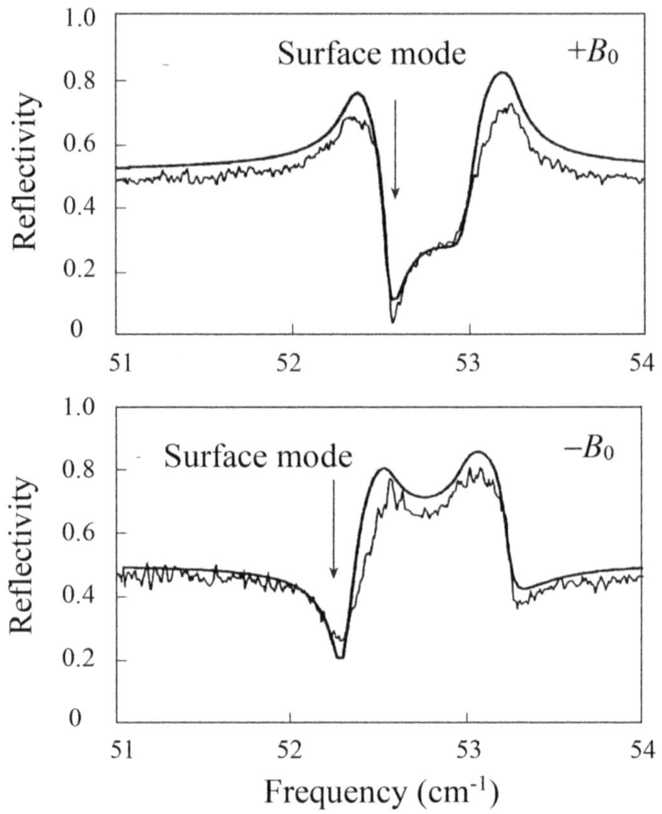

Figure 6.12. ATR spectra of FeF_2 at 1.7 K with resolution 0.02 cm^{-1} for two signs of the magnetic field $B_0 = \pm 0.3$ T. Experiment and theory are shown by the thinner and thicker lines, respectively. (Reproduced with permission from [42].)

6.4.2 Multiple-interface geometries

We now extend some of the preceding calculations for magnon-polaritons in the one- and two-interface geometries to a two-component periodic superlattice using the same geometry as depicted in figure 4.9, except that in the present context the magnetic layers (previously the constituent material B) can be considered as being either ferromagnetic or antiferromagnetic. Also, it turns out to be straightforward to treat the general case where both the A and B constituent materials can be magnetic, as we described next.

The theory can be carried out using a transfer-matrix method as a direct extension of the magnetostatic case in section 4.6 to include the retardation effects (see e.g. [43, 44]. Thus we deal with expressions for the spatial dependence of the electric and magnetic-field components in the layers, instead of using the magnetostatic potential. Specifically, we consider s-polarized polaritons in the Voigt geometry, so the **E** field has only a single component E_z. In medium $J(=A,\ B)$ of the lth cell, defined as before, the field is given by an expression of the form

$$E_{Jz}^{(l)} = [a_J^{(l)} \exp(\kappa_J[x + (l-1)D]) + b_J^{(l)} \exp(-\kappa_J[x + (l-1)D])] \exp(iq_{\|}y - i\omega t). \quad (6.41)$$

where

$$\kappa_J = \left[q_{\|}^2 - \epsilon_J \mu_V^{(J)}(\omega)(\omega^2/c^2)\right]^{1/2}, \quad (J = A, B). \quad (6.42)$$

The analogous expressions for the components of the **H** field can also be written down. The Voigt permeability $\mu_V^{(J)}$ of each material is defined as in equation (6.36).

After using the electromagnetic boundary conditions at each interface of cell l and following through with the calculation of the transfer matrix as described in section 1.5.1, the Rytov-type equation for the bulk magnon-polaritons of the superlattice becomes

$$\cos(QD) = \cosh(\kappa_A d_A)\cosh(\kappa_B d_B) + g \sinh(\kappa_A d_A)\sinh(\kappa_B d_B), \quad (6.43)$$

where

$$g = \left(\frac{q_{\|}^2 \chi_b^{(A)} \chi_b^{(B)}}{\kappa_A \kappa_B}\right) - \left(\frac{q_{\|}\mu_V^{(A)}}{i\kappa_A}\right)\Gamma_B - \left(\frac{q_{\|}\mu_V^{(B)}}{i\kappa_B}\right)\Gamma_A \quad (6.44)$$

and

$$\Gamma_J = \frac{\kappa_J^2(1 + \chi_a^{(J)})^2 - q_{\|}^2(\chi_b^{(J)})^2}{iq_{\|}\kappa_J \mu_V^{(J)}(1 + \chi_a^{(J)})^2}, \quad (J = A, B). \quad (6.45)$$

This general result is algebraically fairly complicated, but there are certain special cases for which the expression for g simplifies. One such case of interest is when one of the materials (say, A) is nonmagnetic, since we then have $\chi_a^{(A)} = \chi_b^{(A)} = 0$ and $\mu_V^{(A)} = 1$. A numerical example of the magnon-polariton dispersion relationships for this case is shown in figure 6.13, where the magnetic material (medium B) is the antiferromagnet MnF_2. Results are included here both for the bulk modes (shaded regions) and for the surface modes (dashed lines) of the superlattice [45]. In figure 6.13(a), where $B_0 = 0$, the surface mode exhibits reciprocal propagation; this contrasts with the situation in figure 6.13(b) where $B_0 \neq 0$ and the surface-mode behaviour is seen to be nonreciprocal.

If the wave-vector propagation is different from the Voigt configuration, meaning that the angle $\phi \neq 90°$ in figure 4.1, the use of a generalized 4×4 transfer-matrix formalism [46] is necessary, since the x-dependence of the electric and magnetic fields is more complicated than in equation (6.41). Another approach, which simplifies the analysis, is to use an effective-medium approximation analogous to that described in section 6.1.3 for phonon-polaritons. This approach was explored in [47] (see also [8]).

6.4.3 Graphene magnonics

Recent calculations for magnetic polaritons have shown that the presence of graphene at one (or more) of the surfaces of a ferromagnetic or antiferromagnetic film modifies the excitations [48]. This occurs through changes to the boundary conditions at any of

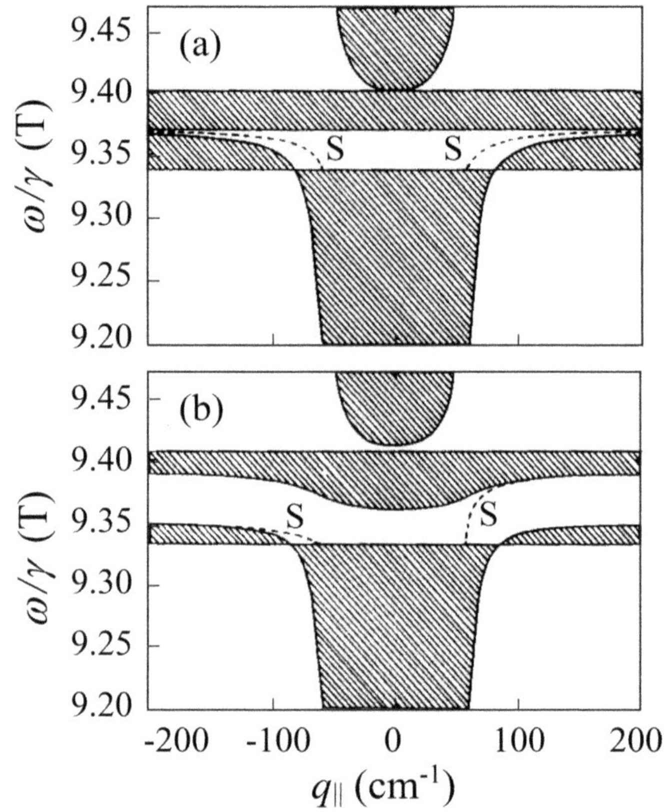

Figure 6.13. Calculated frequencies versus in-plane wave vector for magnon-polaritons in a semi-infinite superlattice of alternating layers of antiferromagnetic MnF_2 (medium B) and a nonmagnetic material (medium A), showing bulk modes (shaded) and surface modes (dashed lines with label S) when $d_A = d_B = 500\ \mu\text{m}$. The two panels correspond to (a) $B_0 = 0$; (b) $B_0 = 0.02$ T. (Reproduced with permission from [45].)

the graphene sheets and applies, in particular, when the magnetization (or sublattice magnetization in the case of an antiferromagnet) is parallel to the surfaces.

As an example, we will assume a simple geometry with a graphene monolayer on a thick (effectively semi-infinite) layer of a ferromagnet with in-plane magnetization $\mathbf{M}_0$ and applied magnetic field $\mathbf{B}_0$ (see figure 6.14). We may then set about generalizing the magnetostatic modes calculation given in section 4.1.1 to include retardation effects (as in section 6.4.1) and the graphene monolayer. The main difference concerns the boundary condition across the $x = 0$ interface, which now requires $\mathbf{H}_2 - \mathbf{H}_1 = \mathbf{j} = \sigma(\omega)\mathbf{E}$. The magnetic and electric-field components can be written down by analogy with results earlier in this chapter, and the conductivity for graphene may be approximated for the low frequencies of the magnetic excitations by a Drude-like formula [49], as

$$\sigma(\omega) = \frac{\mathrm{i}e^2 E_F}{\pi\hbar^2(\omega + \mathrm{i}\Gamma)}, \tag{6.46}$$

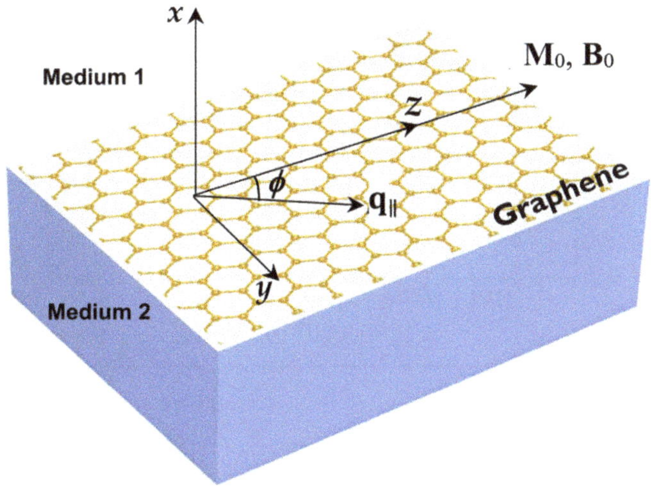

Figure 6.14. The assumed geometry and coordinate axes for graphene magnonics. For the region $x > 0$ we have labelled the nonmagnetic medium as 1, while for $x < 0$ we have a semi-infinite gyromagnetic medium (labelled as 2). In the plane $x = 0$ we have a graphene monolayer. Reprinted with permission from [48].

instead of the more accurate form with contributions given in equation (5.55). Here E_F is the Fermi energy and Γ is the phenomenological damping associated with the electron scattering. After some straightforward algebra, it can be shown that the implicit dispersion relation for the surface magnetic polariton [48] is equivalent to

$$\kappa_a\mu_V + \kappa_b = -\mu_V\left(\frac{\sigma\omega}{\epsilon_0 c^2}\right) + \mathrm{i}\frac{q_{\|}\mu_{xy}}{\mu_{xx}}. \tag{6.47}$$

The notation here is that μ_{xx} and μ_{xy} are the permeability components for the magnetic material, as defined in equation (6.26), and μ_V is its Voigt permeability as given by equation (6.36). Also, κ_a and κ_b are the complex wave vectors in the x-direction, defined in section 6.4.1. It can now be seen that equation (6.47) reduces to the previous result due to Marchand and Caillé [39] in equation (6.37) when the limit is taken of $\sigma \to 0$ for the graphene sheet.

A numerical example of the magnetic polariton dispersion relations is given in figure 6.15, taking the case of the Voigt geometry so $q_y = \pm q_{\|}$. The bulk modes (shown shaded) are the same as in the absence of a graphene sheet, but it can be seen from the figure that the nonreciprocal surface branch depends on the Fermi energy E_F of the graphene sheet, which modifies the conductivity. The frequencies are scaled relative to $\omega_0 = \gamma B_0 = g\mu_B B_0$ and the limiting frequencies ω_S and ω_B represent the surface and bulk magnetostatic modes discussed earlier in chapter 4.

The calculations described here have been extended to finite-thickness magnetic films with graphene layers at one or both surfaces [48] and to the case of multiple graphene layers embedded at the interfaces of magnonic crystals (magnetic superlattices) [50]. In all cases, it is found that the graphene layers provide an excellent means to control the surface magnetic excitations.

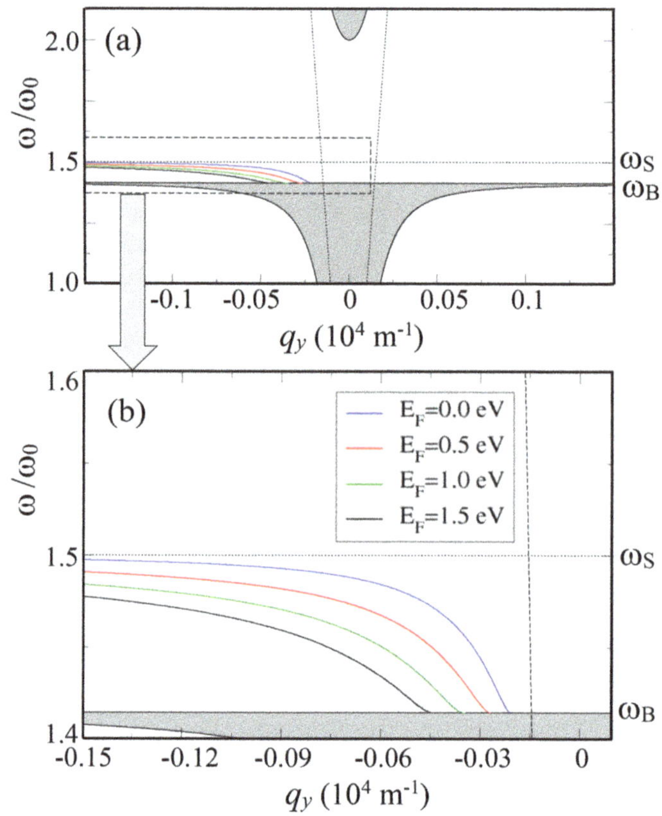

Figure 6.15. Dispersion relations for the bulk (shaded areas) and surface (continuous lines) magnetic polaritons when a graphene layer is included (see red, green and black lines for different E_F) or there is no graphene layer (see blue line) at the interface with a semi-infinite ferromagnet. The magnetic parameters (typical for a ferrite medium) are $B_0 = \mu_0 M_0 = 0.175$ T. The nearly vertical dotted lines are the vacuum light lines. The surface-mode region within the dashed rectangle in panel (a) is shown enlarged in panel (b). See the text for other notation and parameter values. (Adapted with permission from [48].)

6.5 Other types of polaritons

In the previous parts of this chapter we have covered the most important types of polaritons, i.e. those involving the coupling of light (photons) to phonons, plasmons and magnons. Other cases to be considered more briefly in this section are magnetoplasmon-polaritons and exciton-polaritons. From the theoretical perspective, both cases involve the use of more complicated expressions for the dielectric functions. Looking ahead, piezoelectric materials are one of the topics in chapter 7, where further discussion of polaritons will be given.

6.5.1 Magnetoplasmon-polaritons

As part of the discussion of the dielectric function of a charge plasma in section 4.3, we mentioned the modifications that could arise if a static field of magnitude B_0 is applied. In particular, the scalar ϵ becomes replaced by a gyrotropic tensor form defined by equations (5.37)–(5.39). This brings in another characteristic frequency for the material, the cyclotron frequency ω_c, which is given by equation (5.40), as well as the plasma frequency ω_p.

The electromagnetic modes associated with this modified dielectric tensor are known as *magnetoplasmon-polaritons*, and their properties at interfaces were discussed by Wallis *et al* [51] and Albuquerque *et al* [52], among others. The polariton results depend on whether the applied field is oriented parallel or perpendicular to the interface planes, but the former case is usually adopted. The topic is comprehensively reviewed in [4], where a 2×2 transfer-matrix calculation is described for a two-component *ABABAB*... superlattice and for a four-component n–i–p–i superlattice.

Following [51] some bulk and surface magnetoplasmon-polariton dispersion relations are shown in figure 6.16 for a semi-infinite two-component superlattice. Here the axes are chosen such that $\mathbf{B}_0$ is in the y-direction and the z-axis is normal to the interfaces. The propagation of modes along the x-direction is studied, so essentially we have the Voigt configuration. The superlattice parameters correspond to $\omega_{pA} = 2\omega_{pB}$ and $d_A = 0.5d_B$, while the magnitude of the applied field is chosen so that the cyclotron frequency $\omega_c = 0.5\omega_{pA}$ in both materials. In the absence of an applied magnetic field it can be shown (e.g. from the formalism of section 6.2) that there would be two surface plasmon-polariton branches, doubly degenerate and corresponding to propagation along the $\pm x$-directions. The effect of $\mathbf{B}_0$ is to remove this degeneracy, leading to the four surface magnetoplasmon-polariton branches labelled with the plus and minus signs in the figure. One of these branches occurs close to, but just below, the lower bulk band at larger q_x and merges into it when $cq_x/\omega_{pA} \simeq 3$. The other three surface modes lie between the two bulk bands, and one of these has the unusual property that it ceases to exist (because it no longer satisfies the localization condition) for an intermediate q_x range when it meets the $\kappa_B = 0$ line, which is shown as a dashed line in the figure.

Magnetoplasmon-polaritons of a different kind have arisen in recent studies by Vasconcelos and Cottam [53], who extended earlier work on semiconductor superlattices with graphene layers at interfaces to include effects of a strong applied magnetic field. Specifically, they considered a magnetic field perpendicular to the graphene layers and hence perpendicular to the interfaces. In these circumstances, with applied fields greater than about 1 T, it is known that the scalar conductivity of graphene, as quoted in equation (5.55) for the zero field case, becomes a tensor (or matrix) with large off-diagonal components (see e.g. [54, 55]) due to the role of the Landau levels. As a consequence, it was shown [53] that new types of surface plasmon modes, controlled by the magnetic field, can arise in the semiconductor multilayers.

6.5.2 Exciton-polaritons

The concept of excitons was introduced in section 5.3, where it was explained that the spatial dispersion effects could become large due to the kinetic energy term for

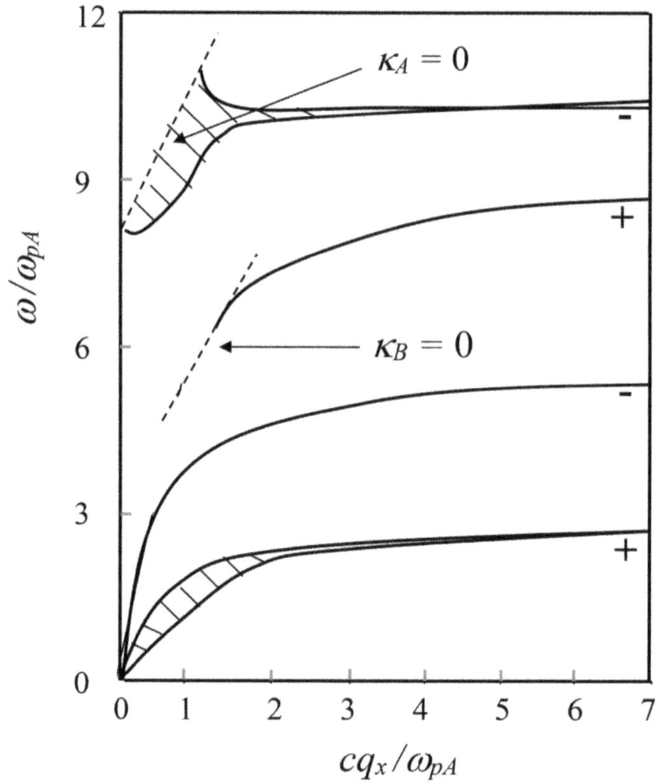

Figure 6.16. Dispersion curves giving the frequencies of bulk and surface magnetoplasmon-polaritons in a semi-infinite two-component superlattice as a function of in-plane wave vector q_x. Bulk-mode regions are shown shaded and the four surface modes are indicated by solid lines with plus or minus labels. See the text for explanation and for parameter values. (Adapted with permission from [51]. Copyright 1987 American Physical Society.)

the bound electron–hole system, as seen from equation (5.47). As a result, the dielectric function $\epsilon(q, \omega)$ takes the form given by equation (5.48). When this dielectric function is substituted into equation (1.20) we obtain the bulk exciton-polariton dispersion relation. It is found to consist of two branches, by analogy with the phonon-polariton case in figure 6.1, but has novel features due to the q-dependence of $\epsilon(q, \omega)$.

In particular, if the bulk frequencies are plotted as a function of q the exciton-polariton analogue of the lower branch in the previous phonon-polariton case will not simply flatten out at large q as in figure 6.1, but instead it will continue to curve upwards. The outcome is that, for any chosen frequency above the exciton frequency ω_e defined in section 5.3, there will be *two* possible values of q (see e.g. [8]). This property has some important consequences when a boundary with another medium is introduced. We may emphasize this behaviour through an example. Suppose s-polarized light with $\omega > \omega_e$ is incident from vacuum onto the planar surface of an excitonic material such as CdS or GaN. There will be a reflected wave and two transmitted transverse waves at different wave vectors. Maxwell's equations, however, provide us with just two boundary conditions, so an additional boundary condition (or ABC) is required. This need for an ABC is somewhat similar to the situation for spin waves in the macroscopic dipole-exchange theory (see section 4.3.1). For exciton-polaritons a phenomenological ABC is usually adopted with the form [56]

$$\frac{\partial \mathbf{P}}{\partial z} + \xi \mathbf{P} = 0, \tag{6.48}$$

where $\mathbf{P}$ is the excitonic polarization, z is the coordinate axis perpendicular to the interface and ξ is a constant. This expression has some formal similarities with the ABC quoted in equation (4.34) for the magnetic case. Another simple form was $\mathbf{P} = 0$ for the excitonic polarization at the surface ($z = 0$) as proposed in [57]. Later work has helped to put the above forms of ABC on a firmer theoretical basis (see discussion in [4]).

A numerical calculation for the surface and guided exciton-polaritons in a GaN thin film [58] is shown in figure 6.17. The $\mathbf{P} = 0$ form for the ABC was used at both interfaces, and damping was ignored in the dielectric function. Among the techniques for observing these modes are photoreflectance (PR) and photoluminescence (PL), and the above theory was compared with experiments by Gil *et al* [59] giving reasonable agreement. The full calculations for exciton-polaritons in superlattices become extremely complicated, but in practice an effective-medium approach similar to that described earlier in this chapter can often be employed (see e.g. [17, 60]).

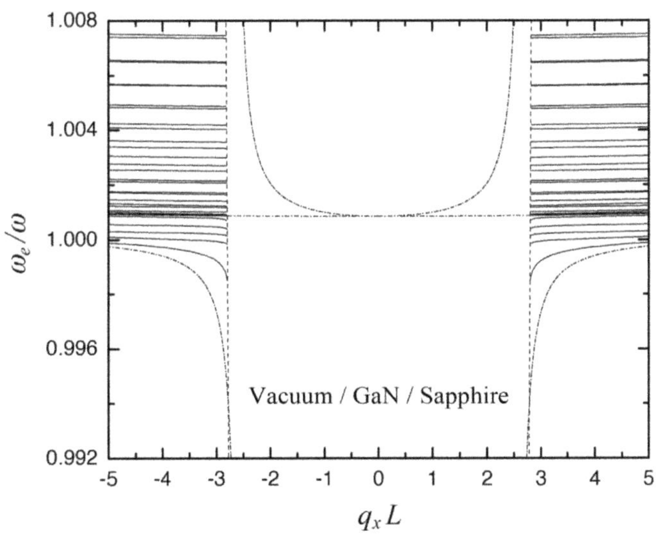

Figure 6.17. Dispersion curves for p-polarized exciton-polaritons in a 50 nm GaN film bounded by vacuum and sapphire. The full lines represent the calculated surface and guided modes. The near-vertical lines are the light lines and the chain-dashed lines are two transverse and one longitudinal bulk mode. (Reproduced with permission from [58]. Copyright 2002 Elsevier.)

References

[1] Burstein F and de Martini F (ed) 1974 *Polaritons* (New York: Pergamon)
[2] Agranovich V M and Mills D L (ed) 1982 *Surface Polaritons* (Amsterdam: North-Holland)
[3] Boardman A D (ed) 1982 *Electromagnetic Surface Modes* (Chichester: Wiley)
[4] Albuquerque E A and Cottam M G 2004 *Polaritons in Periodic and Quasiperiodic Structures* (Amsterdam: Elsevier)
[5] Otto A 1976 *Optical Properties of Solids: New Developments* ed B O Seraphim (Amsterdam: North-Holland) p 678
[6] Kawata S (ed) 2001 *Near-Field Optics and Surface Plasmon Polaritons* (Berlin: Springer)
[7] Kittel C 2004 *Introduction to Solid State Physics* 8th edn (New York: Wiley)
[8] Cottam M G and Tilley D R 2005 *Introduction to Surface and Superlattice Excitations* 2nd edn (Bristol: IOP Publishing)
[9] Oliveira F A, Cottam M G and Tilley D R 1981 *Phys. Stat. Solidi* b **107** 737
[10] Fukui M, So V C Y and Normandin R 1979 *Phys. Stat. Solidi* b **91** K61
[11] Sarid D 1981 *Phys. Rev. Lett.* **47** 1927
[12] Wendler L and Haupt R 1986 *Phys. Stat. Solidi* b **137** 269
[13] Marschall N and Fischer B 1972 *Phys. Rev. Lett.* **28** 811
[14] Sambles J R, Bradberry G W and Yang F 1991 *Contemp. Phys.* **32** 173
[15] Evans D J, Ushioda S and McMullen J D 1973 *Phys. Rev. Lett.* **31** 372
[16] Raj N and Tilley D R 1985 *Solid State Commun.* **55** 373
[17] Agranovich V M and Kravstov V E 1985 *Solid State Commun.* **55** 85
[18] Liu W M, Eliasson G and Quinn J J 1985 *Solid State Commun.* **55** 533
[19] Raj N, Camley R E and Tilley D R 1987 *J. Phys. C: Solid State Phys.* **20** 5203
[20] Maslin K A, Parker T J, Raj N, Tilley D R, Dobson P J, Hilton D and Foxon C T B 1986 *Solid State Commun.* **60** 461
[21] Bergman D J and Stroud D 1992 *Solid State Physics* vol 46 ed H Ehrenreich and D Turnbull (New York: Academic) p 147
[22] Dumelow T, Parker T J, Smith S R P and Tilley D R 1993 *Surf. Sci. Rep.* **17** 151
[23] Constantinou N C and Cottam M G 1986 *J. Phys. C: Solid State Phys.* **19** 739
[24] Qin G, Giuliani G F and Quinn J J 1983 *Phys. Rev.* B **28** 6144
[25] Farias G A, Auto M M and Albuquerque E L 1988 *Phys. Rev.* B **38** 12540
[26] Vollath D 2008 *Nanomaterials* (Weinheim: Wiley)
[27] Johnson B L and Camley R E 1988 *Phys. Rev.* B **38** 3311
[28] Chang Y-C, Liu C-H, Liu C-H, Zhang S, Marder S R, Narimanov E E, Zhong Z and Norris T B 2016 *Nat. Comm.* **7** 10568
[29] Huang S, Song C, Zhang G and Yan H 2017 *Nanophotonics* **6** 1191
[30] Liu J M and Lin I T 2018 *Graphene Photonics* (Cambridge: Cambridge University Press)
[31] Zhan T, Shi X, Dai Y, Liu X and Zi J 2013 *J. Phys.: Condens. Matter* **25** 215301
[32] Vasconcelos M S and Cottam M G 2019 *J. Phys. D: Appl. Phys.* **52** 285104
[33] Hartstein A, Burtstein E, Maradudin A A, Brewster R and Wallis R F 1973 *J. Phys. C: Solid State Phys.* **6** 1266
[34] Jackson J D 1998 *Classical Electrodynamics* 3rd edn (New York: Wiley)
[35] Griffiths D J 2012 *Introduction to Electrodynamics* 4th edn (Boston: Addison-Wesley)
[36] Auld B A 1960 *J. Appl. Phys.* **31** 1642
[37] Sarmento E F and Tilley D R 1975 *J. Phys. C: Solid State Phys.* **10** 795
[38] Karsono A D and Tilley D R 1978 *J. Phys. C: Solid State Phys.* **11** 3487

[39] Marchand M and Caillé A 1980 *Solid State Commun.* **34** 827
[40] Abraha K and Tilley D R 1996 *Surf. Sci. Rep.* **24** 125
[41] Jensen M R F, Parker T J, Abraha K and Tilley D R 1995 *Phys. Rev. Lett.* **75** 3756
[42] Jensen M R F, Feiven S A, Parker T J and Camley R E 1997 *J. Phys.: Condens. Matter* **9** 7233
[43] Barnás J 1988 *J. Phys. C: Solid State Phys.* **21** 1021
[44] Raj N and Tilley D R 1989 *The Dielectric Function of Condensed Systems* ed I V Keldysh, D A Kirzhnitz and A A Maradudin (Amsterdam: Elsevier)
[45] Barnás J 1990 *J. Phys.: Condens. Matter* **2** 7173
[46] Barnás J 1991 *Phys. Stat. Solidi* b **165** 529
[47] Raj N and Tilley D R 1987 *Phys. Rev.* B **36** 7003
[48] Vasconcelos M S, Cottam M G and Anselmo D H A 2021 *J. Phys.: Condens. Matter* **33** 055801
[49] Fan Y, Wei Z, Li H, Chen H and Soukoulis C M 2013 *Phys. Rev.* B **88** 241403
[50] Vasconcelos M S and Cottam M G 2021 *J. Phys.: Condens. Matter* **33** 315802
[51] Wallis R F, Szenics R, Quinn J J and Giuliani G F 1987 *Phys. Rev.* B **36** 1218
[52] Albuquerque E L, Fulco P, Farias G A, Auto M M and Tilley D R 1991 *Phys. Rev.* B **43** 2032
[53] Vasconcelos M S and Cottam M G 2020 *J. Phys. D: Appl. Phys.* **53** 135101
[54] Mikhailov S A and Ziegler K 2007 *Phys. Rev. Lett.* **99** 016803
[55] Gusynin V P, Sharapov S G and Carbotte J P 2006 *J. Phys.: Condens. Matter* **19** 026222
[56] Agranovich V M and Ginzburg V L 1962 *Sov. Phys.—Usp.* **5** 323 and 675
[57] Pekar S I 1958 *Sov. Phys.—JETP* **7** 813
[58] Vasconcelos M S, Albuquerque E L, Farias G A and Freire V N 2002 *Solid State Commun.* **124** 109
[59] Gil B, Clur S and Briot O 1997 *Solid State Commun.* **104** 267
[60] Albuquerque E L and Fulco P 1994 *Z. Phys.* B **100** 289

IOP Publishing

Dynamical Properties in Nanostructured and Low-Dimensional Materials (Second Edition)

Michael G Cottam

Chapter 7

Mixed excitations

Throughout the last several chapters we have covered the main types of excitations in solids, with the emphasis being on their properties at surfaces and interfaces. Nevertheless, there are some important areas not yet touched upon or only mentioned briefly, and so it is useful to provide an account here in terms of a general survey with illustrative examples.

Most of the examples covered here will be for 'mixed' types of excitation that involve coupled waves of two or more different kinds. This is apart from the coupling of excitations to photons, which was the topic of the previous chapter 6 on polaritons. The mixed excitations to be considered here include *magnetoelastic waves*, where there is a coupling between magnetic and elastic (or vibrational) modes, and *piezoelectric waves*, where the electric-field properties are coupled to crystal deformations. There will also be discussion of excitations in *ferroelectrics*, i.e. materials that may exhibit a spontaneous electric-dipole moment at low temperatures, just as ferromagnets have a spontaneous net magnetization below the Curie temperature T_C. These topics lead us into important classes of materials known as *ferroics* and *multiferroics*, where ferromagnetism and ferroelectricity (for example) are strongly coupled together and may be exhibited simultaneously in the same material.

7.1 Magnetoelastic waves

Magnetoelastic waves arise from a coupling between the magnetic and vibrational (phonon) modes when the latter are treated in terms of an elastic continuum approach, as was done in section 2.3. The form taken by the magnetoelastic interaction was first considered by Turov and Irkhin [1] and Kittel [2] using phenomenological arguments, but a detailed derivation for the interaction Hamiltonian was given shortly afterwards by Schlömann [3] who found a combination of four terms in general:

doi:10.1088/978-0-7503-3903-2ch7

$$\mathcal{H}_{m-e} = \frac{B_1}{M_0^2}\sum_{\mu} M_\mu^2 \bar{u}_{\mu\mu} + \frac{B_2}{M_0^2}\sum_{\mu,\nu(\neq\mu)} M_\mu M_\nu \bar{u}_{\mu\nu} + \frac{A_1}{M_0^2}\sum_{\mu,\nu,\eta(\neq\nu)} \frac{\partial M_\mu}{\partial\mu}\frac{\partial M_\mu}{\partial\eta}\bar{u}_{\nu\eta} + \frac{A_2}{M_0^2}\sum_{\mu,\nu}\left(\frac{\partial M_\mu}{\partial\mu}\right)^2 \bar{u}_{\nu\nu}. \tag{7.1}$$

The notation here is consistent with that used in earlier chapters, with $\bar{u}_{\mu\nu}$ and M_μ being Cartesian components of the strain tensor (see section 2.3) and total magnetization (see section 3.2.2), respectively. The summations are taken over x-, y- and z-coordinates. The first two terms, proportional to B_1 and B_2, represent the dependence of the magnetic dipole–dipole and spin-orbit interactions, respectively, on the elastic strain. These coefficients are expected to be comparable in magnitude; in fact, it can be deduced that $B_1 = B_2$ for a ferromagnet with assumed isotropic elastic properties. In practice these coefficients may be different, e.g. measurements for the ferrimagnet YIG have shown that $B_1 \simeq 0.5B_2$. The last two terms in equation (7.1), proportional to A_1 and A_2, are due to modulations of the Heisenberg exchange term by the strain. Their main effect is to describe magnon-phonon processes in which two magnons interact with a phonon, and they are more appropriately considered along with other nonlinear effects (see chapters 8 and 9). All of the terms appearing in equation (7.1) can also provide contributions to the effective magnetic anisotropy at the bulk and at the surface of the material (see e.g. [4]).

At sufficiently long excitation wavelengths, as discussed in chapters 3 and 4, the exchange effects become relatively unimportant in the dynamics. It can then be shown that the main effect of the magnetoelastic coupling in the dynamical properties comes from the B_2-term in equation (7.1). The dominant contributions in the summation are those involving products like $M_z M_x \bar{u}_{zx}$, which yields a term like $M_0 m_x \bar{u}_{zx}$ after a linearization approximation. Hence a simplified form for the magnetoelastic interaction when the anisotropy energy is neglected can be taken as

$$\mathcal{H}_{m-e} = \frac{B_2}{M_0}\left[m_x(\bar{u}_{xz} + \bar{u}_{zx}) + m_y(\bar{u}_{yz} + \bar{u}_{zy})\right]. \tag{7.2}$$

This is just the same as the expression introduced phenomenologically by Kittel [2]. Apart from the references already given, some other introductory accounts of magnetoelastic interactions and waves can be found in [5–8].

7.1.1 Bulk waves

We can derive the dispersion relation(s) for the bulk magnetoelastic waves by writing down the appropriate equations of motion satisfied by the four coupled quantities m_x, m_y, u_x and u_y. The total Hamiltonian consists of $\mathcal{H}_{m-e}$ together with the equations for the noninteracting elastic waves (phonons) and magnons. The calculation will be done first for a bulk medium where we seek solutions with a plane-wave form like $\exp(i\mathbf{q}\cdot\mathbf{r} - i\omega t)$, as usual. Using equation (7.2) and taking the case of the propagation wave vector $\mathbf{q}$ lying along the z-direction of magnetization in a cubic ferromagnet, we find

$$
\begin{gathered}
\mathrm{i}\omega m_x + \omega_{\mathrm{mag}}(q) m_y = \mathrm{i}qB_2 u_y, \quad \mathrm{i}\omega m_y - \omega_{\mathrm{mag}}(q) m_x = -\,\mathrm{i}qB_2 u_x, \\
\omega^2 \rho u_x - q^2 \lambda_{44} u_x = \mathrm{i}q \frac{B_2}{M_0} m_x, \\
\omega^2 \rho u_y - q^2 \lambda_{44} u_y = \mathrm{i}q \frac{B_2}{M_0} m_y.
\end{gathered} \tag{7.3}
$$

Here the noninteracting magnon and phonon dispersion relations are $\omega_{\mathrm{mag}}(q) = \omega_0 + Dq^2$ and $\omega_{\mathrm{ph}} = v_T q = (\lambda_{44}/\rho)^{1/2} q$, respectively. The notation follows that in chapters 2 and 3, and we note in particular that v_T corresponds to the velocity of transverse acoustic phonons with λ_{44} denoting one of the elastic constant of a cubic material [9]. Instead of the transverse m_x and m_y variables we could alternatively have employed the circular combinations $m^+ = m_x + \mathrm{i}m_y$ and $m^- = m_x - \mathrm{i}m_y$ by analogy with chapter 3. The mode coupling occurs when $B_2 \neq 0$, leading to the extra terms on the right-hand side of each equation in (7.3). We could add another equation, the wave equation for u_z, to those in equation (7.3), but it is uninteresting because it does not couple to the magnons in the linear approximation.

From the condition for the above four coupled equations to have consistent and nontrivial solutions, we may deduce that ω satisfies

$$
\{\omega \mp \omega_{\mathrm{mag}}(q)\}\{\omega^2 - \omega_{\mathrm{ph}}^2(q)\} \mp bq^2 = 0, \tag{7.4}
$$

where $b = B_2^2/\rho M_0$ is an effective coupling coefficient due to the magnetoelastic interaction. The upper and lower signs in equation (7.5) refer to the waves with right-

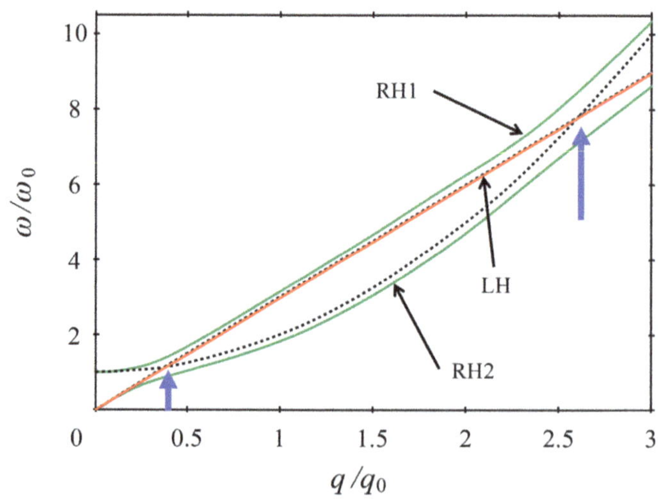

Figure 7.1. Dispersion relations (full lines) for transverse-polarized bulk magnetoelastic modes in a cubic ferromagnet for the case of propagation along the magnetization direction. Here $q_0 = (\omega_0/D)^{1/2}$ is a characteristic wave vector and we have assumed $v_T q_0 = 3\omega_0$ and $b = D\omega_0^2$ for illustration. The right-hand magnetoelastic modes (green lines) are labelled RH1 and RH2, while the left-hand mode (red line) is LH. The noninteracting magnons and phonons correspond to the dashed lines, and they intersect at the two points indicated by the vertical blue arrows.

hand and left-hand sense of circular polarization, respectively. The form of the dispersion relations can be explained using figure 7.1. In the absence of magnetoelastic coupling ($b = 0$) it is clear that the unperturbed modes at positive frequency are simply $\omega = \omega_{mag}(q)$ and $\omega = \omega_{ph}(q)$; these are the dashed lines in the figure. For small enough applied magnetic field these lines have two points of intersection as indicated by the vertical blue arrows. The solid lines in figure 7.1 show the magnetoelastic modes obtained when $b \neq 0$. For the right-handed polarization there are two solutions (labelled as RH1 and RH2) which exhibit a typical 'mode repulsion' behaviour near the intersection points. For the left-handed polarization there is just one solution (labelled as LH) which lies very close to the transverse phonon line.

The bulk magnetoelastic modes can similarly be found for other propagation directions, but they are slightly more complicated because the magnetic dipole–dipole terms contribute. For example, suppose we now take the case of the propagation wave vector **q** perpendicular to the magnetization direction and along one of the cubic symmetry axes. After deriving a set of coupled equations analogous to those in equation (7.3) we find that the magnetoelastic spectrum consists of just one branch with its implicit dispersion relation given by

$$\left[\omega^2 - \omega_{mag}^2(q)\right]\left[\omega^2 - \omega_{ph}^2(q)\right] - b\omega_M q^2 = 0, \tag{7.5}$$

where b is the same as before. Here ω_M is an angular frequency related to the magnetization as in equation (4.3) and the unperturbed magnon dispersion is now given by the modified expression

$$\omega_{mag}(q) = (\omega_0 + Dq^2)^{1/2}(\omega_0 + \omega_M + Dq^2)^{1/2}, \tag{7.6}$$

which is equivalent to the bulk dipole-exchange result obtained in equation (4.33). Discussion of the solution of equation (7.5), along with properties of magnetoelastic waves propagating in lower-symmetry directions, can be found in [8].

7.1.2 Surface and superlattice waves

Next we consider the magnetoelastic surface wave in the single-interface case of a semi-infinite ferromagnetic material. In principle, there are now several possible schemes whereby the unperturbed magnetic and elastic modes could be coupled when $B_2 \neq 0$. The relevant modes are those discussed in chapters 2 and 4, which include the surface and bulk magnetostatic or dipole-exchange modes on the magnetic side together with the bulk and surface (Rayleigh) elastic waves on the other. Typically, the 'mixed' magnetoelastic surface modes will exhibit nonreciprocal propagation for geometries where the magnetization is parallel to the interface, since this is a characteristic of the unperturbed surface magnetic modes, as discussed in chapter 4.

For an example in the single-interface case, we will briefly discuss a calculation by Scott and Mills [10]. They studied a ferromagnetic material in the geometry of figure 4.1, but with film thickness $L \to \infty$ for the semi-infinite case. The magnetization was parallel to the interface plane and surface acoustic-type waves, as well as magnetostatic waves, could propagate in the material. The other half-space with

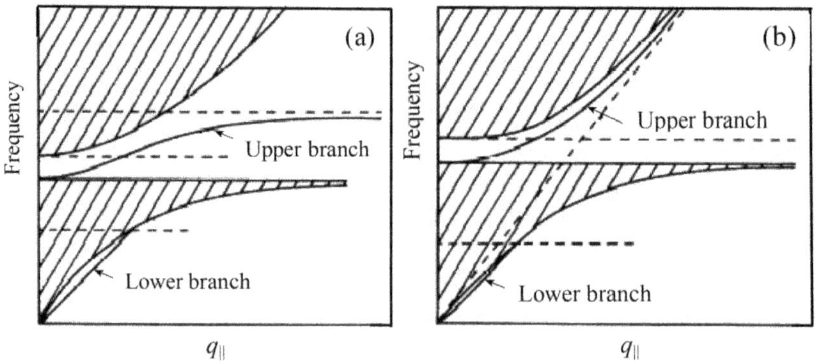

Figure 7.2. Frequency of shear-polarized magnetoelastic surface waves (solid lines) as a function of the in-plane wave vector $q_{\parallel}$. The Voigt geometry is assumed, taking two different propagation directions corresponding to (a) $\phi = 90°$ and (b) $\phi = -90°$. The bulk continuum regions are shown shaded. (Reproduced with permission from [10]. Copyright 1977 American Physical Society.)

$x > 0$ was considered to be vacuum. The authors assumed a form for the magnetoelastic coupling $\mathcal{H}_{m-e}$ equivalent to equation (7.2). Essentially, by combining the same theoretical techniques and the boundary conditions as used earlier in sections 2.3.1 and 4.1.1, they carried out calculations for two generalizations of the surface acoustic waves in the presence of the magnetoelastic coupling. One calculation was for the Rayleigh surface wave which becomes modified by the $\mathcal{H}_{m-e}$ term and the other was for a shear-like magnetoelastic wave that exists only when $\mathcal{H}_{m-e} \neq 0$. Some results for the dispersion properties are given in figure 7.2, where we show two cases for the angle ϕ (see figure 4.1) corresponding to reversing the in-plane propagation wave vector $\mathbf{q}_{\parallel}$ in the Voigt geometry. The nonreciprocity properties of the surface modes (with two branches) are evident by comparing the two cases in figure 7.2. In each case for ϕ the lower surface branch is analogous to the Rayleigh surface wave, while the upper surface branch is a distinctively new feature. Scott and Mills concluded that the magnetoelastic coupling could provide a sensitive experimental technique for probing the magnetic modes (e.g. via the Rayleigh waves).

Experimental studies of the surface and bulk magnetoelastic waves in thin films of the ferrimagnet YIG were, for example, made by Temiryazev *et al* [11]. As well as requiring a generalization of the preceding theory to apply to a finite film thickness L, their work also brought in new effects due to the strong exchange interactions in YIG. Other calculations have been reported for magnetoelastic waves in ferromagnetic films (see e.g. [12]) and at planar interfaces between two different ferromagnets [13].

There have been numerous extensions, largely from a theoretical perspective, to study magnetoelastic waves in other geometries, including waveguides and multilayer systems. For example, Vanderveken *et al* [14] have recently studied the confined magnetoelastic waves propagating in ferromagnetic CoFeB waveguides, taking the form of stripes with a rectangular cross section where the width (~200 nm) is much larger than the thickness (~20 nm). The confined magnetic modes are essentially the dipole-exchange SWs described in section 4.4 with quantization effects in the width direction. The modifications of these modes due to the magnetoelastic coupling to the confined acoustic waves were analysed.

For multilayer magnetic systems with magnetoelastic coupling, the emphasis was initially on using the effective-medium approximation, which was described for other excitations in earlier chapters. As mentioned before, it is a useful approximation if the periodic length D of the superlattice is sufficiently small compared with any of the characteristic wavelengths for the excitation. Detailed calculations applicable under these conditions were made by Tarasenko and coworkers (see e.g. [15] and references therein). For two-component superlattices with an alternating $ABABAB\ldots$ structure and larger periodic lengths, Belykh *et al* [16] have employed a transfer-matrix approach analogous to that described in section 1.5 and subsequently. Their formalism took account of the superlattice surface modes in finite and semi-infinite structures, as well as the superlattice bulk modes of an infinitely extended structure. More recent work for multilayers was reviewed by Shah *et al* [17], who emphasized the large nonreciprocity acquired by the surface acoustic waves due to their magnetoelastic coupling to the nonreciprocal SWs. Finally, as another example involving multiple interfaces between materials, we mention calculations by Graczyk and Krawczyk [18] in 1D photonic-magnonic crystals consisting of alternating nanowire stripes of a ferromagnetic material with a nonmagnetic material. These authors used the coupled mode technique familiar from optical wave studies to calculate the effects of magnetoelastic coupling between the SWs and the acoustic waves.

So far our attention has been directed towards magnetoelastic waves in the context of ferromagnetic materials, but there has also been interest regarding antiferromagnetic and paramagnetic materials. The extension to antiferromagnets is rather straightforward in principle, because the main difference in the theory is that the magnetic susceptibility terms χ_a and χ_b (see section 4.1), which are used in obtaining the coupled equations in (7.3), have to be replaced by the corresponding susceptibility expressions for antiferromagnets. One noteworthy consequence of this, however, is that in the special case of zero applied magnetic field we have the simplification that $\chi_b = 0$ and so the antiferromagnetic magnetoelastic surface modes have reciprocal propagation properties. The case of magnetoelastic modes in paramagnetic materials (such as rare-earth compounds) is rather simpler, because there is no static magnetization or sublattice magnetization. Hence the unperturbed magnon branch is simply the dispersionless $\omega = \omega_0$. The magnetoelastic interactions arise from the coupling between strain and rotational terms in the deformation tensor of the unfilled 4*f* electron shell. Typically, this results in the elastic properties of the materials showing a strong dependence on temperature and applied magnetic field. These effects can be probed very effectively through the properties of the Rayleigh wave using a single-interface geometry. Experimental observations of the behaviour of the Rayleigh wave in paramagnetic $CeAl_2$ and SmSb were reported by Lingner and Lüthi [19], giving good agreement with the magnetoelastic mode theory.

7.2 Piezoelectric waves

Piezoelectrics form a class of materials with the property that, when they are deformed, they give rise to an electric field. Conversely, when an electric field is applied to a piezoelectric material a deformation takes place. To describe a wave propagating in such a material we require a combination of the elastic equations of motion in the material together with Maxwell's equations of electromagnetism. Both of these have to be augmented by terms that describe the piezoelectric coupling. This type of coupling has a tensorial character that depends on the symmetry of the material, as we will describe shortly

Quartz is a well-known example of a piezoelectric material; others include materials with non-centrosymmetric lattices such as lead titanate ($PbTiO_3$), barium titanate ($BaTiO_3$), and lead zirconate titanate ($PbZr_xTi_{1-x}O_3$ with $0 < x < 1$, usually called PZT). The electric field produced when a surface acoustic wave propagates on a piezoelectric material may be employed to couple to an external circuit or to other surface waves that have an electromagnetic component or to the charge carriers in a semiconductor. Hence, together with the inverse effect, there are many possibilities for practical applications of piezoelectricity in the general area of acousto-electronics. Some references for examples and further details of applications are found in [20, 21].

The basic results for the development of piezoelectric waves are well known from the standard text books (see e.g. [22, 23]). First there is the generalization of Hooke's law in elasticity (see section 2.2) to give an expression for the stress σ_{ij} as

$$\sigma_{ij} = \sum_{kl} \lambda_{ijkl}\bar{u}_{kl} - \sum_{k} e_{kij}E_k, \tag{7.7}$$

where λ_{ijkl} and $\bar{u}_{kl}$ are elements of the elasticity tensor and strain, respectively, as before. The last term in equation (7.7) describes the coupling to the electric field where e_{kin} is a component of the piezoelectric tensor. The summations are over Cartesian components x, y and z. The inverse effect is that components of the electric displacement **D** are now given by

$$D_i = \epsilon_0 \sum_{j} \epsilon_{ij}E_j + \sum_{jk} e_{ijk}\bar{u}_{kl}, \tag{7.8}$$

which again involves the piezoelectric tensor. In addition to the above relationships, when solving for surface- and interface-wave properties, we require the use of appropriate electromagnetic and elastic boundary conditions. Some review accounts of surface piezoelectric waves and applications are to be found in [24–26].

7.2.1 One-interface geometry

The first calculations made for the surface waves at the surface of a semi-infinite piezoelectric medium were essentially modifications of surface acoustic waves (e.g. see [27] and references therein). An interesting development, leading to a new type of surface mode, was reported independently by Bleustein [28] and Gulyaev [29].

This so-called Bleustein–Gulyaev (or BG) wave has the unusual property that there is no direct analogue in the purely elastic system. Other surface-wave calculations were later made by Albuquerque [30, 31] in a more general context to include other crystal symmetries and other orientations of the surface plane relative to the crystal axes. A surface response function or Green's function (GF) method, which enabled the relative intensities of the various piezoelectric modes to be deduced, was outlined in [24].

A simplification incorporated into the majority of calculations for piezoelectric waves involves the use of the *static field approximation*, which is based on the fact that that the electromagnetic-wave component of the mixed wave has a velocity that is several orders of magnitude larger than that of the acoustic-wave (or elastic-wave) component. Hence it is reasonable for the approximation to be made that the electric field can be expressed in terms of an electrostatic scalar potential V by writing $\mathbf{E} = -\nabla V$. Then the coupled equations of motion for piezoelectric waves that are consistent with the constitutive relations in equations (7.7) and (7.8) take the form (see e.g. [32])

$$\rho\frac{\partial^2 u_j}{\partial t^2} - \sum_{ikl}\lambda_{ijkl}\frac{\partial^2 u_k}{\partial i\partial l} = \sum_{ik} e_{kij}\frac{\partial^2 V}{\partial i\partial k}, \quad \sum_{ik}\epsilon_{ik}\frac{\partial^2 V}{\partial i\partial k} = \sum_{ikl} e_{ikl}\frac{\partial^2 u_k}{\partial i\partial l}. \tag{7.9}$$

As before, the summations and partial derivatives are with respect to the coordinates x, y and z. Both parts of equation (7.9) have been arranged so that the piezoelectric coupling terms result in a nonzero right-hand side. Depending on the orientation of the surface relative to the crystal axes, the boundary conditions might typically consist of requiring some component(s) of the stress tensor to vanish at the surface and for the potential V to satisfy a surface condition. Additionally, there would be a condition for mode localization (i.e. for the wave to decay as the distance into the material becomes large).

For a specific example, we start with the situation that gives rise to the BG wave mentioned above. In this case the surface is taken to be the plane $y = 0$. The piezoelectric material, chosen to have symmetry in the class C_{6v} (or $6mm$) in standard group-theoretical notations [33], fills the half-space $y \geqslant 0$. The sixfold axis is taken to coincide with the z-axis, and the external medium (where $y < 0$) is air or vacuum. The relevant equations, which are a subset of those in equation (7.9), involve the component u_z of the displacement and its coupling to V. They lead to

$$\rho\frac{\partial^2 u_z}{\partial t^2} - \lambda_{44}\nabla_2^2 u_z = e_{15}\nabla_2^2 V, \quad \epsilon_{xx}\nabla_2^2 V = e_{15}\nabla_2^2 u_z, \tag{7.10}$$

where ∇_2^2 is shorthand for the 2D Laplacian operator $\partial^2/\partial x^2 + \partial^2/\partial y^2$. Also λ_{44} and e_{15} denote elements of the elastic and piezoelectric tensors in a conventional contracted notation (see e.g. [9, 32]).

Bleustein [28] considered the planar $y = 0$ surface of the piezoelectric material to be coated with an infinitesimally thin, perfectly conducting electrode which was grounded. The problem is therefore to find surface-wave solutions of equation (7.10) subject to the set of boundary conditions that: (i) $\sigma_{yx} = \sigma_{yy} = \sigma_{yz} = 0$ at $y = 0$

assuming stress-free surface conditions; (ii) $V = 0$ at the grounded surface $y = 0$; and (iii) both $u_z \to 0$ and $V \to 0$ as $y \to \infty$ for localization.

The method of solution is facilitated if we define a transformation by $W = V - (e_{15}/\epsilon_{xx})u_z$, which allows equation (7.10) to be recast as

$$\rho\frac{\partial^2 u_z}{\partial t^2} - \Lambda_{44}\nabla_2^2 u_z = 0, \qquad \nabla_2^2 W = 0. \tag{7.11}$$

Here $\Lambda_{44} = \lambda_{44} + (e_{15}^2/\epsilon_{xx})$ is known as the 'piezoelectrically stiffened' elastic constant. It now follows from these equations that the surface-wave solutions for u_z and W must take the form

$$\begin{aligned} u_z &= A_1 \exp(-\alpha y)\exp(\mathrm{i}q_x x - \mathrm{i}\omega t),\\ W &= A_2 \exp(-|q_x|y)\exp(\mathrm{i}q_x x - \mathrm{i}\omega t), \end{aligned} \tag{7.12}$$

where A_1 and A_2 are constants, and the positive quantity α satisfies

$$\rho\omega^2 = \Lambda_{44}(q_x^2 - \alpha^2). \tag{7.13}$$

The result of applying the boundary conditions gives us the extra relationship that $\alpha = |q_x|e_{15}^2/\epsilon_{xx}\Lambda_{44}$, from which it may be deduced that the phase velocity v_{BG} of the BG surface wave satisfies

$$v_{\mathrm{BG}}^2 = \frac{\Lambda_{44}}{\rho}\left(1 - \frac{e_{15}^2}{\epsilon_{xx}\Lambda_{44}}\right). \tag{7.14}$$

If we examine the limit in which the piezoelectric coupling becomes vanishingly small ($e_{15} \to 0$) we see that $\alpha \to 0$ and so we see that there is no longer a localized mode in this case.

Other types of surface modes were analysed by Albuquerque [30, 31] and reviewed in [25]. He examined the dispersion relations and scattering properties of the surface waves, again for a semi-infinite piezoelectric material, assuming either cubic or hexagonal symmetries. Attention was given to different polarizations, leading to mixed waves with other components of the displacement being involved by comparison with the BG case.

7.2.2 Superlattices

Superlattices where at least one of the multilayer components is a piezoelectric material have also received attention. Often, the static field approximation has been employed, as described in the previous subsection, and it is generally well justified since the focus has mainly been on modifications to the acoustic (elastic) modes. By contrast, the case of piezoelectric-polaritons, where electromagnetic retardation effects play a role, has been less studied.

Some of the earlier superlattice calculations, such as those reported by Akcakaya and coworkers [34, 35], also made use of an effective-medium approximation analogous to that described for superlattice polaritons in chapter 6. These authors concentrated on semi-infinite, two-component superlattices where both component materials were piezoelectric. They were concerned with modifications to the acoustic-type modes. The validity of their approach in using effective elastic and piezoelectric constants for the superlattice required the acoustic wavelength to be large compared with the periodic length of the superlattice.

By contrast, Albuquerque and Cottam [36] later employed a transfer-matrix approach as described below, which is a generalization of that technique introduced in section 1.5. They were concerned with superlattices that could be either periodic or quasiperiodic (following a Fibonacci sequence in this case) and were composed of two materials A and B, where only the latter was taken to be piezoelectric. They also made use of the static field approximation, so that the basic relationships involve the electrostatic scalar potential V and the components of the displacement $\mathbf{u}$ as in equation (7.9).

Here we adopt the same geometry and choice of coordinate axes as used for the superlattice optical waves and acoustic waves in chapters 1 and 2, respectively, i.e. the z-axis is perpendicular to the interfaces and the wave propagation is along the x-direction. For the cubic and hexagonal classes of piezoelectric materials being studied in [36], it turns out that the coupled expressions in equation (7.9) lead to the pairs (u_x, u_z) and (u_y, V) satisfying separate equations, with the coupling between the latter pair coming specifically from the piezoelectric terms. Then, we may seek wavelike solutions for u_y and V in the piezoelectric layers that can be cast in the form $\exp(\mathrm{i}q_z z)\exp(\mathrm{i}q_x x - \mathrm{i}\omega t)$, where q_z may be complex in general. It can next be shown that there are actually two solutions for q_z, so a superposition has to be formed, for generality. It is eventually found [36], on dropping the common $\exp(\mathrm{i}q_x x - \mathrm{i}\omega t)$ factor, that we have

$$\begin{aligned} u_y &= \{B_1 \exp(\mathrm{i}qz) + B_2 \exp(-\mathrm{i}qz)\}, \\ V &= (e_{x5}/\epsilon_{xx})[B_1 \exp(\mathrm{i}qz) + B_2 \exp(-\mathrm{i}qz)] + B_3 \exp(-q_x z) + B_4 \exp(q_x z), \end{aligned} \tag{7.15}$$

taking the case of hexagonal symmetry, e.g. as in ZnO. The B-coefficients can be determined later through application of the boundary conditions, and the two values of the effective wave vector q_z are found to be $\mathrm{i}q_x$ and q, where

$$q = \left[(q_{Tz}^2 - q_x^2 p)/(1 + p)\right]^{1/2} \tag{7.16}$$

and $p = e_{x5}^2/\lambda_{44}\epsilon_{xx}$ represents a dimensionless piezoelectric coupling coefficient. The z-component of the transverse acoustic wave vector is

$$q_{Tz} = \left[(\omega/v_T)^2 - q_x^2\right]^{1/2}, \quad \text{with } v_T = (\lambda_{44}/\rho)^{1/2}. \tag{7.17}$$

When the analogous expressions to equations (7.15) and (7.16) are written down for piezoelectric materials with cubic symmetry (e.g. as in GaAs), it is again found that there are combinations of four terms in u_y and V, although the results are slightly more complicated [36]. The consequence for the transfer-matrix method in both symmetry cases, however, is that 4×4 matrices must be employed instead of 2×2 matrices as previously. The appropriate 4×4 transfer matrices were

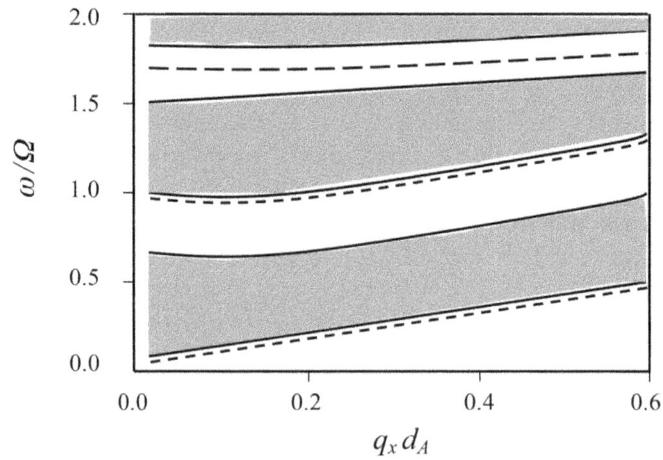

Figure 7.3. Dispersion relations for piezoelectric waves in a semi-infinite periodic superlattice with alternating layers of A (a non-piezoelectric material, taken as fused silica with thickness $d_A = 20$ nm) and B (a hexagonal piezoelectric material, taken as ZnO with thickness $d_B = 40$ nm). The bulk and surface modes are shown as the shaded areas and the dashed lines, respectively. (Reproduced with permission from [36]. Copyright 1992 Elsevier.)

calculated in [36] and employed to discuss the piezoelectric modes (see also [4] for a review account).

A numerical example of these results is given in figure 7.3 for a periodic semi-infinite superlattice, showing the bulk modes (shaded) and surface modes (dashed lines). The piezoelectric layers are assumed to be composed of the hexagonal material ZnO and the non-piezoelectric layers are fused silica. The characteristic frequency Ω, which scales the frequencies, is defined as v_{TA}/d_A where v_{TA} is the transverse acoustic wave velocity in material A (the silica layers). The boundaries of the bulk-mode regions correspond to $QD = 0$ and π for the Bloch wave vector Q, which is similar to the situations described for superlattice polaritons in chapter 6.

In some of the more recent work on piezoelectric waves in superlattices, there has been attention paid to novel aspects due to quasiperiodicity in the sequencing of layers. For example, in [37] the transverse acoustic (TA) waves of a Fibonacci quasiperiodic superlattice where the two components were both piezoelectric, having the wurtzite structure (e.g. ZnO and MgO), were analysed. The dispersion relations of the modes and their spatial localizations were calculated for ultrathin layers (less than ~5 nm in thickness). Further calculations for the TA waves in a Fibonacci superlattice were given in [38] for the case where one component is piezoelectric (e.g. PZT) and the other is a metal (e.g. Fe). The fractal scaling associated with the bands for high Fibonacci generation number were investigated.

7.3 Ferroelectric materials

Ferroelectrics are materials that undergo a phase transition below a critical temperature to a state characterized by a spontaneous electric-dipole moment. Some general references on ferroelectricity are the books by Blinc and Zeks [39] and

Lines and Glass [40]. As a prelude to tackling the problem of the modes or excitations in ferroelectrics, it is necessary to have an understanding of the static properties in the vicinity of a surface (i.e., to understand if there are modifications to the equilibrium electric-dipole moments). This was notably studied by Kretschmer and Binder [41], using Landau theory which will be described later.

The dynamical properties in the bulk and at surfaces of ferroelectrics will be discussed here using three different approaches, starting with the treatment for surface excitations at a single interface in ferroelectrics, following [42]. One study is based on formal analogies with the transverse Ising model of ferromagnetism and is generally assumed to be appropriate for hydrogen-bonded ferroelectrics such as potassium dihydrogen phosphate (KDP). Another approach is an adaptation of the Landau theory of phase transitions where an expansion of the free energy is made in terms of a chosen order parameter. Finally a theory can be developed using a polariton-type formalism (as in chapter 6) where now the dielectric function is given an appropriate anisotropic and temperature-dependent matrix form.

7.3.1 The Ising model in a transverse field

As mentioned in section 3.4, the Ising model is well known as an exchange model of ferromagnetism in which the isotropic $\mathbf{S}_i \cdot \mathbf{S}_j$ exchange interaction of the Heisenberg model, as in equation (3.3), is replaced by the anisotropic product $S_i^z\, S_j^z$, where the preferred magnetization direction is along the z-axis. If there is a transverse applied magnetic field acting in the x-direction the Hamiltonian becomes

$$\mathcal{H} = -\frac{1}{2}\sum_{i,j} J_{ij} S_i^z\, S_j^z - \sum_i \Omega_i S_i^z, \tag{7.18}$$

where the notation is as before, but we have introduced the parameter Ω_i as being proportional to the strength of a transverse field that may be site dependent. The applicability of this Hamiltonian as a pseudo-spin model for hydrogen-bonded ferroelectrics is extensively discussed in the literature (see e.g. [39, 40]). In these materials a spontaneous electric-dipole moment occurs when there is ordering that can be associated with the protons in the hydrogen bonds. Each proton has two possible positions, so it can be considered as being in a double potential well and mathematically represented by the two eigenstates $S^z = \pm\frac{1}{2}$ of an effective spin $S = \frac{1}{2}$. The transverse-field term is then interpreted as a measure of the ability of the proton to tunnel between the two eigenstates.

In an infinitely extended system described by equation (7.18) a phase transition is predicted at a temperature T_C, which is given in mean-field theory (see [39]) by

$$\tanh\,(\Omega/2k_B T_C) = 2\Omega/n_0 J. \tag{7.19}$$

To obtain this result, we take $\Omega \equiv \Omega_i$ at all sites and J denotes the exchange coupling between any magnetic site and its n_0 nearest neighbours. It can be shown that for temperature $T < T_C$ the static spin average at any site has nonzero components R^x and R^z in the x- and z-directions, respectively, but for $T \geqslant T_C$ only $R^x \neq 0$.

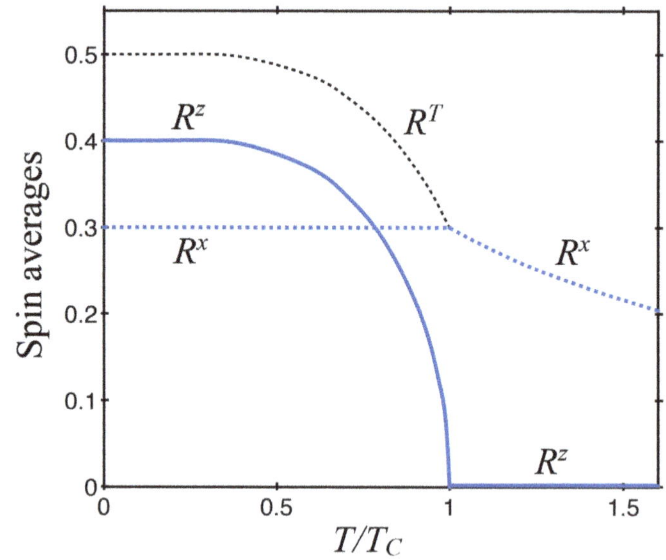

Figure 7.4. The spin averages R^x (blue dashed line) and R^z (blue solid line), calculated in mean-field theory for the Ising model in a transverse field, as a function of T/T_C. It has been assumed here that $\Omega/n_0J = 0.3$. The transition temperature T_C is deduced using equation (7.19). Also, we show (black dashed line) the temperature dependence of $R^T = \sqrt{(R^x)^2 + (R^z)^2}$ representing the total magnitude of the spin average.

The behaviour for the spin averages is sketched in figure 7.4. In fact, R^x is constant at temperatures below T_C with the value Ω/n_0J, and then it monotonically decreases at higher temperatures. In the context of the transverse Ising model it is R^z that behaves as the order parameter for ferroelectricity, being proportional to the polarization. The bulk modes (essentially the magnons or SWs) that correspond to equation (7.18) for an infinite system were calculated in [39, 43].

Now, using methods similar to those described in section 3.2.2 for Heisenberg ferromagnets, we may deduce the dispersion relation for the surface modes in a semi-infinite ferroelectric. Thus, taking the coordinate axes as before and writing

$$\mathbf{S}_j = \mathbf{R}_j + \mathbf{s}_j, \tag{7.20}$$

we may employ the torque equation of motion as in (3.18) and a functional derivative for the effective field as in equation (3.21). A set of three linearized equations is then obtained for the components of the fluctuating spins $\mathbf{s}_j$ as

$$\begin{aligned} -\mathrm{i}\omega s_j^x &= s_j^x \underset{i}{\,} J_{ij}R_i^z, \\ -\mathrm{i}\omega s_j^y &= \Omega_j s_j^z - R_j^x \underset{i}{\,} J_{ij}s_i^z - s_j^x \underset{i}{\,} J_{ij}R_i^z, \\ -\mathrm{i}\omega s_j^z &= \Omega_j s_j^y. \end{aligned} \tag{7.21}$$

The quantities such as $\mathbf{R}_j$ and Ω_j in a semi-infinite system may now depend on the site label j in relation to the surface. The solution of these equations is complicated in general because a mix of different spin components is involved [42], but here we will consider for simplicity the solution for the mode in the high-temperature phase where $T > T_C$. This assumption allows us to put $R_j^z = 0$ in this phase, and the nonzero component is $R_j^x = (1/2)\tanh(\Omega_j/2k_BT)$ according to mean-field theory. The equations in (7.21) then combine to give

$$-(\omega^2 - \Omega_j^2)s_j^z + \Omega_j R_j^x \sum_i J_{ij} s_i^z = 0. \tag{7.22}$$

The solution of the above equation (involving only the z-components) is now straightforward to obtain by methods described earlier in this book and in the appendix. We assume that the exchange J_{ij} couples only nearest neighbours, with the value J_S if both spins are in the surface and the value J otherwise (by analogy with the theory in section 3.2). Also we take Ω_j to be Ω_S if j is in the surface and Ω otherwise. The latter assumption means that R_j^x has only two values, which we denote as R_S^x if j is in the surface layer and R^x otherwise. If we now seek solutions for s_j^z with a dependence like $s_n \exp(i\mathbf{q}_\parallel \cdot \mathbf{r}_\parallel - i\omega t)$, where the notation is the same as in section 3.2 with n (= 1, 2, 3, ...) being the layer index, we deduce from equation (7.22) that

$$\begin{aligned} &\{\omega^2 - \Omega_S^2 + 4\Omega_S R_S^x J_S \gamma(\mathbf{q}_\parallel)\}s_1 + \Omega_S R_S^x J s_2 = 0, \\ &\{\omega^2 - \Omega^2 + 4\Omega R^x J\gamma(\mathbf{q}_\parallel)\}s_n + \Omega R^x J(s_{n-1} + s_{n+1}) = 0 \quad (n > 1). \end{aligned} \tag{7.23}$$

The definition of $\gamma(\mathbf{q}_\parallel)$ is the same as given in equation (3.10). We may now observe that the above equations are formally similar to those encountered in chapters 2 and 3 and can be solved by the same methods as used there, i.e. either by seeking travelling-wave and attenuated solutions or by the tridiagonal-matrix (TDM) method described in the appendix. Here we will just quote the final results.

The bulk modes correspond to $\omega = \pm\omega_B(\mathbf{q})$ and have the dispersion relation

$$\omega_B(\mathbf{q}) = \left[\Omega^2 - 2\Omega R^x J\{2\gamma(\mathbf{q}_\parallel) + \cos(q_z a)\}\right]^{1/2}, \tag{7.24}$$

assuming an sc crystal structure. The surface-mode solutions are more complicated, but in the special case of $\Omega_S = \Omega$ we have $\omega = \pm\omega_S(\mathbf{q}_\parallel)$ where

$$\omega_B(\mathbf{q}) = \left[\Omega^2 - \Omega R^x\left\{4J_S\gamma(\mathbf{q}_\parallel) + \frac{J}{4\sigma\gamma(\mathbf{q}_\parallel)}\right\}\right]^{1/2}. \tag{7.25}$$

Here we have denoted $\sigma = (J_S - J)/J$ and equation (7.25) is applicable only if $J_S < (3/4)J$ or $J_S > (5/4)J$ as the localization requirements for a surface mode. The bulk and surface dispersion relations are illustrated with numerical examples in figure 7.5.

In later work the analysis of ferroelectrics on the basis of the Ising model in a transverse field was extended to films (see [44]) and then to two-component periodic superlattices where both materials could be ferroelectrics (see e.g. [45, 46]).

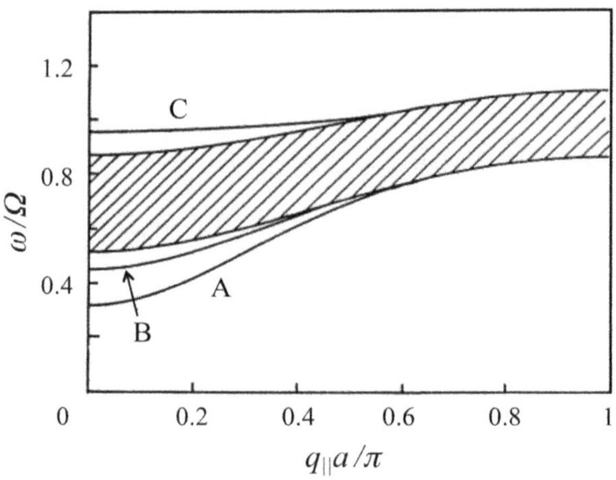

Figure 7.5. Reduced frequency ω/Ω plotted against $q_{\|}a/\pi$ for bulk excitations (shaded region) and surface excitations (lines) for the Ising model in a transverse field when $\mathbf{q}_{\|} = (q_{\|}, 0)$. We have assumed $T = 1.5T_C$ and $\Omega_S = \Omega = 2J$. The J_S/J values for the three surface branches are: A, 1.75; B, 1.5; C, 0.25. (Reproduced with permission from [42].)

7.3.2 Landau theory for ferroelectrics

The basic concept in the Landau theory of phase transitions, e.g. as discussed in the books by Landau and Lifshitz [47] and Plischke and Bergersen [48], is that in the vicinity of the critical point the free energy of the system can be expanded in a power series involving a chosen order parameter.

As in [42], we may consider a simple ferroelectric material filling the half-space $z \geqslant 0$ and having just one component (denoted by P) of the polarization as being the order parameter. It is assumed that the free energy $\mathcal{F}$ per unit area (in the xy-plane) has an expansion in terms of P of the form

$$\mathcal{F} = \int_0^\infty \mathrm{d}z \left[\frac{1}{2}AP^2 + \frac{1}{4}BP^4 + \frac{1}{2}C(\nabla P)^2 - E_{\mathrm{ext}}P\right] + \frac{1}{2}\xi C\{P(z=0)\}^2, \quad (7.26)$$

where A, B and C are the usual phenomenological expansion parameters employed in this type of theory. They depend on temperature in general, and A has the required properties that it vanishes at the critical temperature, so we postulate that $A(T_C) = 0$, and $A(T) < 0$ for $T < T_C$. The quantity E_{ext} is an applied static electric field that couples linearly to the polarization. The final term in equation (7.26), with ξ denoting a constant, is introduced as a phenomenological boundary condition at $z = 0$.

The equilibrium polarization is found by applying the variational condition $\delta\mathcal{F}/\delta P = 0$ to equation (7.26), which gives

$$AP + BP^3 - C\nabla^2 P = E_{\mathrm{ext}}(z > 0), \quad (7.27)$$

with the boundary condition being

$$\frac{dP}{dz} - \xi P = 0, \; (z = 0). \tag{7.28}$$

We remark in passing that equation (7.28) has the same general form as the exchange boundary condition in section 3.2, where ξ was a pinning parameter. We may note also that the effects of ξ are broadly analogous to the roles played by J_S/J and Ω_S/Ω in the Ising-model approach.

The mode frequencies for fluctuations in the polarization P can next be investigated by postulating its equation of motion to have a damped oscillator form (see [42, 49]) as

$$m\frac{\partial^2 P}{\partial t^2} + \gamma\frac{\partial P}{\partial t} + \frac{\delta \mathcal{F}}{\delta P} = 0. \tag{7.29}$$

Here m is an inertial factor, γ is a damping factor, and the next term represents an effective force. Next we assume a time dependence like $\exp(-i\omega t)$ and linearize the equation of motion by writing $P = P_{eq} + p$, where P_{eq} is the equilibrium value of the polarization. The result is

$$-C\frac{d^2p}{dz^2} + \left(-m\omega^2 - i\gamma\omega + A + 3BP_{eq}^2 + Cq_{\|}^2\right)p = 0, \tag{7.30}$$

where $q_{\|}$ is the wave vector parallel to the surface. A useful mathematical device to solve this equation is to define an eigenvalue E corresponding to the linear differential equation $-C(d^2p/dz^2) + 3BP_{eq}^2\, p = E\, p$, whereupon the implicit dispersion relation can be expressed formally as

$$m\omega^2 + i\gamma\omega = A + E + Cq_{\|}^2. \tag{7.31}$$

In solving for E, and thus for the dependence of the mode frequency ω on $q_{\|}$ or T, formal results due to Lubensky and Rubin [50] may be used. To illustrate the predictions of the theory we take $A = A_0[(T/T_C) - 1]$ as a simple choice, while B and C are assumed to be independent of temperature. A numerical example of the T dependence for the surface mode in the case of $q_{\|} = 0$ is shown in figure 7.6 for $T < T_C$ and $\xi > 0$. In this case the surface mode, which becomes a 'soft' mode (meaning that its frequency tends to zero) at T_C, lies below the bulk continuum region. Here the surface mode is well split off from the bottom of the bulk band, but if ξ is larger it moves closer. When $\xi < 0$ a modified surface mode can still exist under some conditions [42].

The type of approach described above for a semi-infinite ferroelectric material can readily be generalized to apply to thin films and to alternating periodic superlattices composed of two different ferroelectric materials. Some useful references to specific calculations and applications are [51–54], while comprehensive reviews have been given by Ishibashi [55] and Chew [56], including applications to many different materials.

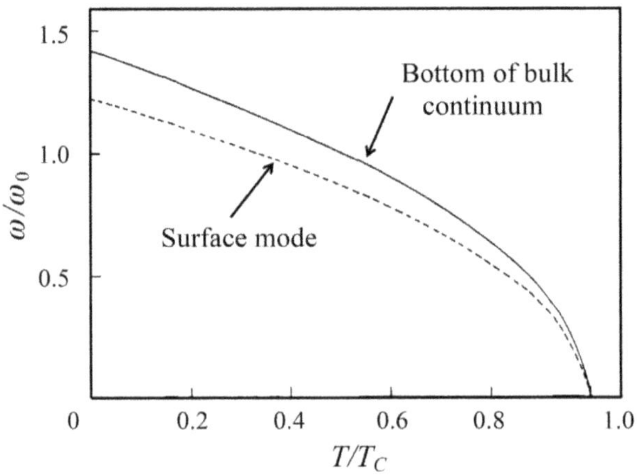

Figure 7.6. Predicted temperature dependence for the surface-mode frequency (dashed line) at $q_{\|} = 0$ for a semi-infinite ferroelectric material obtained using Landau theory. For comparison the bottom of the bulk-mode band is shown as the solid line. It is assumed that $\xi = 0.1(A_0/C)^{1/2}$ and $\omega_0 = (A_0/m)^{1/2}$. (Reproduced with permission from [42].)

7.3.3 Polariton theory for ferroelectrics

In the electromagnetic regime of small wave vectors, such that $q \sim \omega/c$, another alternative is to follow a polariton-like description to model the ferroelectric excitations. This has been applied both to the hydrogen-bonded ferroelectrics (such as KDP) and to the displacive ferroelectrics (such as $BaTiO_3$).

Specifically, the polariton approach involves using an *anisotropic* form here for the effective dielectric function, which is typically written as a uniaxial matrix

$$\bar{\epsilon} = \begin{pmatrix} \epsilon_\perp & 0 & 0 \\ 0 & \epsilon_\perp & 0 \\ 0 & 0 & \epsilon_\| \end{pmatrix}. \tag{7.32}$$

Typically, it is only the longitudinal component $\epsilon_\|$ that displays a significant dependence on frequency ω and temperature T characteristic of a ferroelectric [42]. By contrast, in most of the examples for nonmagnetic polaritons considered in chapter 6, we employed an isotropic dielectric function. The exceptions there were for magnetoplasmon-polaritons and where the effective-medium approximation was used in certain superlattices. The theory for polaritons in anisotropic media is well established, mainly due to work by Borstel and Falge [57, 58], using a more general matrix form than given in equation (7.32), so their results can be utilized here. We restrict attention to TM modes and assume that the principal axes for the ferroelectric material are aligned in directions parallel and perpendicular to the surface. Then, we find that the previous calculations in chapter 6 generalize to give

$$q_x^2 = \frac{\omega^2}{c^2}\left(\frac{\epsilon_A \epsilon_\perp[\epsilon_A - \epsilon_\|]}{\epsilon_A^2 - \epsilon_\| \epsilon_\perp}\right). \tag{7.33}$$

Here A refers to the inactive medium with isotropic dielectric constant ϵ_A in the half-space $z > 0$, while the ferroelectric medium (labelled as B) fills $z < 0$. We note that the above expression is just a generalization of the previous equation (6.7), which is recovered when we set $\epsilon_{\|} = \epsilon_{\perp} \equiv \epsilon_B$.

A suitable choice for the specific expression to be adopted for $\epsilon_{\|}$ in the case of a ferroelectric material takes the same form as in equation (5.44) for optic phonons in an ionic crystal. In the present context we make the replacement $\omega_L \to \omega_0$ for the ferroelectric mode. In the absence of damping ($\Gamma = 0$) this frequency ω_0 is assumed to have the soft-mode property that $\omega_0 \to 0$ as $T \to T_C$. At temperatures sufficiently close to T_C, however, the damping cannot realistically be ignored, and so either the full form of equation (5.44) should be employed, or a relaxational form like

$$\epsilon_{\|}(\omega) = \frac{\epsilon_{\|}(0)}{1 + \mathrm{i}\omega\tau} \tag{7.34}$$

may be appropriate, where τ denotes a relaxation time.

The existence conditions for surface polaritons can now be deduced by following the same formal steps as in chapter 6, i.e. by taking account of the signs of $\mathrm{Im}(q_{Az})$ and $\mathrm{Im}(q_{Bz})$ needed for localization. If we employ the modified equation (5.44), putting $\Gamma = 0$ and adapting results from [57, 58] for anisotropic materials, we find that necessary requirements are that

$$(\epsilon_A\omega^2/c^2) - q_x^2 < 0 \quad \text{and} \quad \epsilon_{\|}\epsilon_{\perp}\left[(\epsilon_{\perp}\omega^2/c^2) - q_x^2\right] < 0. \tag{7.35}$$

In analysing the predictions of the above theory it is useful, when using equation (5.44) for $\epsilon_{\|}$ with $\Gamma = 0$, to define a characteristic frequency ω_S by $\epsilon_{\|}(\omega_S) = \epsilon_A^2/\epsilon_{\perp}$ and to recall the property that $\epsilon_{\|}(\omega_L) = 0$. It can then be deduced that the surface polaritons have the properties

$$\begin{aligned} q_x &= \epsilon_A^{1/2}\omega/c \quad \text{when} \quad \omega = \omega_0, \\ q_x &= \epsilon_{\perp}^{1/2}\omega/c \quad \text{when} \quad \omega = \omega_L, \end{aligned} \tag{7.36}$$

along with $\omega \to \omega_S$ as $q_x \to \infty$. When the damping is nonzero but still small, these surface-polariton results become modified and there is also some mixing with another mode that has a purely bulk-like character when $\Gamma = 0$.

A numerical example of the surface-polariton dispersion relation is displayed in figure 7.7, where a relatively weak damping is assumed with $\Gamma/\omega_0 = 0.2$. Here ω is taken as real, and so q_x is complex in the presence of damping (which makes the dielectric function $\epsilon_{\|}$ complex). The real and imaginary parts of q_x are plotted in the figure for real ω, which is the representation that would be relevant for an experimental ATR scan (see section 6.1.2). In this context, the imaginary part Im (q_x) can be interpreted as the reciprocal of the decay length for the surface polariton. The main features in the behaviour for both $\mathrm{Re}(q_x)$ and $\mathrm{Im}(q_x)$ are resonances around $\omega \simeq \omega_S$. Some polariton calculations for the heavily damped regime, obtained using equation (7.34) for $\epsilon_{\|}$, are given in [42] and are typically non-resonant in their form.

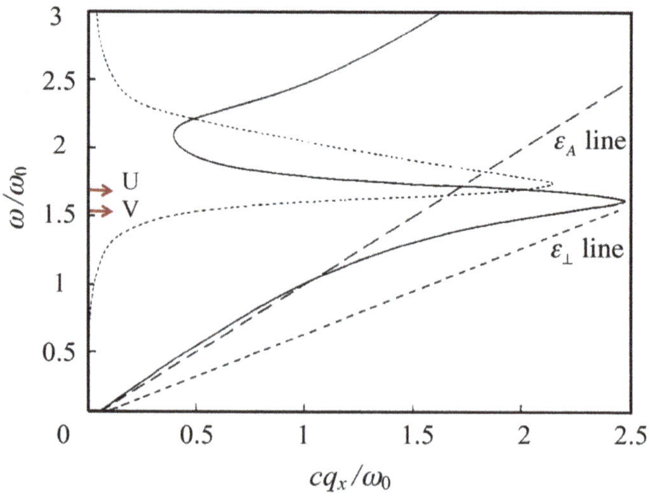

Figure 7.7. Dispersion properties of a TM surface polariton on an anisotropic medium representing a ferroelectric, as calculated using equation (7.33). The mode frequency is plotted against $\mathrm{Re}(q_x)$ (solid curves) and $\mathrm{Im}(q_x)$ (dotted curves). The straight lines marked as ϵ_A line and $\epsilon_\perp$ line refer to the conditions in equation (7.36). The frequency intercepts labelled as U and V refer to ω_S and ω_L, respectively. The parameters are $\Gamma/\omega_0 = 0.2$, $\epsilon_A = 1$, $\epsilon_\perp = 2.5$, $\epsilon_\infty = 1.5$ and $\omega_L/\omega_0 = 1.53$. (Reproduced with permission from [42].)

7.4 Multiferroic materials

In this final section we will provide a brief account of *multiferroics*, which can be defined (see, e.g., [59, 60]) as materials that can simultaneously in the same phase exhibit at least two of the following four types of 'ferroic' behaviour: *ferromagnetism*, *ferroelectricity*, *ferroelasticity* and *ferrotoroidicity*. The addition of the fourth of these phenomena is in accordance with recent trends. On the magnetic side, more complicated spin-ordering schemes such as found in antiferromagnetism and ferrimagnetism are generally included within the definition, and likewise for the other ferroics (e.g. antiferroelectric materials possess ordered electric-dipole moments that cancel each other within each unit cell).

7.4.1 Ferroelastic and ferrotoroidal materials

We have already discussed the properties of ferromagnetic and ferroelectric materials individually in this book, but the other two categories of ferroic materials just mentioned, specifically ferroelastic and ferrotoroidal materials, require some brief explanation, which is given in the following paragraphs.

First, ferroelastic materials may broadly be defined as those which display a spontaneous strain which is stable below a phase transition temperature T_C (by analogy with the appearance of a spontaneous magnetization or electric polarization in the previous cases). Also the strain can be switched, accompanied by a hysteresis effect, using an applied stress. An example of a ferroelastic material having a pronounced hysteresis loop is $Pb_3(PO_4)_2$. A review account of ferroelasticity can be found in the article by Salje [61], who describes experimental studies of the domain walls, principally by x-ray diffraction, as well as the theoretical interpretation of the data by a Landau-type theory (similar to that given in section 7.3.2 for ferroelectrics). The Landau theory was found to account successfully for the temperature dependence of the domain wall thickness. More recently, there has been interest in the ferroelastic properties of monolayers of certain transition metal dichalcogenides, such as WTe_2 and ReS_2, in their 1T and 1T′ phases (see e.g. [62]). These 2D and quasi-2D materials are studied because of their extraordinary mechanical flexibility and their potential as memory devices using the switching capabilities between the distorted phases. A theoretical analysis has been carried out in [62] using techniques of density functional theory (DFT).

Next, ferrotoroidic materials are those for which the order parameter can be taken as the curl of a magnetization (for a spin system) or the curl of a polarization (for electric dipoles). This parameter has a stable value and occurs spontaneously below a transition temperature. Alternatively, a toroidal-like moment **T**, defined in general by

$$\mathbf{T} = \sum_{i=1}^{N} \mathbf{r}_i \times \mathbf{S}_i, \tag{7.37}$$

may be introduced for an assembly of N (>1) spins $\mathbf{S}_i$ arranged at positions $\mathbf{r}_i$ to characterize the ferrotoroidicity, as discussed by Zimmermann *et al* [63]. Loosely, one way to picture ferrotoroidicity is as an array of magnetic vortices. A material typically given as an example of ferrotoroidicity is the orthophosphate $LiCoPO_4$. Other orthophosphates, along with silicates and germanates with the pyroxene crystal structure, are also promising candidates as ferrotoroidic materials [64].

7.4.2 Excitations in multiferroics

Turning now to the phenomenon of multiferroics, it has to date been the simultaneous occurrence of magnetic and ferroelectric behaviour that has attracted the most interest. These types of multiferroics (sometimes called *magnetoelectric* multiferroics) will therefore be the focus of our brief discussion here. Typically, the problem is to achieve a magnetoelectric coupling between the coexisting magnetization **M** and polarization **P** in a material, so that the magnetic excitations may in principle be controlled by an electric field or conversely the polarization dynamics controlled by a magnetic field.

The earlier work in the 1960s to 1980s on dynamical processes in multiferroics notably included calculations by Bar'yakhtar and Chupis [65, 66] and later by Maugin [67]. They examined the role of the magnetoelectric coupling in modifying the magnon frequencies in ferroelectric ferromagnets. In effect, they studied electric-dipole-active magnons, sometimes referred to as *electromagnons*. The methods employed were analogous to those described for magnons in chapter 3 but with an extra coupling term that was quadratic in both the polarization and the magnetization components. The main difference in the two works cited was that Maugin included magnetostatic effects that were ignored in the previous papers.

At around the same time, Tilley and Scott [68] used a Landau-type theory to study the high-frequency modes in the antiferromagnet $BaMnF_4$ with an assumed magnetoelectric coupling. This material is a two-sublattice antiferromagnet with a complicated phase behaviour whereby the sublattice magnetization directions are canted slightly, so that they are not inclined exactly at 180° to one another. Hence $BaMnF_4$ behaves also as a *weak ferromagnet* and there can be a low-frequency magnon associated with precession about this weak magnetic moment, as well as the higher-frequency magnon branch typical of an antiferromagnet.

Since the predicted, as well as the observed, effects of the magnetoelectric coupling were so small, interest in multiferroics declined. It is only fairly recently that the discovery of new materials with more complicated magnetic and electric ordering has led to a great revival in this field. In particular, we mention that an issue of *Nature Materials* (volume 6, number 1) in 2007 was devoted to this topic. It included review articles by Cheong and Mostovoy [69] and Ramesh and Spaldin [70] detailing some of the advances for thin-film multiferroics and the potential applications for electronic devices and sensors.

Developments on the theoretical side were also spurred on by these advances, and we now outline as an example some studies by de Sousa and Moore [71]. As in [68], they considered a weak ferromagnetic (canted antiferromagnetic) material where the polarization is perpendicular to the weak ferromagnetic moment. The basic geometry and choice of coordinate axes are as depicted in figure 7.8, taking the equilibrium polarization to be perpendicular to the plane of the sublattice magnetization vectors (and hence also the weak ferromagnetic moment). The model is appropriate to multiferroics such as $BiFeO_3$ where the canting angle θ is of order 0.1°.

The free energy per unit volume for the system is assumed, following [71], to have the form $\mathcal{F} = \mathcal{F}_E + \mathcal{F}_M + \mathcal{F}_{\rm ME}$, where $\mathcal{F}_E$ comprises the Landau expansion terms

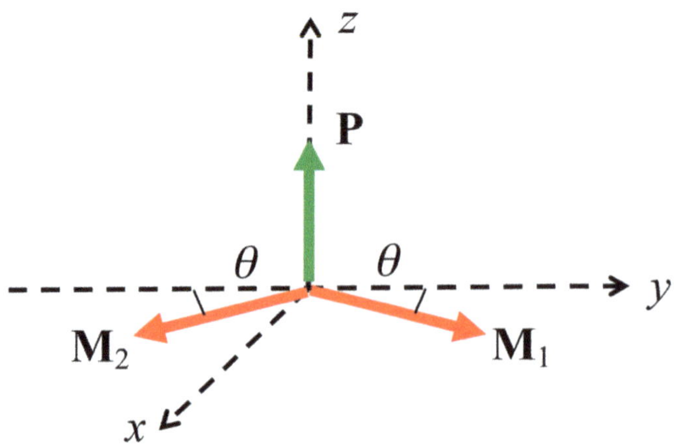

Figure 7.8. Geometry for a ferroelectric canted antiferromagnet (weak ferromagnet). The equilibrium polarization is along the z-direction, while the equilibrium sublattice magnetizations are in the xy-plane with a small canting angle θ away from the y-axis. The weak ferromagnetic moment is along the x-direction.

(analogous to those in equation (7.26) but omitting the surface term for simplicity) for the ferroelectric properties in terms of **P**. Next $\mathcal{F}_M$ involves similar expansions for the magnetic system in terms of the sublattice magnetizations $\mathbf{M}_1$ and $\mathbf{M}_2$. The coefficients chosen are such as to make the z-axis the preferred direction for **P** and the $\pm y$-axes the preferred directions for $\mathbf{M}_1$ and $\mathbf{M}_2$, respectively. Finally, the magnetoelectric contribution is written as

$$\mathcal{F}_{\mathrm{ME}} = (J + \alpha\, P^2)\, \mathbf{M}_1 \cdot \mathbf{M}_2 + \beta\, \mathbf{P} \cdot (\mathbf{M}_1 \times \mathbf{M}_2), \tag{7.38}$$

where α and β are coupling constants. The first term on the right-hand side describes the nearest-neighbour exchange between opposite sublattices, by analogy with results in section 3.2.2, except that now a term proportional to P^2 is introduced into the effective exchange interaction, which becomes $(J + \alpha\, P^2)$. The second term, being proportional to the combination $\mathbf{M}_1 \times \mathbf{M}_2$, is an example of a Dzyaloshinskii–Moriya (or DM) antisymmetric exchange interaction [72, 73], which was introduced in section 3.5.2. This term is allowed by symmetry in $BiFeO_3$. It provides the mechanism for the canting of the sublattice magnetizations, and here its strength is proportional to the polarization. Minimization of the total free energy $\mathcal{F}$ leads to the result [71] for the canting angle θ that

$$\tan\theta = \frac{\beta P_0}{[J + \alpha P_0^2] + [(\beta P_0)^2 + (J + \alpha P_0^2)^2]^{1/2}}, \tag{7.39}$$

where P_0 is the magnitude of the equilibrium polarization.

The dynamics for the coupled fluctuations in the magnetization vectors and in the polarization were studied by de Sousa and Moore [71] by using the Landau–Lifshitz torque equation of motion, by analogy with equation (3.17), together with Maxwell's equations of electromagnetism in the magnetostatic approximation. The authors found, on examining a limit where fluctuations in the polarization were small, that there are two SW branches (or electromagnons), each with a dependence on P_0. The low-frequency SW mode corresponds to spin fluctuations that leave the overall canting angle unchanged, since the spins precess about their equilibrium directions in phase. For a propagation wave vector **q** in the xz-plane this mode is gapless in the sense that its frequency is proportional to q. By contrast, for propagation along the y-axis (the direction of the weak ferromagnetic moment) the frequency is nonzero and proportional to βP_0 as $q \to 0$. This gap arises due to a magnetostatic correction in the presence of the DM interaction. The high-frequency SW mode is less interesting; it changes the overall canting angle but otherwise is broadly similar to the usual antiferromagnetic SW in a simple antiferromagnet. A fuller description of these modes, including their **q** dependence and the magnetic polariton regime, can be found in [71]. The experimental detection of electromagnons in $BiFeO_3$ was reported in [74].

Further calculations for the electromagnons, using broadly similar methods and again with a model of a DM canted antiferromagnet, were presented by Livesey and Stamps [75]. They focussed on obtaining the dynamic susceptibility over a wide frequency range, including that corresponding to the high-frequency branch for the electromagnons. They also used an effective-medium approach applied to a heterostructure superlattice of alternating layers of $BiFeO_3$ (the multiferroic component) and permalloy (the ferromagnetic component). Experimental work on other types of multiferroic superlattices has been reported, for example, by Yang *et al* [76]. More recent studies for electromagnons, mainly from a theoretical perspective, are described in [77] and references therein.

Finally we mention that there have been recent studies of multiferroic behaviour in other low-dimensional systems, in particular for cylindrical nanowire and nanotube geometries by Eliseev *et al* [78]. These authors considered the mixed compound $Eu_xSr_{1-x}TiO_3$ for various concentrations x of Eu and calculated the phase diagrams under various conditions of stress in these low-dimensional geometries. Recent advances in multiferroics, particularly in thin-film heterostructures are reviewed in the book edited by Cano *et al* [79], which also gives a good introduction to the fundamentals.

References

[1] Turov E A and Irkhin Y P 1956 *Fiz. Met. Metalloved.* **3** 15

[2] Kittel C 1958 *Phys. Rev.* **110** 836

[3] Schlömann E 1960 *J. Appl. Phys.* **31** 1647

[4] Bland J A C and Heinrich B (ed) 1994 *Ultrathin Magnetic Structures* vol 1 (Berlin: Springer)

[5] Akhiezer A I, Bar'yakhtar V G and Peletminskii S V 1968 *Spin Waves* (Amsterdam: North-Holland)

[6] Keffer F 1966 *Handbuch Phys.* **18** 1

[7] L'vov V S 1991 *Wave Turbulence under Parametric Excitation* (Berlin: Springer)

[8] Gurevich A G and Melkov G A 1996 *Magnetization Oscillations and Waves* (Boca Raton, FL: CRC Press)

[9] Landau L D and Lifshitz E M 1970 *Theory of Elasticity* (Oxford: Pergamon)

[10] Scott R Q and Mills D L 1977 *Phys. Rev.* B **15** 3545

[11] Temiryazev A G, Tikhomirova M P and Zilberman P E 1996 *Nonlinear Microwave Signal Processing: Towards a New Range of Devices* ed M Marcelli and S A Nikitov (Dordrecht: Kluwer) p 165

[12] Gulyaev Y V and Zil'berman P E 1988 *Russ. Phys. J.* **31** 860

[13] Gorobets Y I, Reshetnyal S A and Khomenko T A 2010 *Acta Phys. Pol.* A **117** 211

[14] Vanderveken F, Mulkers J, Leliaert J, Van Waeyenberge B, Sorée B, Zografos O, Ciubotaru F and Adelmann C 2021 *Phys. Rev.* B **103** 054439

[15] Tarasenko O S, Tarasenko S V and Yurchenko V M 2006 *Crystallogr. Rep.* **51** 296

[16] Belykh N V, Lapteva T V, Tarasenko O S and Tarasenko S V 2007 *Crystallogr. Rep.* **52** 864

[17] Shah P J, Bas D A, Lisenkov I, Matyushov A, Sun N X and Page M R 2020 *Sci. Adv.* **6** eabc5648

[18] Graczyk P and Krawczyk M 2017 *Phys. Rev.* B **96** 024407

[19] Lingner C and Lüthi B 1981 *Phys. Rev.* B **23** 256

[20] Royer D and Dieulesaint E 2000 *Elastic Waves in Solids* vols 1 and 2 (Berlin: Springer)

[21] Manbachi A and Cobbold R S C 2011 *Ultrasound* **19** 187

[22] Cady W G 1946 *Piezoelectricity* (New York: McGraw-Hill)

[23] Yang J 2005 *An Introduction to the Theory of Piezoelectricity* (Berlin: Springer)

[24] Cottam M G and Maradudin A A 1984 *Surface Excitations* ed V M Agranovich and R Loudon (Amsterdam: North-Holland) p 1
[25] Albuquerque E A and Cottam M G 2004 *Polaritons in Periodic and Quasiperiodic Structures* (Amsterdam: Elsevier)
[26] Fu Y Q *et al* 2017 *Prog. Mater. Sci.* **89** 31
[27] Tseng C C 1967 *J. Appl. Phys.* **38** 4281
[28] Bleustein J L 1968 *Appl. Phys. Lett.* **13** 412
[29] Gulyaev Y V 1969 *JETP Lett.* **9** 37
[30] Albuquerque E L 1979 *Phys. Stat. Solidi* b **96** 475
[31] Albuquerque E L 1981 *Phys. Stat. Solidi* b **104** 667
[32] Auld B A 1990 *Acoustic Fields and Waves in Solids* vols 1 and 2 2nd edn (Malabar, FL: Krieger)
[33] Wherrett B S 1986 *Group Theory for Atoms, Molecules and Solids* (Englewood Cliffs, NJ: Prentice-Hall)
[34] Akcakaya E and Farnell G W 1988 *J. Appl. Phys.* **64** 4469
[35] Akcakaya E, Farnell G W and Adler E L 1990 *J. Appl. Phys.* **64** 4469
[36] Albuquerque E A and Cottam M G 1992 *Solid State Commun.* **83** 545
[37] Martínez-Gutiérrez D and Velasco V R 2014 *Surf. Sci.* **624** 58
[38] Alami M, Quotane I, El Boudouti E H and Djafari-Rouhani B 2019 *Mater. Today: Proc.* **13** 541
[39] Blinc R and Zeks B 1974 *Soft Modes in Ferroelectrics and Antiferroelectrics* (Amsterdam: North-Holland)
[40] Lines M E and Glass A M 1979 *Principles and Applications of Ferroelectrics and Related Materials* (Oxford: Oxford University Press)
[41] Kretschmer R and Binder K 1979 *Phys. Rev.* B **20** 1065
[42] Cottam M G, Tilley D R and Zeks B 1984 *J. Phys. C: Solid State Phys.* **17** 1793
[43] Stinchcombe R B 1973 *J. Phys. C: Solid State Phys.* **6** 2459
[44] Salman A and Cottam M G 1994 *Surf. Rev. Lett.* **1** 23
[45] El Aouad N, Laaboudi B, Kerouad M and Saber M 2001 *J. Phys.: Condens. Matter* **13** 797
[46] Saber M, Ainane A and Bärner K 2007 *Surf. Sci.* **601** 4274
[47] Landau L D and Lifshitz E M 1969 *Statistical Physics* (Oxford: Pergamon)
[48] Plischke M and Bergersen B 2006 *Equilibrium Statistical Physics* 3rd edn (Singapore: World Scientific)
[49] Ginzburg V L, Levanyuk A P and Sobyanin A A 1980 *Phys. Rep.* **57** 151
[50] Lubensky T C and Rubin M H 1975 *Phys. Rev.* B **12** 3885
[51] Tilley D R and Zeks B 1984 *Solid State Commun.* **49** 823
[52] Schwenk D, Fishman F and Schwabl F 1990 *Ferroelectrics* **104** 349
[53] Li S, Eastman J A, Vetrone J M, Newnham R E and Cross L E 1997 *Phil. Mag.* B **76** 47
[54] Ong L H, Osman J and Tilley D R 2001 *Phys. Rev.* B **63** 144109
[55] Ishibashi Y 2005 *Ferroelectric Thin Films: Basic Properties and Device Physics for Memory Applications* ed M Okuyama and Y Ishibashi (Berlin: Springer)
[56] Chew K H 2012 *Solid State Phenom.* **189** 145
[57] Borstel G and Falge H J 1977 *Phys. Stat. Solidi* b **83** 11
[58] Borstel G and Falge H J 1978 *J. Appl. Phys.* **16** 211
[59] Schmid H 1994 *Ferroelectrics* **162** 317
[60] Eerenstein W, Mathur N D and Scott J F 2006 *Nature* **442** 759

[61] Salje E K H 2012 *Annu. Rev. Mater. Res.* **42** 1.1
[62] Li W and Li J 2016 *Nat. Commun.* **7** 10843
[63] Zimmermann A S, Meier D and Fiebig M 2014 *Nat. Commun.* **5** 4796
[64] Gnewuch S and Rodriguez E E 2019 *J. Solid State Chem.* **271** 175
[65] Bar'yakhtar V G and Chupis I E 1969 *Sov. Phys.—Solid State* **10** 2818
[66] Bar'yakhtar V G and Chupis I E 1970 *Sov. Phys.—Solid State* **11** 2628
[67] Maugin G A 1981 *Phys. Rev.* B **23** 4608
[68] Tilley D R and Scott J F 1982 *Phys. Rev.* B **25** 3251
[69] Cheong S-K and Mostovoy M 2007 *Nat. Mater.* **6** 13
[70] Ramesh R and Spaldin N A 2007 *Nat. Mater.* **6** 21
[71] de Sousa R and Moore J E 2008 *Appl. Phys. Lett.* **92** 022514
[72] Dzyaloshinskii I E 1958 *J. Phys. Chem. Solids* **4** 241
[73] Moriya T 1960 *Phys. Rev.* **120** 9
[74] Cazayous M, Gallais Y, Sacuto A, de Sousa R, Lebeugle D and Colson D 2008 *Phys. Rev. Lett.* **101** 037601
[75] Livesey K L and Stamps R L 2010 *Phys. Rev.* B **81** 094405
[76] Yang Z, Ke C, Sun L L, Zhu W, Wang L, Chen X F and Tan O K 2010 *Solid State Commun.* **150** 1432
[77] Morimoto T and Nagaosa N 2019 *Phys. Rev.* B **100** 235138
[78] Eliseev A A, Glinchuk M D, Khist V V, Lee C-W, Deo C S, Behera R K and Morozovska A N 2013 *J. Appl. Phys.* **113** 024107
[79] Cano A, Meier D and Trassin M (ed) 2021 *Multiferroics: Fundamentals and Applications* (Berlin: De Gruyter)

Michael G Cottam

Chapter 8

Nonlinear dynamics for nonmagnetic excitations

The preceding chapters have been mainly devoted to the linear aspects of the dynamics, which have been the subject of detailed studies. We introduced, however, the concept of nonlinear excitations in chapter 1 (see section 1.1.2) where we described nonlinear aspects of the dielectric function and outlined a calculation for a nonlinear surface wave that had no counterpart in the linear regime. Also, in chapter 4 we indicated briefly how nonlinear effects would arise, in principle, for spin waves or magnons in magnetic systems in terms of boson operator methods.

The aim in this and the next chapter is to develop, in particular cases, the nonlinear dynamical properties for the types of excitations covered earlier in the book, in particular for low-dimensional structures and for surface/interface geometries. This is not intended as a comprehensive treatment of the nonlinear dynamics, since that is a major topic in its own right and there are already textbooks dealing with specialized aspects. Instead, it is included as a concise survey with examples chosen to relate to the material of the previous chapters and to recent developments. As a general reference, the book by Cottam and Tilley [1] includes several chapters on nonlinear waves summarizing work in this field up to about 2005. There has been a stronger emphasis on 2D and 1D materials in the more recent work, and we give specialized references later.

In the present chapter we cover different *nonmagnetic* excitations, following approaches developed in nonlinear optics and for the nonlinear properties of phonons using many-body theoretical techniques, for example. The next chapter 9 focuses on the nonlinear aspects of magnetic excitations.

8.1 Nonlinearities in optics

We start by summarizing here some of the basic results for nonlinear waves in optics, including the origins of the nonlinearities and the different approaches that may be

doi:10.1088/978-0-7503-3903-2ch8

followed for the excitations in either calculating their properties or probing them experimentally. It will be convenient to consider separately the approaches conventionally followed in nonlinear optics and for dielectric media from those used in magnetic systems. In the former case, the development is typically in terms of an expansion of the nonlinear dielectric function, or its corresponding susceptibility, as already mentioned briefly in section 1.1.2. In the latter case (see the next chapter), there is either a macroscopic method adopted in terms of a nonlinear magnetic susceptibility (the analogue of the nonlinear optics approach) or a microscopic dipole-exchange or exchange-dominated method is used, depending in part on the magnetic regime for the excitation wave vector.

Some general nonlinear optics references are the established books by Yariv [2] and Mills [3], both of which provide detailed accounts of the theoretical formalism. More recent accounts are given in the books by Powers and Haus [4] and Boyd [5]. Already, in chapter 1, we discussed the response of a dielectric material to a position- and time-dependent electric field $\mathbf{E}$. The important relationships are those involving the polarization $\mathbf{P}$ and the dielectric function, written as $\epsilon(\mathbf{q}, \omega)$ if spatial dispersion is included. They are given in equations (1.13)–(1.17). We may also add the definition of the electrical susceptibility $\chi(\mathbf{q}, \omega)$ as expressed by $\mathbf{P}(\mathbf{q}, \omega) = \epsilon_0\chi(\mathbf{q}, \omega)\mathbf{E}(\mathbf{q}, \omega)$, and so from equation (1.17) we have the identity that

$$\epsilon(\mathbf{q}, \omega) = 1 + \chi(\mathbf{q}, \omega). \tag{8.1}$$

The conventional approach, as followed in nonlinear optics, is to expand the polarization, or alternatively the susceptibility or the dielectric function, in terms of Cartesian components of the electric-field vector, taking account of symmetry. Hence the expressions for the linear (L) and nonlinear (NL) parts of the Fourier component of the polarization, chosen at any single reference frequency (denoted by ω_0), take the form

$$P_i^{\mathrm{L}}(\omega_0) = \sum_j \chi_{ij}^{(1)}(-\omega_0; \omega_1)\, E_j(\omega_1), \tag{8.2}$$

$$\begin{aligned} P_i^{\mathrm{NL}}(\omega_0) = &\sum_{jk} \chi_{ijk}^{(2)}(-\omega_0; \omega_1, \omega_2)\, E_j(\omega_1)E_k(\omega_2) \\ &+ \sum_{jkl} \chi_{ijlk}^{(3)}(-\omega_0; \omega_1, \omega_2, \omega_3)\, E_j(\omega_1)E_k(\omega_2)E_l(\omega_3) + \cdots. \end{aligned} \tag{8.3}$$

We have dropped the $\mathbf{q}$ labels here for simplicity of the notation. On the right-hand side there are terms involving Fourier components of the electric-field vectors at the indicated frequencies ω_1, ω_2, etc. Each of these frequencies can be positive or negative (or zero in special cases). In equation (8.2) the linear susceptibility is just the second-order tensor (or matrix) with elements $\chi_{ij}^{(1)}$, which we have already discussed with examples (in terms of the linear dielectric function) in chapter 5 and elsewhere. The nonlinear susceptibility terms in equation (8.3) start with $\chi_{ijk}^{(2)}$ and $\chi_{ijkl}^{(3)}$ in lowest orders. These are just the elements of third- and fourth-order tensors, respectively. The allowed nonzero elements of all these susceptibility tensors can be deduced by symmetry arguments based on the crystal lattice structure (see e.g. [6, 7]). It is important to note that these elements may have different values at surfaces and interfaces compared with the bulk of a material due to the reduction in symmetry there.

The notation that we are using for the frequency dependence of the susceptibility terms follows that in [8], where the general term with the form $\chi^{(n)}_{ij\ldots m}(-\omega_0; \omega_1, \omega_2, \ldots, \omega_n)$ has the conserving property that

$$\omega_0 = \omega_1 + \omega_2 + \cdots + \omega_n. \tag{8.4}$$

In the linear case, no new frequencies are produced since we simply have $\omega_0 = \omega_1$ due to this conservation property. Various different situations arise, however, with the higher-order terms and in the following paragraphs we will briefly consider the second- and third-order nonlinear effects.

The possibilities for the overall frequency coming from the second-order $\chi^{(2)}$ nonlinearities are the sum and difference terms such as $|\omega_1| \pm |\omega_2|$. This can lead to a frequency doubling of the initial frequency in the sense that if the input frequencies of the electric fields are $\omega_1 = \omega_2 \equiv \omega$, then the output signal for the nonlinear polarization involves a frequency $\omega_0 = 2\omega$. This effect is known as *second-harmonic generation* (SHG). Likewise, there will be a difference-frequency effect coming at $\omega_0 = 0$, and this is *optical rectification.* Another case of interest is when one of the input fields is static (e.g. with $\omega_1 = 0$) and the other has frequency $\omega_2 \equiv \omega$. The output frequency for the nonlinear polarization is then also at the value ω. This is the same value as in the linear effect but the total polarization now has an additional contribution related to $\chi^{(2)}_{ijk}(-\omega; 0, \omega)$ and known as the *Pockel's effect.*

Further nonlinear effects arise in third-order from the term $\chi^{(3)}_{ijlk}(-\omega_0; \omega_1, \omega_2, \omega_3)$ when various choices are made for the input frequencies ω_1, ω_2 and ω_3. There are now many possibilities, but in particular we will mention the cases of

$$\chi^{(3)}_{ijlk}(-3\omega; \omega, \omega, \omega) \quad \text{and} \quad \chi^{(3)}_{ijlk}(-\omega; \omega, -\omega, \omega). \tag{8.5}$$

The first of these corresponds to *third-harmonic generation* (THG), since there is a scaling by a factor three for the frequency. The second example produces output at the same input frequency ω and, together with similar terms, it gives rise to what is known as the *Kerr effect.* It follows that the total effective dielectric function at the frequency ω due to this process is given by

$$\epsilon^{\text{eff}}_{ij} = \epsilon_{ij} + \frac{3}{4}\sum_{lk}\chi^{(3)}_{ijlk}(-\omega; \omega, -\omega, \omega)\, E_k(-\omega)E_l(\omega), \tag{8.6}$$

where the factor of 3/4 comes from a careful consideration of the permutations involved. This simplifies if we assume the case of an isotropic material (such as glass), since the linear term reduces to a scalar (multiplied by the unit matrix) and the only nonzero elements of the NL tensor are the equivalent diagonal terms such as $\chi^{(3)}_{xxxx}$. Hence the modified scalar dielectric function is given by

$$\epsilon^{\text{eff}} = \epsilon + \frac{3}{4}\chi^{(3)}_{xxxx}|\mathbf{E}|^2, \tag{8.7}$$

which is just the same result as was quoted in equation (1.18) with the nonlinear coefficient given by $\rho = (3/4)\chi^{(3)}_{xxxx}$. We now see the origin of the intensity-dependent contribution to the dielectric function described in section 1.1.2 as being the Kerr effect. The field dependence predicted in equation (8.7) for ϵ^{eff} eventually breaks down at large enough values of $|\mathbf{E}|$, because higher-order terms in the previous electric-field expansions become important. Typically a saturation effect at large $|\mathbf{E}|$ is observed, in which ϵ^{eff} tends to a constant value.

We mention that, apart from the specific nonlinear effects described in the previous paragraphs, the $\chi^{(2)}$ and $\chi^{(3)}$ terms can be used in many-body perturbation theories to calculate the contributions to the damping of the excitations in the nonlinear material. Damping effects were introduced phenomenologically into many expressions for the dielectric functions in chapters 5 and 6 (e.g. for plasmons and phonons). There is further discussion of phonon damping in the next section.

One way to enhance the interactions between light and matter is to place the material sample inside an electromagnetic cavity designed to excite specific photon modes. The hybrid cavity modes that result from this process may exhibit strong nonlinear and nonclassical behaviours. These modes have received increasing attention, particularly during the last decade when experimental techniques have advanced. Some general accounts for mainly nonmagnetic materials can be found, for example, in [9–11]. The hybrid magnons (or magnon-polaritons) in cavities will be discussed in section 9.5.

8.2 Nonlinearities for phonons

In this section we will discuss some nonlinear aspects of the phonon dynamical properties. As in chapter 2, where the linear behaviour was the focus, there are two methodologies that can be followed. One approach is to use the lattice dynamics of vibrating atoms (but with generalizations made to the form of the interatomic potential) and the other approach is based on elastic-wave propagation (with the additional effects of higher-order tensors to relate stress and strain in the material). Both approaches will be mentioned here. Some useful general references for the linear and nonlinear properties of phonons are the books by Srivastava [12] and Snoke [13], whereas the book by Rushchitsky [14] gives a detailed treatment of nonlinear elastic-wave theory.

In the quantum-mechanical 1D approach to the lattice dynamic in section 2.1.1 for a monatomic chain of atoms we expressed the Hamiltonian for the total energy as in equation (2.8), which was subsequently rewritten as a sum of simple harmonic oscillators (SHOs) in equation (2.10). The interaction potential was seen to be quadratic in the atomic displacements in the linear regime. Thus for a single SHO in 1D, where the displacement in the z-direction is $u = z - z_0$ and z_0 denotes the equilibrium position, the potential energy would take the quadratic form $U(z) = U_0 + \frac{1}{2}C\,(z - z_0)^2$. Here C is just the force constant introduced in section 2.1. The simplest phenomenological way to generalize this results to include effects of

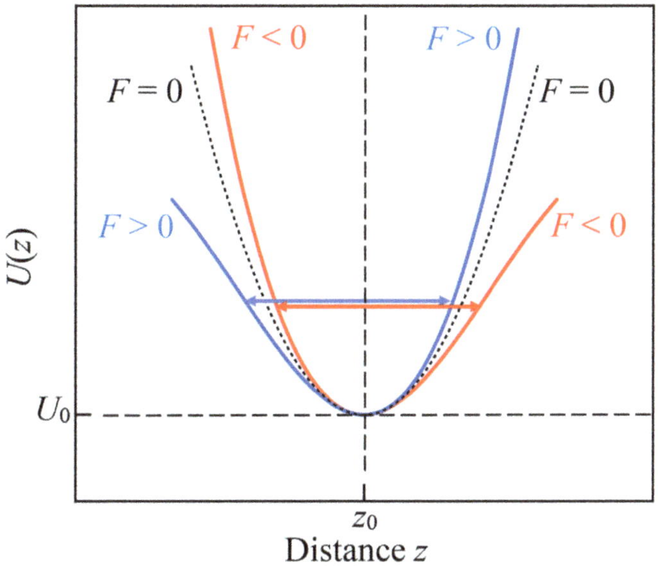

Figure 8.1. Schematic plot of the interatomic potential $U(z)$, given by equation (8.8), as a function of separation distance z. The blue and red curves illustrate the anharmonic cases of $F > 0$ and $F < 0$, respectively, while the case of a parabolic potential well ($F = 0$) for a SHO is shown by the black dotted curve.

anharmonicity at larger displacements would be to add a term with a higher power of $z - z_0$. Postulating a leading-order cubic term, for example, we have

$$U(z) = U_0 + \frac{1}{2}C\,(z - z_0)^2 + F\,(z - z_0)^3, \tag{8.8}$$

where F is an anharmonicity constant. The predicted behaviour for the potential well is illustrated schematically in figure 8.1, showing cases when F is either positive or negative. For any chosen value of the total energy, the range of displacements for a classical particle are indicated by the double-arrowed horizontal lines. It can be inferred from this simple classical picture that the mean displacement at larger energy moves to the right (left) when $F < 0$ ($F > 0$), consistent with a thermal expansion (contraction) due to that mode.

The effective interatomic potentials for specific selected phonon modes in various 2D and 3D solids have recently been investigated by strong-field NL phononics. As a groundbreaking example, we mention work by von Hoegen *et al* [15], who used high-power laser pulses, with electric-field strengths of order 10^6 V cm^{-1}, at THz frequencies to drive the phonon nonlinearities to high orders. Their studies were applied to the A_1 TO-phonon mode in the ferroelectric material $LiNbO_3$. The authors were able to deduce (using DFT and other theoretical techniques) the potential-well profiles, showing them to have the same form as that shown in figure 8.1 when $F > 0$. Their analysis, however, included higher-order terms (up to the fifth power) in the atomic displacements, going beyond those in equation (8.8).

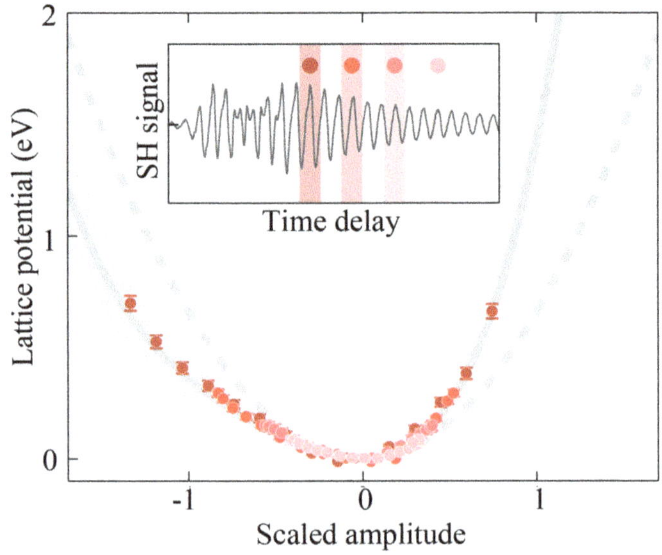

Figure 8.2. Plot of the lattice potential for the A_1 TO-phonon in $LiNbO_3$ versus displacement of the oxygen atoms (in scaled units quoted as Å amu$^{-1/2}$). The experimental points (shown in red symbols) were reconstructed from different cycles of the time-resolved second-harmonic (SH) measurement shown in the inset. The grey solid line is the mode potential obtained by DFT calculations, and the light grey shaded area is an estimate of its systematic uncertainties. The dashed grey curve is the potential in the harmonic approximation. (Reproduced with permission from [15]. Copyright 2018 Nature.)

Some results demonstrating the excellent correspondence between theory and experiment are shown in figure 8.2

To develop our arguments more formally, we may follow a similar approach to that in [13] by expanding the potential energy density U due to the lattice deformations (or atomic displacements) in a Taylor series with the form

$$U = U_0 + \sum_{i,j} C_{ij}^{(1)}\bar{u}_{ij} + \frac{1}{2!}\sum_{i,j}\sum_{i',j'} \bar{u}_{ij}\bar{u}_{i'j'} + \frac{1}{3!}\sum_{i,j}\sum_{i',j'}\sum_{i'',j''} C_{ij}^{(3)}\bar{u}_{ij}\bar{u}_{i'j'}\bar{u}_{i''j''} + \cdots. \tag{8.9}$$

Here the first term U_0 again represents the contribution for the undeformed system and it has a constant value, $\bar{u}_{ij}$ is a component of strain as defined in equation (2.18), and the tensor $C^{(n)}$ is associated with the term in the expansion having n products of the strain ($n = 1, 2, 3, \ldots$). The second term on the right-hand side, which is linear in the strain, vanishes at an equilibrium point. The following term involving $C^{(2)}$ is just the elastic (linear phonon) term, and the term proportional to $C^{(3)}$ gives us the leading-order contributions due to the NL processes. Hence, dropping the constant term and choosing the origin for the displacements as being expressed relative to the equilibrium point, we have more simply

$$U = \frac{1}{2}\sum_{i,j}\sum_{i',j'}\left[C_{ij}^{(2)} + \frac{1}{3}\sum_{i'',j''} C_{ij}^{(3)}\bar{u}_{i''j''} + \cdots\right]\bar{u}_{ij}\bar{u}_{i'j'}. \tag{8.10}$$

The above result for the NL case can be interpreted as representing an effective correction to (or a renormalization of) the elastic constant tensor $C^{(2)}$.

There are now different possibilities regarding how to proceed with the theoretical development. One way is to re-express the terms involving the strain by transforming to boson creation and annihilation operators for the phonons, by analogy with results in section 2.1.1 for a linear chain of elastically coupled atoms. This is the approach that is typical of many-body theory in condensed-matter systems, and we refer to specialist books in this field for details of the techniques (see e.g. [16, 17]). In general, a Hamiltonian of the following form in second quantization is often employed:

$$\mathcal{H} = \sum_{\mathbf{q}} \omega_{\mathbf{q}}\, a_{\mathbf{q}}^{\dagger} a_{\mathbf{q}} + \sum_{\mathbf{q},\mathbf{q}',\mathbf{k}} U(\mathbf{k})\, a_{\mathbf{q}}^{\dagger} a_{\mathbf{q}'}^{\dagger} a_{\mathbf{q}-\mathbf{k}} a_{\mathbf{q}'+\mathbf{k}}, \tag{8.11}$$

where we have the property $[a_{\mathbf{q}}, a_{\mathbf{q}'}^{\dagger}] = \delta_{\mathbf{q},\mathbf{q}'}$ for the commutators between boson operators. Here the first term, which is quadratic in the operators, represents non-interacting phonons, and is the same (apart from an unimportant constant) as the result in equation (2.13). The second term represents the leading effects of phonon–phonon interactions; it is quartic in the operators and contains an interaction potential with Fourier components denoted by $U(\mathbf{k})$ at wave vector $\mathbf{k}$. It incorporates scattering processes in which a phonon with a wave vector of (say) $\mathbf{q}$ interacts with another phonon with wave vector $\mathbf{q}'$, e.g. as excited thermally or by a high-power laser pulse, as in the experiment just referenced. These excitations then scatter off one another to produce an output pair of phonons with wave vectors $\mathbf{q} - \mathbf{k}$ and $\mathbf{q}' + \mathbf{k}$. The situation is broadly analogous to that for the four-magnon interactions to be discussed later in chapter 9 for magnetic systems. Using standard many-body theoretical techniques such as diagrammatic perturbation theory, equation (8.11) may be employed to 'renormalize' the phonons, i.e. to calculate the energy shifts produced in the phonons due to their interactions with one another and to obtain the phonon damping (or reciprocal lifetime). Other effects that depend on phonon scattering, such as the thermal conductivity, may also be studied using either equation (8.11) or by methods involving Fermi's golden rule [13] in quantum mechanics. Details may be found in specialized textbooks on many-body theory (see e.g. [16, 17]).

It is relevant to mention at this this stage that, in addition to the direct phonon–phonon interactions, the NL phonon effects may take place indirectly via other excitations (such as magnons and various electronic excitations) in the same material, as well as with defects in the bulk or at a surface. The role of the electron–phonon interaction in modifying the phonon dynamics has been an active area of study. The interaction mechanism is usually taken to arise due to the *deformation potential* (see e.g. [13]), whereby a propagating phonon distorts the lattice and hence modifies the electronic band(s). In terms of creation and annihilation operators, the electron–phonon interaction Hamiltonian typically contains terms proportional to the operator products such as $a_{\mathbf{q}}c^{\dagger}_{\mathbf{q}+\mathbf{q}'}c_{\mathbf{q}'}$ and its Hermitian conjugate, where the a operators are the boson operators for the phonons, as earlier in this section, and the c operators are the fermion operators for the electronic states, as described in section 5.2. This Hamiltonian describes a process in which a phonon may split into a virtual electron–hole pair, which may subsequently recombine to produce another phonon, thus representing a renormalization process for the original phonon. Raman scattering (RS) was employed in a recent study by Eliel *et al* [18] to investigate both the electron–phonon interactions and the phonons in bilayers of graphene.

Apart from the lattice dynamical methods referenced in the preceding paragraphs, there have been ongoing studies of NL elastic waves. Basically, the method proceeds by taking the Hooke's law expression in equation (2.20) and generalizing the right-hand side with the inclusion of terms that are of higher order (quadratic, cubic, and so forth) in the strain components. Then, with the aid of equation (2.19), we may find a generalization of the linear elastic wave equation given in (2.22). It becomes a higher-order partial differential equation. Typically, solutions are investigated using either finite-element methods or perturbative approaches (see e.g. [14, 19, 20]).

8.3 Introduction to solitons

When nonlinearities are included in the relevant wave equation, some of the partial differential equations that involve position and time as the variables may turn out to exhibit types of solutions known as *solitary waves*, for which there is no counterpart in the linear case. In essence, a solitary wave is a disturbance that travels without change of shape. If there is the further property that after two such solitary waves collide (passing through one another) they emerge without change of their shape, then they are called *solitons* (see e.g. the books by Drazin and Johnson [21], Infeld and Rowlands [22] and Dauxois and Peyrard [23] for general accounts). Sometimes, however, the terms solitary wave and soliton tend to be used interchangeably.

The earliest reported observation of a solitary wave was by John Scott Russell in 1834 as a propagating water wave along a canal [22]. The wave was launched from the bow of a canal boat when the boat suddenly stopped. A 'hump' of water detached itself and travelled a large distance along the canal (followed by Scott Russell) without noticeably changing its profile. Later studies of the hydrodynamics of surface waves showed that the phenomenon is governed by a specific nonlinear differential equation, now known as the *Korteweg–de Vries* (KdV) *equation*, that permits solitary wave solutions. An essential feature, both here and in other examples to be described shortly, is that nonlinearities must be present (playing a role as a 'focussing' effect) to counterbalance the broadening or spreading of the wave that would otherwise take place due to dispersion.

In its standard form, the KdV equation in one spatial dimension can be written as the partial differential equation [23]

$$\frac{\partial \phi}{\partial \tau} + 6\phi \frac{\partial \phi}{\partial \xi} + \frac{\partial^3 \phi}{\partial \xi^3} = 0. \tag{8.12}$$

Here τ and ξ are dimensionless time and spatial variables, respectively, and the nonlinearity (in the amplitude function ϕ) arises from the second term. It can be verified (e.g. by direct substitution into the above equation) that solutions exist with the form

$$\phi(\xi, \tau) = A \operatorname{sech}^2 \{\sqrt{A/2}\,(\xi - 2A\tau)\}, \tag{8.13}$$

where A is a positive constant. This represents a travelling hump-like disturbance with the characteristics seen by Scott Russell.

In the context of excitations in solids, the two nonlinear equations capable of soliton solutions that we shall discuss are the *sine-Gordon equation* and the *nonlinear Schrödinger equation*. Both have been studied mathematically in great detail (for example, by the inverse scattering method [21, 22]). We will now give examples of how these two nonlinear equations provide an approximate physical description for some real systems.

8.3.1 The sine-Gordon equation for solitons

We start with the sine-Gordon (or sG) equation, which exhibits soliton-related solutions that have been applied to a variety of nonlinear physical systems, including (in classical mechanics) an array of coupled restricted pendula. The standard sG equation for a variable $\phi(\xi, \tau)$ as a function of a spatial coordinate ξ and time τ takes the form

$$\frac{\partial^2 \phi}{\partial \xi^2} - \frac{1}{c^2}\frac{\partial^2 \phi}{\partial \tau^2} = m^2 \sin \phi. \tag{8.14}$$

Here c is a characteristic velocity and m is a constant that scales the nonlinearity represented by the term on the right-hand side of the equation. If $m = 0$ we would simply have the linear wave equation (e.g. as for electromagnetic waves or for classical elastic waves in a solid). Three types of solution have been identified for the sG equation; they are spatially unbounded oscillatory solutions, spatially bounded oscillatory solutions (called breathers), and finally solitons. An example of a soliton solution is

$$\phi(\xi, \tau) = \cos^{-1}\left\{1 - 2\operatorname{sech}^2\left[\frac{m(\xi - \xi_0 - u\tau)}{\sqrt{1 - (u/c)^2}}\right]\right\}. \tag{8.15}$$

This equation represents a 'kink'-like feature that is centred around $\xi = \xi_0$ and moves with a speed $u < c$. The quantity ϕ, which may be envisioned as the angular orientation of a vector, undergoes a rotation from 0 to 2π in going from one side of the kink through to the other side.

The specific discussion of physical applications involving the sG equation and equation (8.15) will be deferred until the next chapter, where they notably arise in the context of magnetic solitons associated with domain wall effects in magnetic materials and with excitations in certain 1D (chain) ferromagnets.

8.3.2 The nonlinear Schrödinger equation for solitons

The nonlinear Schrödinger equation (NLSE), which we will describe next, has soliton solutions with a variety of applications in condensed-matter physics, occurring both in nonlinear optics (e.g. for pulses in optical-fibre communications) and in the dynamical properties of magnetic films. A standard (dimensionless) form of writing the NLSE for a variable $\phi(\xi, \tau)$ is

$$\mathrm{i}\frac{\partial \phi}{\partial \tau} = -\frac{1}{2}\frac{\partial^2 \phi}{\partial \xi^2} - |\phi|^2\phi. \tag{8.16}$$

This may be compared with the 1D time-dependent Schrödinger equation of quantum mechanics, which may be stated as $\mathrm{i}\hbar\partial\psi/\partial t = -(\hbar^2/2m)(\partial^2\psi/\mathrm{d}x^2)$, where $\psi(x, t)$ is the wave function, m is the particle mass, and the potential function has been set to zero. We see that the last term of equation (8.16) provides the nonlinearity, while ξ and τ are analogous here to the space-like and time-like variables of Schrödinger's equation.

Solutions of the NLSE are discussed in most textbooks on nonlinear dynamics (see e.g. [21, 22]). The simplest solution is for the excitation of one soliton, and it can be written as

$$\phi(\xi, \tau) = (2\kappa)^{1/2}\,\mathrm{sech}[(2\kappa)^{1/2}\tau]\exp(\mathrm{i}\kappa\tau), \tag{8.17}$$

where κ may take any real positive value. The intensity of the pulse, which is proportional to $|\phi^2|$, has the form of a 'sech-squared' hump when plotted as a function of the spatial variable ξ, as we illustrate in figure 8.3 for two cases, depending on the value of κ. In these dimensionless units the pulse height is 2κ and the pulse width (the full width at half-maximum) is $(2/\kappa)^{1/2}$.

Generalizations of the one-soliton solutions to equation (8.16) were made by Satsuma and Yajima [26] by using a powerful mathematical technique known as the inverse scattering method (see e.g. [21, 22]). The additional solutions are found to differ from that quoted in equation (8.17) in that they are 'breathing' modes, meaning here that although the initial pulse shape changes with distance it re-forms unchanged after a certain interval of the distance variable ξ is traversed.

The NLSE in nonlinear optics

For nonmagnetic low-dimensional structures, the propagation of solitons generated by a tuneable laser in a single-mode quartz fibre has been studied experimentally by Mollenauer *et al* [27]. The nonlinearity that provides the connection to the NLSE in this case arises via the Kerr effect already discussed in sections 1.1 and 8.1. The derivation of the NLSE for this system usually proceeds by starting with Maxwell's

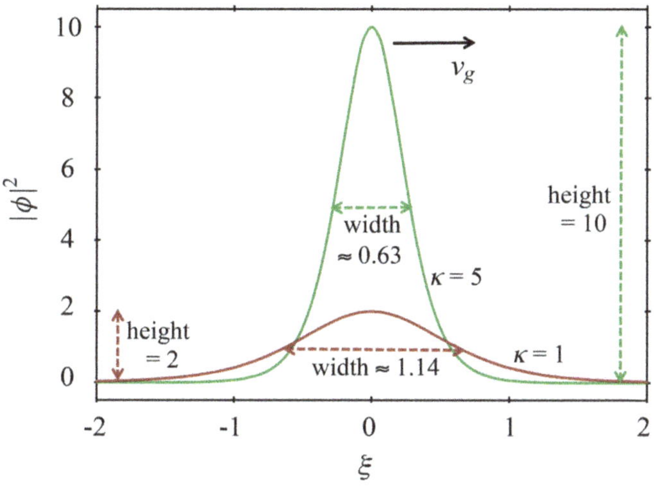

Figure 8.3. Plot to show the dependence of the intensity $|\phi|^2$ on ξ at $\tau = 0$ as given by equation (8.17) for solitons that are described by the nonlinear Schrödinger equation (NLSE) when $\kappa = 1$ and when $\kappa = 5$. The behaviour for the pulse height and width is shown for each case.

equations supplemented by a nonlinear polarization term. Then a wave packet can be formed, which is expanded using the *slowly varying amplitude* (SVA) approximation. The main steps are summarized below and a fuller account can be found in [1].

Maxwell's wave equation for an isotropic nonlinear material can be expressed as (see e.g. [28, 29])

$$\frac{\partial^2 E}{\partial z^2} - \mu_0 \frac{\partial^2 D}{\partial t^2} = \mu_0 \frac{\partial^2 P^{\text{NL}}}{\partial t^2}, \tag{8.18}$$

where the spatial variations are taken, for simplicity, to be in just one propagation direction z. The D field here contains only the linear polarization, and so $D = \epsilon_0 \epsilon(\omega)\, E$ for the Fourier components at frequency ω. The quantity P^{NL} is the nonlinear polarization. For applications to a wave packet (representing a pulse) we may write

$$E(z, t) = F(z, t)\exp(\mathrm{i}q_0 z - \mathrm{i}\omega_0 t), \tag{8.19}$$

with q_0 and ω_0 denoting the central wave number and angular frequency of the packet. The above relationships enable the derivative terms on the left side of equation (8.18) to be formally rewritten in terms of the envelope function F and its derivatives. In particular, it is found that

$$-\mu_0 \frac{\partial^2 D}{\partial t^2} = \exp(\mathrm{i}q_0 z - \mathrm{i}\omega_0 t)\int_{-\infty}^{\infty} (q_0 + k)^2 F(z, \nu)\exp(-\mathrm{i}\nu t)\mathrm{d}\nu, \tag{8.20}$$

where $q_0 + k$ means the wave number corresponding to frequency $\omega_0 + \nu$. The SVA approximation mentioned earlier is basically just the assumption that $|k| \ll q_0$ for the important terms in the integral, so a Taylor series expansion gives $k(\nu) \simeq k' \nu + (1/2)k'' \nu^2$, where we denote for the derivatives

$$k' = \frac{\mathrm{d}k}{\mathrm{d}\omega} = \frac{1}{v_g}, \quad k'' = \frac{\mathrm{d}^2k}{\mathrm{d}\omega^2} = \frac{\mathrm{d}}{\mathrm{d}\omega}\left(\frac{1}{v_g}\right). \tag{8.21}$$

Here the derivatives are evaluated at $\omega = \omega_0$ and v_g denotes the group velocity of the wave packet. With the above approximations and similarly assuming that the frequency spread around ω_0 is small, we find after some straightforward algebra that the envelope function F satisfies

$$\mathrm{i}\left(\frac{\partial}{\partial z} + \frac{1}{v_g}\frac{\partial}{\partial t}\right)F = \frac{k''}{2}\frac{\partial^2 F}{\partial t^2} - k_0\rho|F|^2 F. \tag{8.22}$$

Here we have assumed a Kerr nonlinearity to replace P^{NL}, and so ρ denotes the third-order nonlinear coefficient defined as in equation (1.18). The terms on the left-hand side of the equation describe an envelope travelling at speed v_g, so it is convenient to make a transformation to new set of space and time variables by $(z, t) \rightarrow (Z, T)$ where we have $Z = z - v_g t$ and $T = t$, giving

$$\mathrm{i}\frac{\partial F}{\partial Z} = \frac{k''}{2}\frac{\partial^2 F}{\partial T^2} - k_0\rho|F|^2 F. \tag{8.23}$$

We may now compare the above result with the previous canonical form of the NLSE that was quoted in equation (8.16). Apart from the scaling of the coefficients, we see that the equations are essentially the same with two provisions: on the one hand, the roles of dimensionless ξ and τ in (8.16) are replaced by T and Z, respectively, and on the other hand it is necessary that the coefficients k'' and ρ must have opposite signs. The first of these points (the interchange of space-like and time-like variables) is of no particular significance, but the second point is important in giving us as a necessary condition for solitons, the so-called *Lighthill criterion*, that

$$k''\rho < 0. \tag{8.24}$$

This result ensures the physical property mentioned earlier that the spreading effect of a soliton pulse due to its dispersion must be counterbalanced by the nonlinearity.

The corresponding derivation of the NLSE in the context of magnetic solitons in magnetic thin films and wave guides will be given in chapter 9.

8.4 Some applications to low-dimensional systems

In this section we will focus on three main areas for applications. These are, firstly, nonlinear optics via surface modes, secondly, Kerr effect bistability in layered and multilayered systems, and thirdly, gap solitons in nonlinear optics. These applications involve the third- and fourth-order susceptibility tensors described in section 8.1.

8.4.1 Nonlinear optics via surface modes

In principle, the nonlinear effects mentioned earlier in section 8.1 can lead to an enhancement of the electric field when measuring the optical transmission though a suitable sample, in particular when multiple interfaces are involved. This is because of the higher powers of E appearing in expressions such as equation (8.3). The most basic implementation of these ideas is for transmission through a Fabry–Pérot resonator, in which a film of thickness L is coated with partially reflecting mirrors. The use of a Fabry–Pérot instrument in the *linear* regime is familiar and extensive (see e.g. the general optics text book by Born and Wolf [30]). It is, for example, the essential tool for achieving the high spectral resolution in Brillouin light scattering (BLS) in solids, as mentioned in section 1.3. The calculation of the transmission spectra in a normal-incidence geometry is standard in linear optics, giving repeated peaks that occur when the condition $q_B L = n\pi$ is satisfied, where n is any positive integer and q_B is the optical wave number in the film. The generalization to include the simplest nonlinear effects of SHG through terms like $\chi_{ijk}^{(2)}(-2\omega; \omega, \omega)$ is surprisingly complicated, as a result of all the internal reflections and transmissions in this geometry. A comprehensive account is presented by Hashizume *et al* [31].

Another type of structure that has been studied for the role of SHG on the reflectance and transmission is the periodic superlattice, in which one (or more) of the components is taken to be nonlinear. Notably, calculations have been made by Agranovich and Kravstov [32] and Boyd and Sipe [33]. They were able to take advantage in some cases of a generalized form for the effective medium approximation, which was described in chapters 5 and 6 for the linear regime. Experiments due to Fischer *et al* [34] have provided quantitative confirmation for many of the predictions.

A more interesting illustration of electric-field enhancement, which is also surface specific, is through nonlinear coupling to surface modes such as surface plasmons or surface polaritons in bounded media. For significant nonlinear effects to be observable it is important that the surface mode be deeply penetrating, and for this reason the long-range surface polariton (LRSP) that was mentioned briefly in chapter 6 is suitable. This mode was originally predicted by Sarid [35] to occur for a thin metallic layer (constituting a 2DEG electron gas) with a linear dielectric function, bounded on each side by a non-lossy medium. In later work the theory was extended to nonlinear optics [36]. Experiments to demonstrate nonlinear optical effects using the LRSP were reported by Quail *et al* [37]. An example of their measurements of the SHG reflection coefficient is shown in figure 8.4. The active medium in this case was an Ag film and the experimental set-up is sketched in the inset. The theory line for the variation of the second-harmonic reflection coefficient with angle of incidence on the surface is seen to be in good agreement with experiment.

8.4.2 Kerr effect bistability in multilayers

We now discuss some consequences of the Kerr effect in multilayer geometries, starting with the simple example of a nonlinear surface excitation that was given at

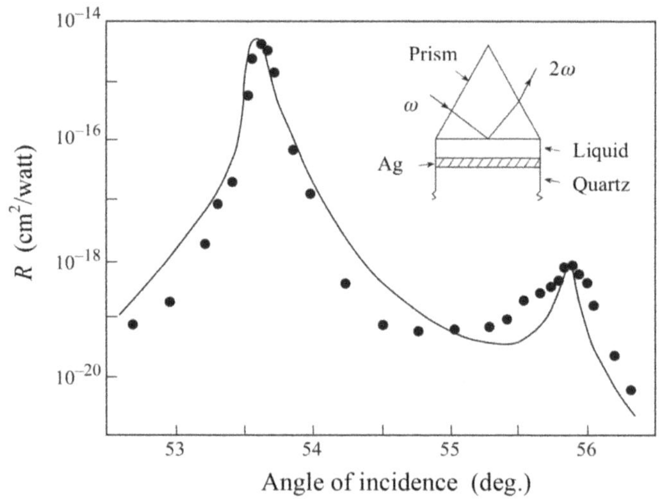

Figure 8.4. Second-harmonic reflection coefficient R for SHG via the long-range surface plasmon in an Ag film plotted versus the angle of incidence. The solid lines are theory and the points represent experimental data. The experimental geometry is shown in the inset. (Reproduced with permission from [37]. Copyright 1983 American Physical Society.)

the beginning of this book in section 1.1.2. There we outlined a calculation for a nonlinear material having a dielectric function with the form of equation (1.18) filling the half-space $z < 0$, while the other half-space ($z > 0$) corresponds to a linear material. By following analogies with self-focussing, we showed that a nonlinear wave is obtained in s-polarization for the electric field localized near the $z = 0$ interface. The form of the solution for the electric field is given explicitly by equations (1.26) and (1.27).

A few further steps are now required in order to arrive at a dispersion relation. First, the usual electromagnetic boundary conditions on the field components can be applied at $z = 0$. These conditions are equivalent to requiring that E and dE/dz are continuous at the boundary, and their implementation leads to the condition

$$\kappa \tanh(\kappa z_0) = \kappa_1, \tag{8.25}$$

where κ and κ_1 were introduced earlier and z_0 is the remaining constant of integration. Equation (8.25) implies $\kappa_1 < \kappa$ (or, equivalently, $\epsilon_1 > \epsilon$) as a necessary inequality. The standard procedure (as discussed fully in [1], for example) is to utilize the time-averaged power flow $\mathcal{P}$ as a basic parameter. The power flow is obtained from the Poynting vector, which is proportional to $\mathbf{E} \times \mathbf{H}$ and may be evaluated in terms of z_0 and other quantities. Eventually the dispersion relation is found from equation (8.25) in the form

$$\mathcal{P} = \mathcal{P}_0\left[1 + \frac{1}{2}\left(\frac{\kappa}{\kappa_1} + \frac{\kappa_1}{\kappa}\right)\right] > 2\mathcal{P}_0, \tag{8.26}$$

where $\mathcal{P}_0 = c^2 q_x \kappa / \mu_0 \rho \omega^3$. This result clearly establishes that there is a critical power level (related to a multiple of $\mathcal{P}_0$) for the nonlinear mode to exist. Further discussion and some minor extensions are provided in [38, 39].

Subsequently the preceding analytic theory of s-polarized nonlinear polaritons was generalized to multilayers (including superlattices) by Vassiliev and Cottam [40], including the situation where all layers in the system may have a Kerr-type dielectric function. Their results are first illustrated in figure 8.5 for a special case of a nonlinear film bounded on each side by the same nonlinear cladding material. It is convenient to plot the results in terms of showing an effective refractive index $n = cq_x/\omega$ versus $p^{1/2}$, where $p \equiv \rho E_1^2/2$ is a scaled nonlinearity coefficient. Here E_1 denotes the magnitude of the electric field at either one of the interfaces with the film. It is the additional parameter to be related to the power flow (just as we had the parameter z_0 in the previous calculation). For illustration, the film thickness has been chosen as one half of the vacuum wavelength λ_0 of the light. The modes labelled Z are branches to the dispersion that occur even in the limit of a linear cladding material, whereas X and Y refer to the additional branches that are found when the cladding material is nonlinear. These results exhibit the property of multistability due to the nonlinear effects, i.e. n is a multivalued function of the nonlinearity parameter $p > 0$ for certain ranges of the materials parameters.

A further example, this time for the case of a two-component periodic superlattice, is given in figure 8.6 where both constituent materials A and B have nonlinear Kerr-type dielectric functions. Although there is no rigorous analogue of Bloch's theorem in the nonlinear regime, we can nevertheless examine the behaviour for particular values of the Bloch factor exp(iQD), following our previous notation, while other types of modes are not necessarily precluded. The assumed parameters in

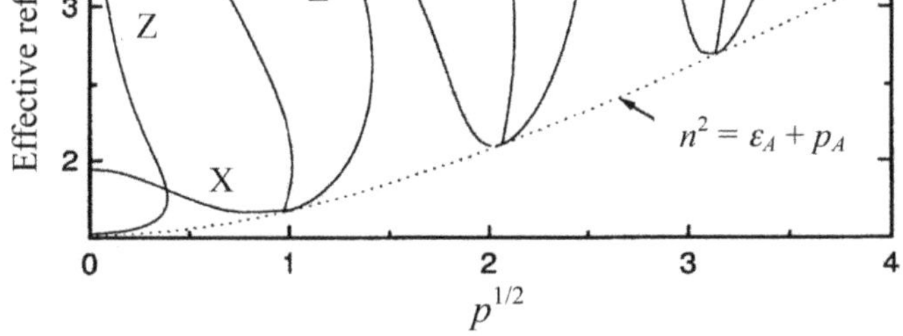

Figure 8.5. Dispersion curves for s-polarized polaritons in the case of a nonlinear film bounded symmetrically by two nonlinear cladding materials. The plots are shown as an effective refractive index n for the film versus $p^{1/2}$, taking $\epsilon_A = 1$ for the bounding material and $p/p_A = 2$ for the ratio of nonlinear coefficients. Other labels and notation are explained in the text. (Reproduced with permission from [40]. Copyright 2000 World Scientific.)

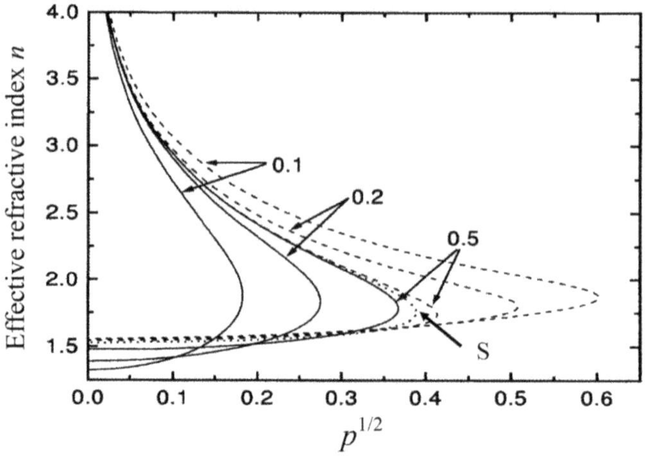

Figure 8.6. Dispersion curves for s-polarized polaritons in the case of a two-component superlattice when both constituents A and B have nonlinear dielectric functions. The numerical values correspond to d_A/λ_0 and other parameters are given in the text. Curves for the Bloch factor equal to -1 and 1 are shown by solid and dashed lines, respectively. The plots are displayed as an effective refractive index n versus $p^{1/2}$. Curve S indicates a corresponding single-layer branch for comparison. (Reproduced with permission from [40]. Copyright 2000 World Scientific.)

this case are $\epsilon_A = 2.3$, $\epsilon_B = 2.45$, $\rho_A/\rho_B = 0.5$ and $d_B/\lambda_0 = 0.5$. Here the nonlinearity parameter p is defined as $\rho_B E_1^2/2$, where E_1 is the magnitude of the electric field at an $A{-}B$ interface.

In recent years there have been continuing studies of various types of polaritons in layered structures, e.g. taking characteristic frequency dependences of the linear and nonlinear dielectric functions into account and studying other electric-field polarizations (see [41] and references therein). For example, the utilization of the Kerr effect for multilayer dielectric coatings was described in [42].

8.5 Gap solitons in nonlinear optics

We described in section 8.3.2 how the Lighthill criterion $k''\rho < 0$ provides a necessary condition for the formation and evolution of solitons in terms of the NLSE. Usually the nonlinearity coefficient ρ is positive, so typically it is the curvature term in the dispersion, $k'' = d^2k/d\omega^2$, that must be negative for optical solitons. One way in which this can be engineered relates to wave propagation in periodic structures such as superlattices. For example, we have seen in several of the previous chapters that band gaps (or stop bands) open up at the mini Brillouin-zone boundaries and that the curvature in the dispersion can have opposite signs just below and just above the gaps, where it changes rapidly. We may notice this property, for example, in figure 1.10 for electromagnetic waves, figures 2.13 and 2.17 for phonons, and figures 4.12 and 4.13 for magnons. It follows, therefore, that the conditions for soliton propagation might potentially be realized near to these band gaps, which leads us to the concept of *gap solitons*.

The conditions for gap solitons in nonlinear optics were first investigated by Mills and co-workers [43, 44]. They assumed a periodic two-component superlattice with a large (but finite) number of layers. One constituent material was taken to be linear while the other was assumed to have a Kerr-type nonlinear dielectric function with the same form as discussed in previous sections of this chapter. The formation of a soliton wave packet was confirmed, and the envelope function was found to be similar to a sech-squared NLSE profile as plotted in figure 8.3. Theoretical studies were also carried out to show the possibility of gap solitons in more complicated layered structures, such as those based on the Fibonacci growth rules (see figure 2.15a). In particular, Dutta Gupta and Ray [45] calculated the transmission through 55-layer and 233-layer sequences taking both the A and B building blocks to have a Kerr-type nonlinearity. They found results that were qualitatively rather similar to those obtained in [43, 44].

Following shortly after the calculation applied to the periodic superlattices, the first experimental evidence was published by Mohideen *et al* [46]. Their measurements were made on a grating that was produced by UV laser interference on a GeO_2-doped SiO_2 glass fibre. By varying the longitudinal strain in the fibre they were able to adjust the grating period and hence achieve tunability. Further discussion of these results can be found in [1]. The review article on gap solitons in nonlinear optics by Maimistov [47] covers the recent work in this field, including developments that have been mainly on the experimental side and with regard to the materials technology.

References

[1] Cottam M G and Tilley D R 2005 *Introduction to Surface and Superlattice Excitations* 2nd edn (Bristol: IOP Publishing)
[2] Yariv A 1989 *Quantum Electronics* 3rd edn (New York: Wiley)
[3] Mills D L 1998 *Nonlinear Optics* 2nd edn (Berlin: Springer)
[4] Powers P E and Haus J W 2017 *Fundamentals of Nonlinear Optics* (Boca Raton, FL: CRC Press)
[5] Boyd R W 2020 *Nonlinear Optics* 4th edn (Boston: Academic)
[6] Nye J F 1985 *Physical Properties of Crystals* (Oxford: Clarendon)
[7] Wherrett B S 1986 Group theory for atoms *Molecules and Solids* (Englewood Cliffs, NJ: Prentice-Hall)
[8] Butcher P N and Cotter D 1990 *The Elements of Nonlinear Optics* (Cambridge: Cambridge University Press)
[9] Weiner J and Ho P T 2008 *Light–Matter Interaction: Fundamentals and Applications* (New York: Wiley)
[10] Xiang Z-L, Ashhab S, You J Q and Nori F 2013 *Rev. Mod. Phys.* **85** 623
[11] Kurizki G, Bertet P, Kubo Y, Mø lmer K, Petrosyan D, Rabl P and Schmiedmayer 2015 *J. Proc. Natl Acad. Sci.* **112** 3866
[12] Srivastava G P 1990 *The Physics of Phonons* (New York: Taylor and Francis)
[13] Snoke D W 2020 *Solid State Physics: Essential Concepts* 2nd edn (Cambridge: Cambridge University Press)
[14] Rushchitsky J J 2014 *Nonlinear Elastic Waves in Materials* (Berlin: Springer)

[15] von Hoegen A, Mankowsky R, Fechner M, Först M and Cavalleri A 2018 *Nature* **555** 79
[16] Nolting W 2009 *Fundamentals of Many-Body Physics: Principles and Methods* (Berlin: Springer)
[17] Cottam M G and Haghshenasfard Z 2020 *Many-Body Theory of Condensed Matter Systems: An Introductory Course* (Cambridge: Cambridge University Press)
[18] Eliel G S N *et al* 2018 *Nat. Commun.* **9** 1221
[19] Di Bella A, Gliozzi A S, Scalerandi M and Tortello M 2019 *Appl. Sci.* **9** 5332
[20] Jia L, Yan S, Zhang B and Huang J 2020 *J. Acoust. Soc. Am.* **148** EL289
[21] Drazin P G and Johnson R S 1989 *Solitons: An Introduction* (Cambridge: Cambridge University Press)
[22] Infeld E and Rowlands G 1990 *Nonlinear Waves, Solitons, and Chaos* (Cambridge: Cambridge University Press)
[23] Dauxois T and Peyrard M 2006 *Physics of Solitons* (Cambridge: Cambridge University Press)
[24] Mikeska H J 1978 *J. Phys. C: Solid State Phys.* **11** L29
[25] Kjems J K and Steiner M 1978 *Phys. Rev. Lett.* **41** 1137
[26] Satsuma J and Yajima N 1974 *Suppl. Prog. Theor. Phys.* **55** 284
[27] Mollenauer L F, Stolen R H and Gordon J P 1980 *Phys. Rev. Lett.* **45** 1095
[28] Jackson J D 1998 *Classical Electrodynamics* 3rd edn (New York: Wiley)
[29] Griffiths D J 2012 *Introduction to Electrodynamics* 4th edn (Boston: Addison-Wesley)
[30] Born M and Wolf E 1966 *Principles of Optics* (Oxford: Pergamon)
[31] Hashizume N, Ohashi M, Kondo T and Ito R 1995 *J. Opt. Soc. Am.* B **12** 1894
[32] Agranovich V M and Kravstov V E 1985 *Solid State. Commun.* **55** 85
[33] Boyd R W and Sipe J E 1994 *J. Opt. Soc. Am.* B **11** 297
[34] Fischer G L, Boyd R W, Gehr R J, Jenekhe S A, Osaheni J A, Sipe J E and Weller-Brophy L A 1995 *Phys. Rev. Lett.* **74** 1871
[35] Sarid D 1981 *Phys. Rev. Lett.* **47** 1927
[36] Deck R T and Sarid D 1982 *J. Opt. Soc. Am.* **72** 1613
[37] Quail J C, Rako J G, Simon H J and Deck R 1983 *Phys. Rev. Lett.* **50** 1987
[38] Stegeman G I, Seaton C T, Hetherington W M, Boardman A D and Egan P 1986 *Electromagnetic Surface Excitations* ed R F Wallis and G I Stegeman (Berlin: Springer) p 261
[39] Wendler L 1986 *Phys. Stat. Solidi* b **135** 759
[40] Vassiliev O N and Cottam M G 2000 *Surf. Rev. Lett.* **7** 89
[41] Dzedolik I V and Karakchieva O 2013 *Appl. Opt.* **52** 3073
[42] Fedulova E, Trubetskov M, Amotchkina T, Fritsch K, Baum P, Pronin O and Pervak V 2016 *Opt. Express* **24** 21802
[43] Chen W and Mills D L 1987 *Phys. Rev. Lett.* **58** 160
[44] Mills D L and Trullinger S E 1987 *Phys. Rev.* B **36** 947
[45] Dutta Gupta S and Ray D S 1990 *Phys. Rev.* B **41** 8047
[46] Mohideen U, Slusher R E, Mizrahi V, Erdogan T, Kuwata-Gonokami M, Lemaire P J, Sipe J E, de Sterke C M and Broderick N G R 1995 *Opt. Lett.* **20** 1674
[47] Maimistov A 2010 *Quantum Electron.* **40** 756

IOP Publishing

Michael G Cottam

Chapter 9

Nonlinear dynamics for magnetic excitations

In this final chapter we continue with the discussion of nonlinear aspects of the dynamics, with the focus here being on the magnetic properties, particularly the spin waves (or magnons) and magnetic solitons. This will be done with special regard to low-dimensional structures and for surface/interface geometries. Again, this is not intended as a comprehensive treatment of the nonlinear dynamics, since there are other detailed textbooks and review articles for these specialized aspects. Instead, it is included as a concise survey with examples chosen to relate to the material of the previous chapters and to recent developments. As a general reference the book by Cottam and Tilley [1] includes chapters on nonlinear magnetic waves that summarize work in this field up to about 2005, and there have been more recent books on magnonics edited by Demokritov and Slavin [2] and by Gubbiotti [3]. The area of *spintronics*, which broadly refers to the processing and control of the electrons through their spins in a magnetic material, has been particularly active and is also discussed in this chapter.

9.1 Fundamentals of magnetic nonlinearities

We start by summarizing here some of the basic results, including the origins of the nonlinearities and the different approaches that may be followed regarding the excitations, either in calculating their properties or when probing them experimentally. In the magnetic case, the description to be followed depends once again on the regime for the excitation wave vector (see figure 3.1) and the interplay of dipole–dipole and exchange interactions. On the one hand, there is the macroscopic method in terms of a nonlinear magnetic susceptibility, which may be expanded in terms of a field variable (i.e. the analogue of the usual nonlinear optics approach). Alternatively, there is the microscopic exchange and dipole-exchange formulations, where the nonlinearity is inherent to the spin Hamiltonian. Also in a later section we introduce magnetic solitons (and solitary waves), which are nonlinear magnetic excitations having no counterpart in the linear regime.

9.1.1 Nonlinear waves in magnetic systems

In the treatments of magnons in chapters 3 and 4, approximations were introduced that made the equations of motion for any transverse component of the dynamic magnetization (or alternatively a spin operator) become linear in the amplitudes. A consequence was that fluctuations were neglected in the longitudinal z-component of magnetization or spin, where the z-axis defines the direction of the local static magnetization in a ferromagnet. It is important to recognize, however, that the starting equations, e.g. the Heisenberg spin Hamiltonian in equation (3.3) or the torque equation for the total magnetization in equation (3.17), are actually non-linear. The role of magnetic nonlinearities becomes more important at elevated temperatures or when considering experiments at higher power levels (e.g. as in FMR with high-intensity pumping magnetic fields). Similar considerations apply for antiferromagnets, although the frequencies of the excitations are typically higher. Some general references, mainly for the nonlinear properties of bulk magnons, are by L'vov [4], Gurevich and Melkov [5], and Stancil and Prabhakar [6]. Some other books that emphasize surface and interface properties are found in [2, 7–10].

For some initial insights, suppose we start with the torque equation of motion for the total magnetization $\mathbf{M}$ of a ferromagnet as expressed in equation (3.17). It follows that $\mathbf{M} \cdot (\mathrm{d}\mathbf{M}/\mathrm{d}t) = 0$ and hence $\mathrm{d}(\mathbf{M} \cdot \mathbf{M})/\mathrm{d}t = 0$, implying that $|\mathbf{M}|$ is a constant, which we denote by M_0. Hence the components of the fluctuating magnetization $\mathbf{m}$, treated as a classical vector, satisfy

$$m_x^2 + m_y^2 + (M_0 + m_z)^2 = M_0^2. \tag{9.1}$$

This can be rearranged to give

$$m_z \simeq -\frac{m_T^2}{2M_0}. \tag{9.2}$$

Here $\mathbf{m}_T = (m_x, m_y)$ denotes the transverse magnetization and we have assumed $m_T \ll M_0$ with $m_T = (m_x^2 + m_y^2)^{1/2}$. The simple result in equation (9.2) not only provides the justification for neglecting the longitudinal fluctuation m_z in a first approximation (e.g. as was done with the decoupling approximation in section 3.2.1), since it is of second order in the small quantity m_T, but it also points the way to developing an improved theory.

As was the case for the linear theory in chapters 3 and 4, it is possible to formulate the results using either microscopic or macroscopic methods. We first describe the main nonlinear processes in terms of three-magnon and four-magnon interactions. Then we include other (parametric) processes that occur when there is 'pumping' of the electromagnetic field as in FMR at high power levels.

9.1.2 Three- and four-magnon processes

We may straightforwardly calculate the leading-order nonlinear terms within a Hamiltonian formalism by extending the boson method that was used in section 4.3.2 for dipole-exchange waves. At that stage we introduced the H-P transformation from spin operators to boson operators and a Hamiltonian expansion in equation (4.44) where the general $\mathcal{H}^{(m)}$ term (with $m = 2, 3, 4, \ldots$) contains products of m operators. Only the effects due to $\mathcal{H}^{(2)}$ were discussed in chapter 4, since this term accounts for the linearized SWs or magnons. In the context of a thin-film geometry $\mathcal{H}^{(2)}$ was quoted in equation (4.45) in terms of the original boson operators (a and $a^\dagger$) and in a 'diagonalized' form in equation (4.46) after a linear transformation to new operators denoted by b and $b^\dagger$. The corresponding results in the case of an infinite ferromagnet take an analogous but slightly simpler form (see e.g. [11]) as

$$\begin{aligned}\mathcal{H}^{(2)} &= \sum_{\mathbf{q}}[A^{(2)}(\mathbf{q})a^\dagger(\mathbf{q})a(\mathbf{q}) + \{B^{(2)}(\mathbf{q})a^\dagger(\mathbf{q})a^\dagger(-\mathbf{q})+\text{H.c.}\}] \\ &= \sum_{\mathbf{q}}\omega_B(\mathbf{q})\, b^\dagger(\mathbf{q})b(\mathbf{q}),\end{aligned} \tag{9.3}$$

where $\mathbf{q}$ is a 3D wave vector and there is only one magnon branch, which has the dispersion relation $\omega = \omega_B(\mathbf{q})$ with

$$\omega_B(\mathbf{q}) = \{[A^{(2)}(\mathbf{q})]^2 - |B^{(2)}(\mathbf{q})|^2\}^{1/2}. \tag{9.4}$$

Going now beyond the $H^{(2)}$ term in the expansion in order to incorporate the nonlinear dynamical effects, we find that the next two terms for the cases of an infinite ferromagnet involve the coefficients

$$\mathcal{H}^{(3)} = \sum_{\mathbf{q}_1,\mathbf{q}_2}\left[A^{(3)}(\mathbf{q}_1, \mathbf{q}_2)b^\dagger(\mathbf{q}_1)b^\dagger(\mathbf{q}_2)b(\mathbf{q}_1 + \mathbf{q}_2) + \text{H.c.}\right], \tag{9.5}$$

$$\begin{aligned}\mathcal{H}^{(4)} = \sum_{\mathbf{q}_1,\mathbf{q}_2,\mathbf{q}_3} \Big[&A^{(4)}(\mathbf{q}_1, \mathbf{q}_2, \mathbf{q}_3)b^\dagger(\mathbf{q}_1)b^\dagger(\mathbf{q}_2)b(\mathbf{q}_3)b(\mathbf{q}_1 + \mathbf{q}_2 - \mathbf{q}_3) \\ &+ \left\{B^{(4)}(\mathbf{q}_1, \mathbf{q}_2, \mathbf{q}_3)b^\dagger(\mathbf{q}_1)b^\dagger(\mathbf{q}_2)b^\dagger(\mathbf{q}_3)b(\mathbf{q}_1 + \mathbf{q}_2 + \mathbf{q}_3) + \text{H.c.}\right\}\Big].\end{aligned} \tag{9.6}$$

Expressions for the amplitude functions $A^{(3)}$, $A^{(4)}$ and $B^{(4)}$ can be found, for example, in [11], where it is shown that when dipole–dipole interactions are neglected (i.e., in the Heisenberg limit) only the function $A^{(4)}$ is nonzero. A physical interpretation of the terms in the above equation can be made with the aid of figure 9.1, where each of the heavy lines represents the propagation of a magnon. We follow the usual convention that an arrow pointing away from (towards) an interaction vertex indicates a creation (annihilation) operation. Figures 9.1(a) and (b) represent three-magnon *splitting* and *confluence* processes, respectively, with conservation of wave

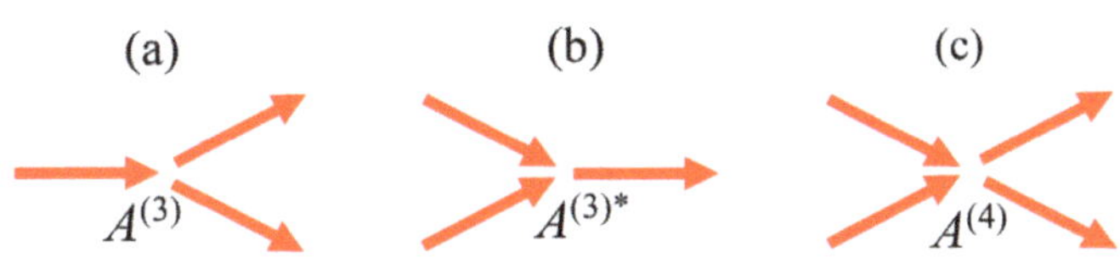

Figure 9.1. Representation of the important interaction processes between magnons corresponding to the terms in equations (9.5) and (9.6): (a) three-magnon splitting process; (b) three-magnon confluence process; and (c) four-magnon scattering process. The heavy lines with arrows represent propagating magnons.

vector. Figure 9.1(c) shows an example of a four-magnon scattering process, again with conservation of wave vector.

The same formal results as described above can also be obtained in the macroscopic (or continuum) limit [11], and there are some limited simplifications for the amplitudes in this case. For example, it is found that the $A^{(2)}$ and $B^{(2)}$ terms of the linear theory can be expressed as

$$A^{(2)}(\mathbf{q}) = \omega_0 + Dq^2 + \frac{1}{2}\omega_M \sin^2\theta, \qquad B^{(2)}(\mathbf{q}) = \frac{1}{2}\omega_M \sin^2\theta \exp(2i\phi), \tag{9.7}$$

where θ and ϕ are the polar and azimuthal angles for the 3D wave vector $\mathbf{q}$ in spherical polar coordinates, and the other notations are the same as in chapters 3 and 4.

9.2 Instability of magnons under microwave pumping

We now generalize the formalism in the preceding paragraphs to incorporate the parametric instabilities that can occur for the magnons when a time-dependent external magnetic pumping field is applied. This effect first attracted attention in some FMR experiments carried out at high power levels by Bloembergen and co-workers [12, 13]. Surprisingly, they observed a saturation of the main FMR absorption peak and the appearance of subsidiary peaks, all with pronounced power thresholds. A theoretical interpretation was provided by Suhl [14] in terms of nonlinear parametric processes. This discovery stimulated related work and shortly afterwards Schlömann *et al* [15] demonstrated an analogous instability effect under parallel pumping with a microwave magnetic field. This observation was remarkable because the conventional FMR configuration is with the microwave field in the perpendicular orientation with respect to the magnetization direction.

The overall effect in the above processes is that one (or more) uniform-precession SWs (meaning magnons with wave vector equal to zero) at the bulk frequency ω_B decays into a pair of magnons with frequencies ω_1 and ω_2 and their corresponding wave vectors $\mathbf{q}_1$ and $\mathbf{q}_2$. The governing conservation conditions for a nth order instability of this type are

$$\omega_1 + \omega_2 = n\omega_B, \quad \mathbf{q}_1 + \mathbf{q}_2 = 0. \tag{9.8}$$

The simplest cases are for $n = 1$ and 2, which correspond to the processes analysed in Suhl's original work.

Within our boson Hamiltonian approach, we may straightforwardly include microwave pumping effects by adding another term to represent the coupling of the magnetization components to a pumping field of the form $\mathbf{h}_p \exp[i\mathbf{q}_p \cdot \mathbf{r} - i\omega_p t]$. Often we may put the pumping wave vector $\mathbf{q}_p = 0$ for the case of spatially homogeneous pumping in a typical FMR experiment. Apart from a constant term, the total effective Hamiltonian has the form

$$\mathcal{H} = \sum_{\mathbf{q}} \omega_B(\mathbf{q}) b^\dagger(\mathbf{q}) b(\mathbf{q}) + \mathcal{H}^{(3)} + \mathcal{H}^{(4)} + \mathcal{H}_p, \tag{9.9}$$

where $\mathcal{H}^{(3)}$ and $\mathcal{H}^{(4)}$ are the previous terms given by equations (9.5) and (9.6), respectively, and there are two cases for the pumping term $\mathcal{H}_p$, which is proportional to the Zeeman energy $-\mathbf{h}_p \cdot \mathbf{m}$. For perpendicular pumping the coupling Hamiltonian $\mathcal{H}_{p,\perp}$ is proportional to the transverse dynamic magnetization $\mathbf{m}_T$, whereas for parallel pumping $\mathcal{H}_{p,\parallel}$ is proportional to the longitudinal dynamic magnetization m_z. Therefore, they are distinct cases in terms of the boson operators involved in the pumping, as we now describe.

Perpendicular pumping occurs when there is a nonzero perpendicular (or transverse) component $h_{p,\perp}$ of the pumping field relative to the static magnetization direction, resulting in

$$\mathcal{H}_{p,\perp} \propto h_{p,\perp}\left[b^{\dagger}(0)\exp(-\mathrm{i}\omega_p t) + \mathrm{H.c.}\right], \tag{9.10}$$

to lowest order in the boson operators, assuming the pumping to be spatially homogeneous. This just describes the usual FMR phenomenon with the excitation of a uniform (wave vector $\mathbf{q} = 0$) magnon. An analogy can be made with a damped harmonic oscillator (as in section 1.4), in which case the mode amplitude becomes expressible in a Lorentzian form [14] as

$$|b(0)| = h_{p,\perp}\frac{U}{\left[(\omega_0 - \omega_p)^2 + \Gamma_0^2\right]^{1/2}}. \tag{9.11}$$

Here Γ_0 is the damping term at zero wave vector (related to the FMR linewidth at low power) and U is an oscillator strength. The *first-order Suhl effect* is illustrated in figure 9.2(a). It involves the pumping (electromagnetic) mode with $\mathbf{q}_p = 0$, shown as

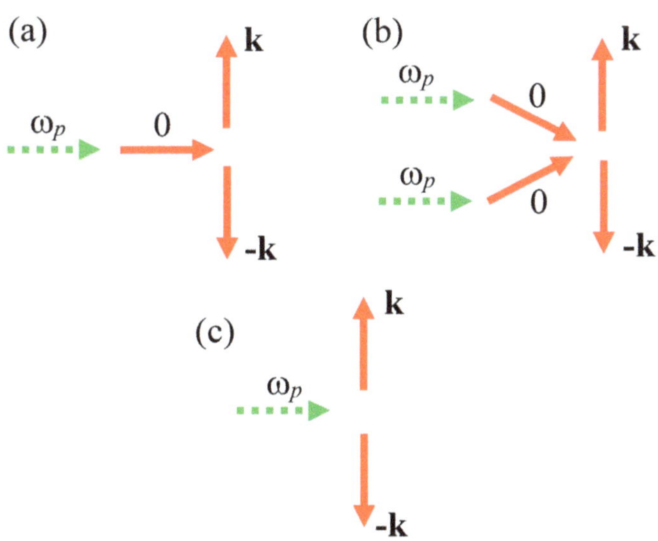

Figure 9.2. Schematic representation of nonlinear processes in perpendicular pumping for (a) a first-order Suhl process and (b) second-order Suhl process and in (c) parallel pumping. The solid (red) and dashed (green) lines with arrows refer to the magnon and electromagnetic modes, respectively. See the text for further explanation.

the dashed line, coupling directly to the $\mathbf{q} = 0$ uniform magnon (solid line), which then decays through the three-magnon splitting term in $\mathcal{H}^{(3)}$ into a pair of magnons with wave vectors $\mathbf{k}$ and $-\mathbf{k}$, each of which must have frequency $\omega_p/2$. From equation (9.5) the part of $\mathcal{H}^{(3)}$ responsible for this process is

$$\sum_{\mathbf{k}\neq 0} A^{(3)}(\mathbf{k}, -\mathbf{k})b^{\dagger}(\mathbf{k})b^{\dagger}(-\mathbf{k})b(0). \tag{9.12}$$

In the macroscopic description for a bulk ferromagnet it may be concluded (see e.g. [5, 11]) that the amplitude term $A^{(3)}(\mathbf{k}, -\mathbf{k})$ is approximately proportional to $\omega_M \sin(2\theta)\exp(-i\phi)$. This is largest when the polar angle $\theta = \pi/4$ for the propagating magnons.

The *second-order Suhl effect*, by comparison, is shown in figure 9.2(b). Two uniform $\mathbf{q} = \mathbf{0}$ magnons are produced from two pumping (electromagnetic) modes, each with the frequency ω_p. In this case they can decay via the four-magnon scattering process into a magnon pair with wave vectors $\mathbf{k}$ and $-\mathbf{k}$. The last step corresponds to terms that have the form

$$\sum_{\mathbf{k}\neq 0} A^{(4)}(\mathbf{k}, -\mathbf{k}, 0)b^{\dagger}(\mathbf{k})b^{\dagger}(-\mathbf{k})b(0)b(0). \tag{9.13}$$

as part of equation (9.6). In a macroscopic bulk ferromagnet this effect is greatest when the magnons, which each have frequency equal to ω_p, are excited at the bottom of the band, i.e. with angle $\theta = 0$.

Finally for *parallel pumping* it is found that the lowest-order term (apart from a constant) in the coupling Hamiltonian $\mathcal{H}_{p,\parallel}$ is quadratic in the boson operators, by contrast with equation (9.10). The important term has the form

$$\mathcal{H}_{p,\parallel} \propto h_{p,\parallel} \sum_{\mathbf{k}} \left[b^{\dagger}(\mathbf{k})b^{\dagger}(-\mathbf{k})\rho(\mathbf{k}) \exp(-i\omega_p t) + \text{H.c.}\right]. \tag{9.14}$$

Here $\rho(\mathbf{k})$ is a weighting factor, which in the macroscopic model for a bulk system is given by $\omega_M \sin^2\theta \exp(-2i\phi)/\omega_B(\mathbf{k})$. This has its maximum when $\theta = \pi/2$ corresponding to the top of the magnon band. This corresponds to a direct excitation of the magnon pair by the pumping mode, and is illustrated in figure 9.2(c).

From the preceding discussion for bulk magnetic systems it is seen that the three types of decay processes all lead to a magnon pair, but modes from different parts of the magnon band are preferentially excited. We will return to this topic of microwave pumping in section 9.4 in the context of low-dimensional magnetic structures.

9.3 Magnetic solitons

We recall that, in section 8.3 of the previous chapter, a general introduction was given to the nonlinear sine-Gordon (sG) equation and the nonlinear Schrödinger equation (NLSE) for soliton excitations in condensed-matter systems. Here we will describe some soliton applications specifically for the case of low-dimensional magnetic structures.

9.3.1 The sine-Gordon equation for magnetic solitons

We start with the sG equation, which has found applications in the context of domain-wall behaviour in magnetic materials and to excitations in certain 1D (chain) ferromagnets. The standard mathematical form of the sG equation was provided in equation (8.14), and one of its simplest soliton solutions was quoted in equation (8.15). As a physical illustration for the use of the sG equation we first mention soliton calculations by Mikeska [16], which were applied to a model of a 1D (chain) ferromagnet with exchange interactions and an easy-plane single-ion anisotropy. The assumed spin Hamiltonian can be expressed in a simple form as

$$\mathcal{H} = -J\sum_n \mathbf{S}_n \cdot \mathbf{S}_{n+1} + D\sum_n (S_n^z)^2 - g\mu_B B_0 \sum_n S_n^x. \tag{9.15}$$

Here $\mathbf{S}_n$ is the spin at position $z = na$ along the z-axis (with n denoting an integer), a is the lattice parameter, and J (>0) is the nearest-neighbour ferromagnetic exchange interaction. When the single-ion anisotropy coefficient D is positive, as will be assumed in this calculation, the spins prefer an orientation in the xy-plane. The transverse applied magnetic field B_0 along the x-direction in this spin Hamiltonian tends to stabilize the spin alignment along this same axis. The spin vectors, if treated classically, can be written in spherical polar coordinates in the form $\mathbf{S}_n = S(\sin\theta_n \cos\phi_n, \sin\theta_n \sin\phi_n, \cos\theta_n)$, so the equilibrium state corresponds simply to $\theta_n = \pi/2$ and $\phi_n = 0$ at all sites. The linear SW excitations for small oscillations in the angles about these values could readily be studied by the methods described in chapters 3 and 4.

In order to show, however, that there are additional solutions of equation (9.15) for which the angles undergo large variations, Mikeska formed the equations of motion satisfied by the angles and then rewrote them by taking a continuum limit, in which $\phi_n(t)$ is replaced by $\phi(z, t)$, etc. He found that this led to a sG equation for $\phi(z, t)$ that has the same form as equation (8.14) with the coefficient definitions that $c = aS(2DJ)^{1/2}$, $m^2 = \mu_B B_0/SJa^2$ and $\theta = (1/2SD)(\partial\phi/\partial t)$ in relation to our model system. The relevant soliton solution quoted in equation (8.15) can therefore be rewritten here as

$$\cos[\phi(z, t)] = 1 - 2\text{sech}^2[m\gamma_0(z - ut - z_0)], \tag{9.16}$$

where z_0 denotes the centre position at $t = 0$. The propagation velocity is u ($<c$) and we denote $\gamma_0 = [1 - u^2/c^2]^{-1/2}$. In figure 9.3 a plot is given to illustrate the dependence of the azimuthal angle ϕ (through the $\cos\phi$ factor) on $(z - z_0)$ at a fixed time $t = 0$. There are large oscillations in ϕ, which switches from being close to zero ($\cos\phi \sim 1$) well away from $z = z_0$ to the opposite direction with $\phi = \pm\pi$ at $z = z_0$. It follows that the polar angle θ also undergoes large oscillations

The above formalism was applied by Mikeska [16] to the quasi-1D material $CsNiF_3$, which is fairly well described by the Hamiltonian in equation (9.15). At around the same time neutron-scattering measurements provided good evidence of soliton-like behaviour in $CsNiF_3$ [17]. In other contexts, the sG equation can also be used to analyse magnetic domain-wall dynamics and magnetization switching.

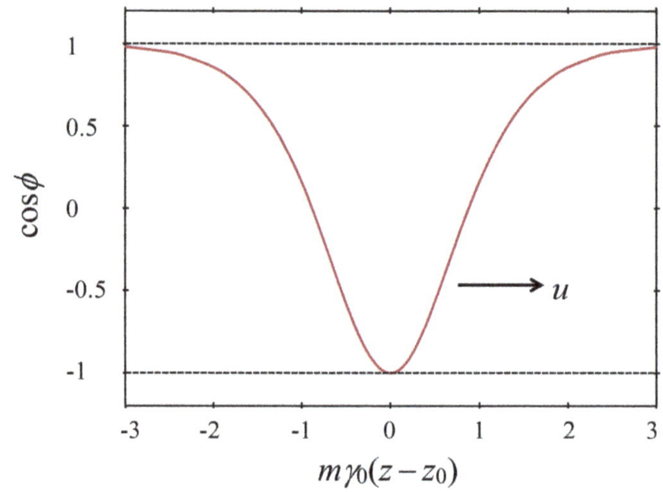

Figure 9.3. Plot to show the dependence of $\cos\phi$ on $(z - z_0)$ at $t = 0$ as given by equation (9.16) for a magnetic-chain soliton as described by the sine-Gordon equation.

We recall that the linear SW properties of quasi-1D magnets were discussed earlier in section 3.4.2.

9.3.2 The nonlinear Schrödinger equation for magnetic solitons

Another important application of the NLSE, apart from the nonlinear optics example already described in chapter 8, involves magnetic systems. Notably, the propagation of envelope solitons in ferromagnetic thin films was first reported as an experimental observation by Kalinikos *et al* [18] for samples of YIG. As in the nonlinear optics case we present just an outline of the main steps that give the connection with the NLSE, while fuller descriptions of the derivation can be found, for example, in [1, 10].

We consider, for simplicity, the situation where the nonlinearities arise only from the four-magnon-interaction processes discussed earlier. In this case the operator equation of motion for $b_\ell(\mathbf{q}_\|)$ is found to be

$$\left(\frac{\partial}{\partial t} + \omega_\ell(\mathbf{q}_\|)\right)b_\ell(\mathbf{q}_\|) + \mathrm{i}\sum_{\ell_1,\ell_2,\ell_3}\sum_{\mathbf{q}_{\|1},\mathbf{q}_{\|2},\mathbf{q}_{\|3}} A^{(4)}_{\ell_1,\ell_2,\ell_3}(\mathbf{q}_{\|1},\mathbf{q}_{\|2},\mathbf{q}_{\|3}) \\ \times\, b^\dagger_{\ell_1}(\mathbf{q}_{\|1})b_{\ell_2}(\mathbf{q}_{\|2})b_{\ell_3}(\mathbf{q}_{\|3})\delta_{\mathbf{q}_\|+\mathbf{q}_{\|1},\,\mathbf{q}_{\|2}+\mathbf{q}_{\|3}} = 0. \tag{9.17}$$

This result is obtained using the quantum-mechanical operator equation (1.39) and we follow the notation for dipole-exchange films in section 4.3.1. Thus, $b_\ell(\mathbf{q}_\|)$ has a band label ℓ and an in-plane 2D wave vector $\mathbf{q}_\|$. The expression used for the $\mathcal{H}^{(4)}$ interaction is analogous to equation (9.6). We now study the evolution of a narrow wave packet of magnons with central in-plane wave vector $\mathbf{q}_{\|0}$ and central frequency $\omega_{\ell\,0}$. The conservation laws for the four-magnon processes within this packet yield $2\omega_{\ell 0} = \omega_{\ell 1} + \omega_{\ell 2}$ for the same SW branch and $2\mathbf{q}_{\|0} = \mathbf{q}_{\|1} + \mathbf{q}_{\|2}$. Approximate forms of the equations of motion of each of the related operators $b_\ell(\mathbf{q}_{\|0})$ and $b_\ell(\mathbf{q}_{\|i})$ (with $i = 1, 2$) can then be obtained by retaining only the dominant terms in equation (9.17).

An important conclusion is that the four-magnon interaction leads to a modification (or a 'renormalization', in the language of many-body theory) of the frequencies for the magnons participating in the above process. Specifically it is found (see e.g. [1, 10]) that we have the replacements $\omega_{\ell 0} \to \Omega_{\ell 0}$ and $\omega_{\ell i} \to \Omega_{\ell i}$, where

$$\begin{aligned} \Omega_{\ell 0} &= \omega_{\ell 0} + A_{00}|b_{\ell 0}|^2, \\ \Omega_{\ell i} &= \omega_{\ell i} + 2A_{i0}|b_{\ell 0}|^2 \quad (i = 1, 2). \end{aligned} \tag{9.18}$$

Here A_{00} and A_{i0} are shorthands for particular matrix elements of the interaction $A^{(4)}$; their definitions are a detail of little concern to us here. What is important is that the nonlinear dispersion relations $\Omega_{\ell i}$ can be expanded in a Taylor series about the central wave vector $\mathbf{q}_{\|0}$ of the wave packet. Thus, putting $\mathbf{q}_{\|i} = \mathbf{q}_{\|0} \pm \mathbf{k}_{\|}$ and assuming $|\mathbf{k}_{\|}|$ is small, we may deduce that

$$\Omega_{\ell 1,2} = \omega_{\ell 1,2} \pm v_g k_{\|} + (1/2)Dk_{\|}^2 + \cdots + 2A_{00}|b_{\ell 0}|^2, \tag{9.19}$$

where the group velocity v_g has been introduced in the same formal manner as for the NLSE derivation in the preceding nonlinear optics case (see equation (8.21)). Also the quantity $D = \partial^2\omega_\ell(\mathbf{q}_{\|})/\partial q_{\|}^2$, which is evaluated at $k_{\|} = 0$, is the dispersion term in the same notation as for chapter 3.

Equations (9.18) and (9.19), taken together with the four-magnon conservation condition, give us a requirement for the parametric instability to occur that

$$Dk_{\|}^2\left(Dk_{\|}^2 + 4A_{00}|b_{\ell 0}|^2\right) < 0. \tag{9.20}$$

By inspection, this can only be true if D and A_{00} are opposite in sign, i.e. we end up with a requirement that the product $A_{00}\,D < 0$ which is just another form of the Lighthill criterion given in equation (8.24). Then, by following the analogous steps for the wave packet as in the nonlinear case earlier, we arrive at an NLSE in the form

$$\left[\mathrm{i}\left(\frac{\partial}{\partial t} + \frac{1}{v_g}\frac{\partial}{\partial \zeta}\right) + \frac{D}{2}\frac{\partial^2}{\partial \zeta^2} - A|\psi_\ell|^2\right]\psi_\ell(\zeta, t) = 0, \tag{9.21}$$

where ζ is an in-plane spatial variable and we define $A = 2M_0A_{00}/\gamma$. The transverse magnetization term ψ_ℓ, which plays an analogous role to the envelope function F in equation (8.22), is formed by writing

$$\psi_\ell(\zeta, t) = \left(\frac{\gamma}{2M_0}\right)^{1/2} \int \mathrm{d}q\; b_\ell(q + q_{\|0}, t)\exp[\mathrm{i}q\zeta - \mathrm{i}\omega_{\ell 0}t]. \tag{9.22}$$

Although we have been able to demonstrate here the formal, qualitative similarities between the envelope solitons in optics and in magnetic films, there are obvious major differences between the two cases at a quantitative level. For example, the pulse width dispersion in optics and in magnetics (taking YIG as a benchmark material) are 1 ps and 100 ns, respectively, and the group velocities are of order 2×10^8 ms^{-1} and 2×10^4 ms^{-1}, respectively.

9.4 Other applications to magnetic systems

By analogy with what was done in the previous chapter for nonlinearities in nonmagnetic systems, we will highlight here some selected topics by choosing just a few applications to nonlinear magnetic low-dimensional systems. The applications all involve the formalism described in sections 9.1 to 9.3, and they comprise the following: decay and instability processes for magnons, resonant frequency multiplication for magnons, the Bose–Einstein condensation in a magnon gas and magnon spintronics. The emphasis in all cases will be on low-dimensional magnetic elements and arrays, such as thin films, nanowire stripes and small spheres.

9.4.1 Decay and instability of magnons

The three- and four-magnon processes that were discussed earlier can give rise to a variety of physical effects, and we start here with some results for the damping (or reciprocal lifetime) of the magnons. Then we follow up with some further discussion and examples of the SW instabilities under intense pumping by a microwave field.

9.4.1.1 SW damping

In principle, the damping may be deduced from the non-instrumental part of the linewidth measured in FMR or inelastic light scattering experiments. From the theoretical perspective, the magnon damping is a finite-lifetime effect that arises due to the interactions between magnons or because magnons can interact with other excitations (such as phonons). The behaviour in bulk materials is fairly well understood at low temperatures (see, for example, the review article by Keffer [11]). In low-dimensional structures the magnon-interaction amplitudes and the overall conservation conditions are modified, but the basic processes, however, are the same.

The simplest contribution to the damping of magnons at low temperatures comes from the three-magnon splitting and confluence that are depicted in figures 9.1(a) and (b). For example, a magnon can split into two other magnons, which can subsequently recombine, representing a decay channel for the initial magnon provided conservation of energy and also (where appropriate) conservation of wave vector are satisfied. In bulk systems the 3D wave vector must always be conserved, but for films or wires only those wave-vector components lying along the directions of translational symmetry are conserved. The damping is of second order in the three-magnon interaction, i.e. proportional to $|A^{(3)}|^2$ in our previous notation. The four-magnon process can similarly mediate a damping process in which there are three intermediate (or 'virtual' magnons), and again it is subject to the conservation of energy and conservation of wave-vector components in the directions of translational symmetry. Typically the simpler three-magnon processes will provide the dominant damping contribution *unless* the magnetic dipole–dipole interactions are negligibly small in that material or when the magnon wave vector is such that conditions for overall wave-vector conservation cannot be satisfied. The latter would be the case at very small wave vectors and/or when the SW energy gap is relatively large.

Magnetic nanowires in stripe geometries form an ideal system for studying the damping, since the magnon spectrum consists of discrete branches due to the lateral spatial quantization (see section 4.4.2) and the wave vectors are conserved only in 1D, i.e. along the length of the wires. Measurements of the damping were reported by Boone *et al* [19] for nanowires (or stripes) with a rectangular cross section. They used a spin torque FMR technique with permalloy nanowire stripes that had a thickness of only 6 nm and different widths in the range 100 – 250 nm. They obtained the linear magnon frequencies as a function of applied magnetic field (along the length of the nanowires), which consisted of discrete (or quantized) branches, and they deduced the behaviour of the damping for the lowest set of magnons.

Some results for the field dependence of the damping of the lowest-frequency magnon are shown in figure 9.4 for nanowires with widths 140 and 125 nm. There were pronounced peaks seen in the damping at an intermediate values of the applied field (as indicated by the arrows in the figure). The damping is presented in terms of scaled dimensionless units after making allowance for inhomogeneous broadening in the experiments [19]. The damping contributions observed by Boone *et al* were ascribed as due mainly to three-magnon processes in conjunction with effects due to roughness at the lateral edges of the wires. The roughness degrades slightly the property of translational symmetry along the length of the wires, and so magnons can decay via direct scattering into another magnon branch with the same energy (but different wave vector). A theory for three-magnon and four-magnon damping in rectangular nanowires was subsequently developed by Nguyen *et al* [20]. This calculation did not include edge roughness effects but could account, at least qualitatively, for the observed field-dependent effects arising from the three-magnon processes.

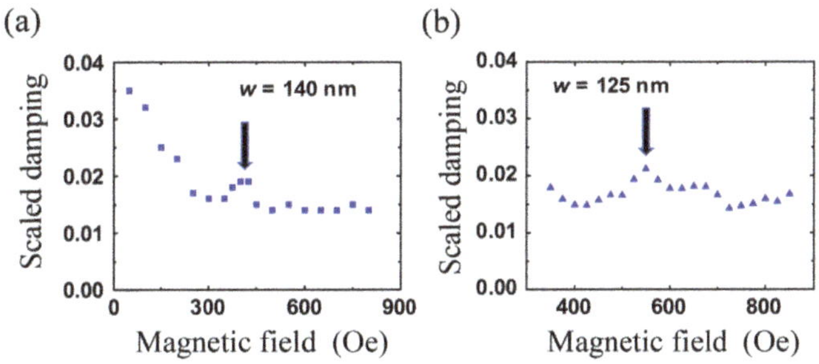

Figure 9.4. Scaled damping parameter measured for the lowest-frequency magnon modes in permalloy nanowire stripes with thickness 6 nm and width w equal to (a) 140 nm and (b) 125 nm, showing the dependence on applied magnetic field (100 Oe is equivalent to 0.01 T). The arrow heads indicate the intermediate peaks in the damping. (Adapted with permission from [19]. Copyright 2009 American Physical Society.)

9.4.1.2 SW instabilities

Next we turn to some examples of SW instabilities under conditions of microwave pumping at elevated power levels. The background theory has already been covered in section 9.2, mainly in the context of macroscopically thick films with in-plane direction of magnetization. It can, however, be extended to other sample geometries where there may be a component of the static magnetization perpendicular to the surface, e.g. as in spheres, flat disks and wires. In figure 9.5 some plots of the threshold microwave field amplitude h for the onset of instability versus the applied magnetic field $H_0 = (B_0/\mu_0)$ are shown, using data from Rezende *et al* [21] for a YIG sphere. Results for parallel pumping and perpendicular pumping for the first-order Suhl effect (or subsidiary resonance) are in panels (a) and (b), respectively. In each case the lower curve shows the instability threshold, while the upper curve correspond to the onset of *auto-oscillations*, which are a further stage of instability acting as a precursor to chaotic behaviour. The instability threshold in figure 9.5(a) for parallel pumping follows the characteristic *butterfly curve* identified in earlier experiments. For applied fields below a critical value at H_c, the excitation of parametric SWs at frequency $\omega_p/2$ becomes energetically allowed, but this is no longer possible for magnon energies in the band of excitations when $H_0 > H_c$ so the instability threshold rises sharply. For perpendicular pumping there is also a critical field as indicated but the shape of the curve is quite different in the case of a sphere rather than a film [21].

For smaller-size samples, with dimensions of the order of several nanometres or micrometres, rather than millimetres in the case just described, the spatial quantization of the magnon modes becomes more pronounced, and consequently the shape of the butterfly curve is expected to take on a quite different form, showing sharp structural features (see e.g. [22–24]). As an example, we show in figure 9.6 some calculations [24] for long ferromagnetic nanowire stripes of permalloy with thickness 10 nm and width 60 nm. The applied magnetic field B_0 is along the nanowire length direction (the easy axis), and so the discrete dipole-exchange SW modes are

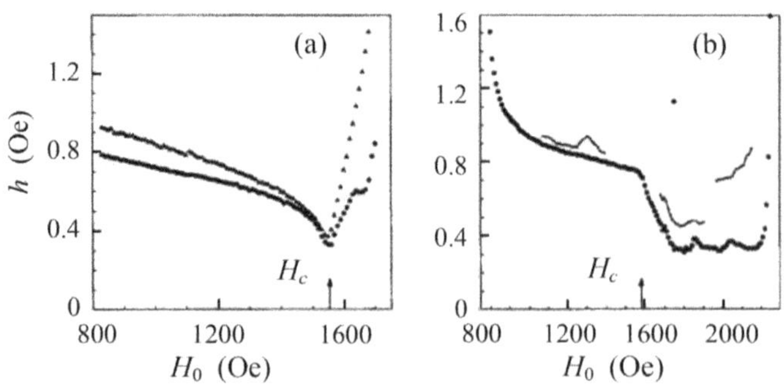

Figure 9.5. Threshold values of the microwave pumping field amplitude h plotted against the static applied field $H_0 = B_0/\mu_0$ in a 1 mm diameter YIG sphere for (a) parallel pumping at 9.5 GHz with H_0 along [111] and (b) first-order Suhl effect at 8.87 GHz with H_0 along [110]. The lower curves refer to the initial instability thresholds and the upper curves (or fragments) refer to auto-oscillations. (Reproduced with permission from [21]. Copyright 1994 World Scientific.)

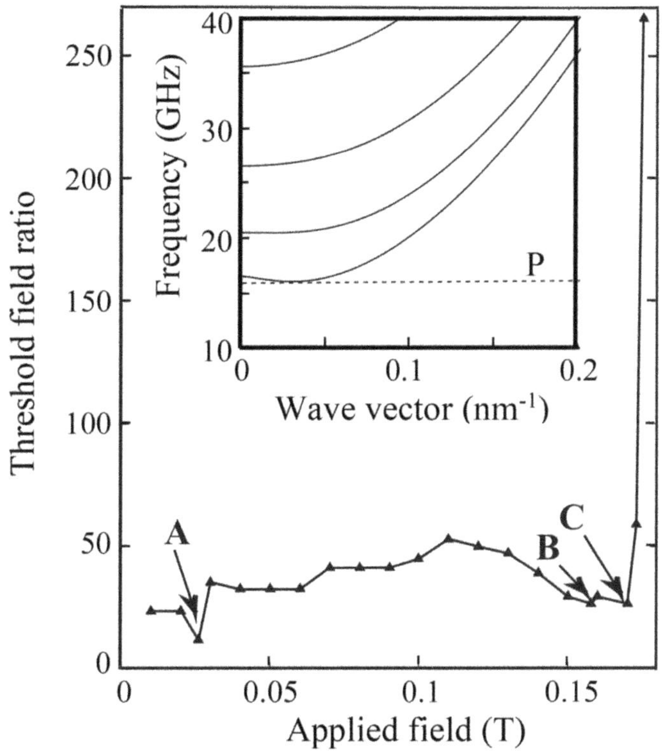

Figure 9.6. SW instability threshold plotted versus applied field B_0 for parallel pumping (taking $\omega_p/2\pi = 32$ GHz) in a permalloy nanowire stripe with cross section 60 nm × 10 nm. The triangles are calculated points and the lines are a guide to the eye. The inset shows the lowest four branches of SW frequencies plotted versus longitudinal wave vector when $B_0 = 0.137$ T. The horizontal line P is drawn corresponding to one-half of the pumping frequency. See the text for other notation. (Adapted with permission from [24].)

characterized by a 1D wave vector q along this same direction. The SW dispersion of frequency versus q for a fixed value of B_0 is shown in the inset for the lowest four modes, where the initial dip for the lowest branch at very small q is an effect analogous to the behaviour of the backward propagating magnetostatic modes described in section 4.1.1 for a complete film. The main part of figure 9.6 shows the calculated value of the threshold field ratio, defined as the field for onset of the instability divided by the half-linewidth in FMR, plotted as a function of the applied field for parallel pumping at a frequency of 32 GHz. The threshold curve no longer has the traditional butterfly wings shape, but instead it shows structural features associated with the discrete SWs of the stripe. The points A, B and C, which correspond to applied field values $B_0 = 0.026$, 0.158 and 0.173 T, respectively, occur when one-half of pumping frequency (as indicated by the horizontal dashed line P in the inset) coincides with either the SW frequency at $q = 0$ or with the frequency minimum for the first two SW branches in the dispersion plot.

Another interesting development arose from experiments by Sandweg *et al* [25] to detect by spin pumping the parametrically excited exchange magnons, rather than the dipole-exchange modes as described earlier. Specifically the magnons were injected into a YIG film by parametric pumping with a microwave field, and the inverse spin-Hall effect was used to detect a voltage. Magnon wavelengths down to the sub-micron level were achievable. This opens new perspectives for applications in magnon spintronics (see section 9.4.4 for general discussion of this topic).

9.4.2 Frequency multiplication for magnons

Nonlinear effects, taking place via the three- and four-magnon processes described earlier, make frequency multiplication for the magnons allowable. For example, if two magnons of frequency ω_0 participate in the confluence process sketched in figure 9.1(b), which corresponds to an interaction amplitude proportional to $[A^{(3)}]^*$ in equation (9.5), a magnon of double the frequency, $2\omega_0$, will be produced in the output. Similarly, the four-magnon interactions can lead to output of magnons with frequency $3\omega_0$ through the amplitude term $[B^{(4)}]^*$ in equation (9.6). It then becomes a matter of how to achieve this in practical terms and how to maximize the NL effect, when this is required.

The effect of resonant frequency multiplication has been demonstrated experimentally by Demidov *et al* [26] by using sub-micron sized permalloy dots which were subjected to an intense microwave magnetic field. They observed harmonics corresponding to output at two and three times the frequency of the applied signal. The magnetic dots were produced starting from a 10 nm thick permalloy film, which was then patterned by electron-beam lithography to form the elliptical dots with semi-major and semi-minor axes of length 500 and 250 nm, respectively. It was found that the efficiency of the signal multiplication is strongly enhanced when the harmonic is arranged to be resonant with the (linear) normal dynamical modes of the dot, which were studied separately using Brillouin light scattering (BLS) at lower power levels. This condition of achieving resonance with the linear modes is different for the cases of double- and triple-frequency harmonics. As a consequence, Demidov *et al* found that their elliptical sample shape was particularly efficient for resonant frequency tripling. Their results may be an important step for the implementation of nanoscale frequency multipliers for integrated microwave technology.

An intriguing new possibility for magnon frequency multiplication, which utilizes the NL dynamics of magnetic *skyrmions*, was recently proposed by Rodrigues *et al* [27]. In this context, magnetic skyrmions refer to small, typically nanometre size, excitations in the otherwise uniform spin texture. They can be stabilized, often by the interfacial DM interaction that was discussed in section 3.5.2, and they have become of immense current interest for information processing and novel spintronics (see e.g., [28, 29] for reviews). There are many different kinds of magnetic skyrmions (and anti-skyrmions), but in many cases they may be envisioned with the spin vectors possessing a swirling form as depicted by the example given in figure 9.7. The theoretical approach followed in [27] relied on the macroscopic torque equation of motion for the dynamical magnetization, plus a phenomenological damping term to yield the nonlinear Landau–Lifshitz–Gilbert (LLG) equation (see equation (3.31) and further discussion in e.g. [5, 6]). The interfacial antisymmetric DM interactions

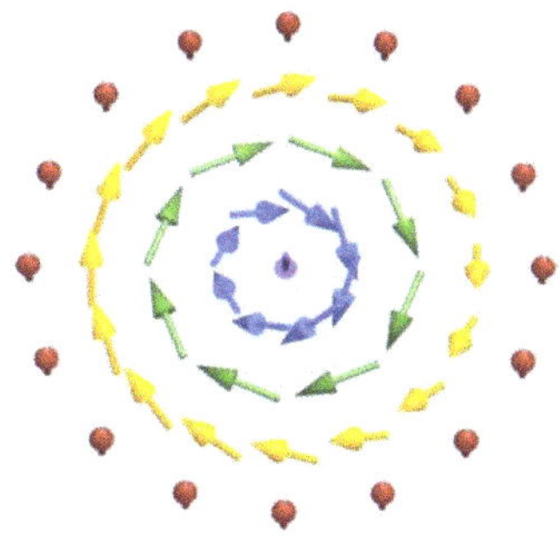

Figure 9.7. Schematic representation of a spin texture known as a Bloch-type magnetic skyrmion. The spiralling spin vectors are drawn in different colours to emphasize that their orientations are tilted upwards or downwards relative to the plane. (Reproduced with permission from [30].)

and stabilizing anisotropy were included in an energy functional, which was studied by analytical methods and micromagnetic simulations to find skyrmion states. From the interaction between a magnon and a skyrmion, it was concluded that frequency multiplication effects would arise.

9.4.3 Bose–Einstein condensation in a magnon gas

The ideas associated with the phenomenon of a Bose–Einstein condensation (BEC) were developed by Bose and Einstein separately in the 1920s. In broad terms, a Bose–Einstein condensate is nowadays taken to refer to a phase or state of matter in which a macroscopically large number of the bosons occupy, or have condensed into, their lowest quantum state (see e.g. the books by Landau and Lifshitz [31] and Pitaevskii and Stringari [32] for general discussions). The occurrence of BEC requires the boson particle density to exceed a critical value, which may be achieved by lowering the temperature or by raising the boson density or by some combination of these effects. Initially superfluids and superconductors provided the only examples of BEC, but more recently it has been achieved for certain diluted atomic gases at ultralow temperatures [33, 34]. This discovery served to suggest that BEC might be a more widespread phenomenon than previously believed and so it spurred on efforts to observe BEC in other physical systems.

A magnon gas in a ferromagnet or ferrimagnet at a temperature well below its Curie temperature T_C is expected to present a particularly promising candidate for BEC formation, since we have already seen that the magnons behave like weakly interacting bosons and that their densities can be manipulated and controlled through microwave pumping techniques. Indeed, the observation of BEC in a magnon gas at room temperature was first reported by Demokritov *et al* [35] in 2006 and consolidated in further work by the same group shortly afterwards (see e.g. [36–38]).

The cited experiments were carried out using thin films of YIG that were magnetized in-plane by a static applied field and had thicknesses in the 2 to 10 μm range. The form of the (linear) magnon dispersion relation in the dipole-exchange regime is in accordance with that discussed in section 4.3. We note in particular that the lowest-frequency magnons are those excited with the in-plane wave vectors $\mathbf{q}_{||}$ parallel to the magnetization direction and that these magnons have their minimum frequency at some value with $q_{||} \neq 0$. This feature occurs due to competition at very small $q_{||} \neq 0$ between the dipole–dipole interactions which decrease the magnon frequency (as in the case of magnetostatic backward bulk modes in section 4.1) and the exchange interactions which cause an increase in frequency. This SW behaviour is apparent also in the inset to figure 9.6. In the BEC experiments the frequency minimum occurred at a wave vector with magnitude $q_{||,m} \simeq 2 \times 10^5$ cm^{-1} as verified by the use of Brillouin light scattering (BLS) spectroscopy.

The magnons are driven by a microwave pumping field at angular frequency ω_p until the stage when a threshold high power level is exceeded and magnon pairs at frequency $\omega_p/2$ and with equal and opposite wave vectors are then produced parametrically, as explained earlier in this chapter. This stage is followed by a rapid redistribution of magnon energies (through scattering events) going down to the frequency minimum $\omega(q_{||,m})$ for the bosons. This redistribution occurs on a time scale that is much shorter than the relatively long spin-lattice relaxation time, so the 'hot' magnon gas state persists. In contrast, if the pumping signal power exceeds another threshold level, typically much larger than the one discussed earlier for the SW instability under parallel pumping, the already elevated population of magnons begins to condense into a state around the minimum magnon frequency and to display the sought-after quantum coherence.

A theory that was successful in explaining many of the observed features of the BEC was developed by Rezende [39]. His model utilizes the four-magnon-interaction terms, as discussed in section 9.1.2, to incorporate the cooperative properties of magnons with frequencies close to the dispersion minimum at $\omega(q_{||,m})$. The theory accounts well for the onset of the BEC as the microwave signal power P_S is increased above a threshold value. Also it predicts coherence for the magnons in the condensate and the separation of the magnon gas above threshold into two parts, one that is in a thermal equilibrium state and one with coherent magnons. Comparisons between theory and two sets of experimental data (the first obtained with coherent pumping and the second with incoherent pumping) are illustrated in figures 9.8(a) and (b). It is seen that there is excellent agreement in both cases. The signal power relates to the mean power radiated by the uniform magnetization precessing about the static field.

In recent BEC work [40] it has been found advantageous experimentally to employ spatially confined samples of YIG, rather than the complete (or extended) films with thicknesses of order several micrometres as before. Specifically, platelets with thickness 85 nm and lateral dimensions of around 1 μm were used both in the experiments and for numerical simulations of the magnetization dynamics based on the LLG equation. This sample geometry with the lateral confinement effects led to a stronger quantization of the magnons and, in particular, to a more pronounced minimum in the dispersion of the lowest branch. An increased flexibility for control of the BEC, including its lifetime, was demonstrated.

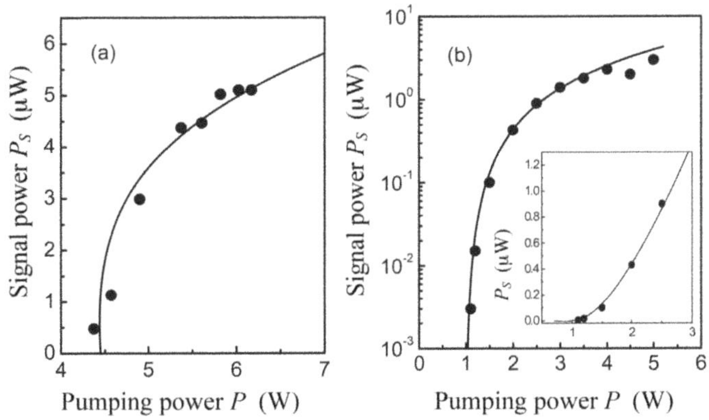

Figure 9.8. Plots of microwave signal power P_S versus microwave pumping power P: (a) theory (solid line) compared with experimental data (black dots) from [37]; (b) the same but using experimental data (see the text) from [38]. Also the inset in (b) shows the behaviour near the threshold value in greater detail. (Reproduced with permission from [39].)

9.4.4 Magnon spintronics

By analogy with electronics as being the manipulation of the charge and charge currents of electrons, the term *spintronics* conventionally refers to the control and processing of the electrons through their spins. For example, the manipulation of spin currents in magnetic nanoelements and arrays of these nanoelements, that are composed of metals and semiconductors, offers numerous applications for magnetic memories and computational devices. The book edited by Wu and Hoffmann [10] contains excellent reviews of the field, covering basics to advanced topics. Other reviews are given in [2, 41]. Spintronics now constitutes a very large field of application that can only be touched upon briefly here. More specifically, the term *magnon spintronics* has come into usage when general concepts from spintronics are used in combination with the SW dynamics as one way to overcome a limitation due to the relatively short distance (referred to as the spin-diffusion length) over which an electron keeps the memory of its spin orientation.

The concept of a spin current density $\mathbf{J}_s$ can be introduced by analogy with the usual definition of a charge current as

$$\mathbf{J}_s = \sum_{\mathbf{q}} S_{\mathbf{q}}^z \, \mathbf{v}_{\mathbf{q}}, \tag{9.23}$$

where $S_{\mathbf{q}}^z$ is the component (along the spin-quantizing axis z) of the spin density at wave vector $\mathbf{q}$ and $\mathbf{v}_{\mathbf{q}}$ is the corresponding velocity for the excitation. This definition can be applied both to the conduction electrons in the system under consideration (giving $\mathbf{J}_s^{\text{el}}$) and to the localized spins that support the SW (giving $\mathbf{J}_s^{\text{sw}}$). In a formalism using creation and annihilation operators, we would choose $S_{\mathbf{q}}^z$ accordingly. Thus, for the spin-$\frac{1}{2}$ conduction electrons we would write $S_{\mathbf{q}}^z = c_{\mathbf{q},\uparrow}^{\dagger} c_{\mathbf{q},\uparrow} - c_{\mathbf{q},\downarrow}^{\dagger} c_{\mathbf{q},\downarrow}$, where the c are fermion operators (see section 5.2) for the spin-up and spin down states. For the

SW we would use boson operators and the HP representation, as before (see section 4.3), to write $S_{\mathbf{q}}^{z} = S - a_{\mathbf{q}}^{\dagger} a_{\mathbf{q}}$.

Apart from the spin currents, the other essential concept in spintronics is the spin-transfer torque (STT) [42]. The typical physical situation is a YIG/Pt layered heterostructure, where the insulating YIG material has exceptionally low damping and so the propagating SWs are relatively long-lived. The Pt metal is a source of conduction electrons, and the STT provides an interfacial coupling between the two systems. Basically, it involves the transfer of spin angular momentum from the conduction electrons in the metal to the magnetization in the magnetic insulator. In effect the magnetization absorbs a spin current and thereby acquires a torque, while the reverse process can also take place. The detailed derivation is complicated and will not be given here. Typically, the theoretical analysis would be carried out using a macroscopic approach in which the LLG equation of motion is supplemented by two extra terms that incorporate the spin-transfer torque and the microwave pumping described earlier. Initially the emphasis in spintronics was on ferromagnetic and ferrimagnetic materials, but more recently there has been increased interest in antiferromagnetic spintronics (see e.g. [43] for a review).

A novel device to utilize concepts in magnon spintronics was proposed by Chumak *et al* [44]. These authors employed a magnonic crystal (MC) that consisted of a thin film of the ferrimagnet YIG into which an array of parallel grooves had been etched into one surface. The partial reflection and transmission of an SW propagating parallel to the film surface and perpendicular to the grooved surface constitutes a periodic magnetic system for which the magnonic bands can be calculated using the methods covered in chapters 3 and 4. The use of this system as a transistor was proposed, with the three terminals represented by a 'source' (a microstrip antenna at one end of the MC providing injection of the SWs), a 'drain' (a second antenna acting as a detector of SWs at the other end) and an intermediate 'gate' (a third antenna used to inject another SW signal directly into the device). In this manner, the spin current through the device can be modified by the spin current from the gate. This occurs through a nonlinear process whereby two interacting SWs scatter into two other SWs (in other words, via the four-magnon interaction process that was discussed in section 9.1.2). The spacing of the grooves (representing the periodicity length D) was of order a few hundred micrometres in the reported prototype device [44], but the potential exists to scale the size down to a few tens of nanometres. The important objective was achieved of having one spin current control another spin current. More recent progress in this evolving field is described, for example, in reviews by Chumak *et al* [45] and by Hirohata *et al* [41].

9.5 Hybrid magnons in cavities

Schemes for controlling the interactions between light and matter on the *quantum level* has attracted considerable attention (see e.g. [46]) because of the fundamental importance of this topic and its potential applications. In the last decade in particular, there have been numerous studies on cavity–magnon systems since they allow for strong coupling between the magnons and cavity photons, often at room temperatures. The strong coupling becomes achievable because of the high spin density in the magnetic sample and the selective excitation of specific cavity electromagnetic modes (photons). Excellent reviews of this burgeoning field have recently been given by Bittencourt *et al* [47], Leo *et al* [48] and Li *et al* [49].

For many experimental purposes, the use of YIG single crystals has provided a remarkably suitable platform for studying enhanced interactions between light and matter, on account of the low damping rate for SWs in this material (as mentioned earlier) in conjunction with the large spin density. A strong coupling between the simplest magnetic mode in a YIG sphere and the microwave cavity photons has been realized [50], leading to new types of quasiparticles called *cavity magnon-polaritons* [51, 52]. Numerous other phenomena that have been investigated in cavity–magnon systems include the observation of magnon-polariton bistability, magnon dark modes and gradient memory, cavity spintronics and the magnon Kerr effect (see [48, 49]).

In particular, the magnons in ferrimagnetic YIG spheres have been used to study nonclassical effects. One of the most common forms of nonclassical phenomena (without any classical analogue) is sub-Poissonian statistics. In quantum optics it has potential applications in imaging and communications, where signals with the least possible noise are needed. Recently, investigations [53] of the statistical properties of magnons in a hybrid ferromagnet-superconductor system were successful in demonstrating a nonclassical sub-Poissonian behaviour. Typically, the evolution of the Mandel parameter $Q(t)$ at time t is used to specify the magnon-counting statistics. It can be defined generally (for any particular magnon mode) by the expression [54]

$$Q(t) = \frac{\langle (b^\dagger(t)b(t))^2\rangle - \langle b^\dagger(t)b(t)\rangle^2}{\langle b^\dagger(t)b(t)\rangle} - 1. \tag{9.24}$$

Here the $b^\dagger$ and b boson operators for the magnons are defined as in section 9.2. If we have $Q(t) = 0$ as the time evolves the statistics is called Poissonian, whereas the cases of $Q(t) > 0$ or $Q(t) < 0$ are referred to as *super-Poissonian* or *sub-Poissonian*, respectively.

As an example to illustrate the time evolution of $Q(t)$ we consider the nonlinear Kerr effect in a cavity–magnon system, where the uniform magnetostatic (or Kittel) mode in a single uniformly magnetized YIG sphere is coupled to a microwave cavity photon. Novel experiments of this type, which employed a sample diameter of order 1 mm were reported in [55]. A detailed theoretical analysis was subsequently given in [56], where the authors studied nonclassical effects and size scalability using the effective Hamiltonian

$$\mathcal{H} = \omega_c a^\dagger a + \omega_m b^\dagger b + K(b^\dagger b)^2 + g(a^\dagger b + ab^\dagger). \tag{9.25}$$

In this case $a^\dagger$ and a denote the boson creation and annihilation operators for the cavity photon at angular frequency ω_c, while $b^\dagger$ and b are the corresponding operators for the magnon at angular frequency ω_m. The coupling between the two subsystems is represented by the final term in equation (9.25) with interaction strength g. The term $K(b^\dagger b)^2$ in equation (9.25) describes the nonlinearity due to the magnetic Kerr effect, which is an analogue of the Kerr effect in quantum optics (see section 8.1) and arises here due to the magnetic anisotropy in the YIG sphere. It is known that K is inversely proportional to the volume [55], so the NL effects scale strongly with sample size (as does the coupling g). Calculations for the Mandel parameter using equation (9.25) have shown that the magnon states, although starting off as Poissonian, may evolve with time to exhibit the nonclassical sub-Poissonian statistics at later times [56] when $K \neq 0$. An example to illustrate this behaviour is given in figure 9.9. The above theory made use of a representation of the magnons in terms of Glauber *coherent states*, as frequently done in quantum optics (see e.g. [54]), and it was shown also that the Kerr nonlinearity in this system led to quadrature squeezing of the magnons.

Another direction taken in recent work on cavity magnonics involves the use of materials other than YIG and similar ferrites, which are limited in their range of frequencies and also cannot be patterned easily or grown as magnetic nanoelements. Thus, transition-metal magnetic materials such as permalloy have been studied in this context. For example, in [57, 58] the strong magnon-photon coupling with a permalloy nanofilm was demonstrated. By utilizing lithographically defined superconducting resonators, the authors reported a coupling strength of around 0.15 GHz for a permalloy thin film with a thickness of 30 nm. These results point to a promising alternative for achieving integrated hybrid quantum systems with nanoscale magnetic materials, where the coupling strength can be tuned by various mechanisms. The above-mentioned experimental developments motivated further theory in which the microscopic dipole-exchange method was employed in a detailed study of the interaction between cavity

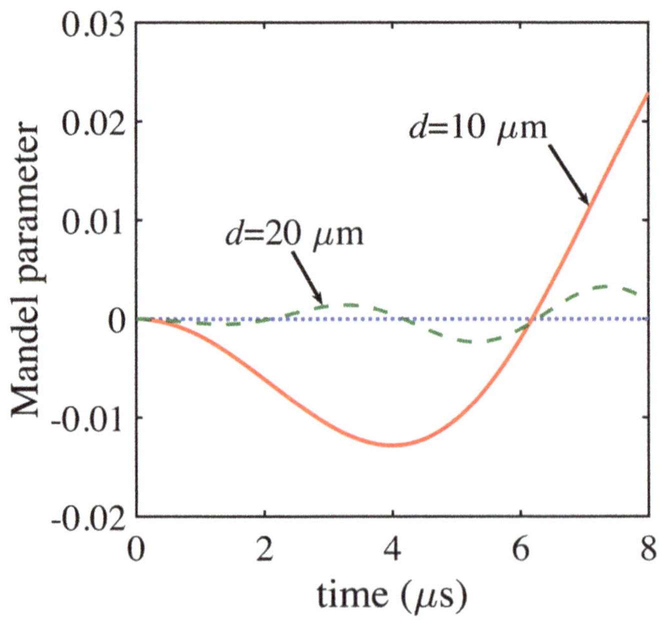

Figure 9.9. The scaled Mandel parameter Q as a function of time, assuming parameters for a coherent magnon state and frequencies $\omega_c/2\pi = \omega_m/2\pi = 10.1$ GHz. The solid (red) line corresponds to the diameter $d = 10\ \mu$m for the YIG sphere (implying $g/2\pi = 0.042$ MHz, $K/2\pi = 3.8 \times 10^{-4}$ MHz). The dashed (green) line corresponds to $d = 20\ \mu$m ($g/2\pi = 0.12$ MHz, $K/2\pi = 0.47 \times 10^{-4}$ MHz). The horizontal line is for the limit of zero Kerr effect. (Reproduced with permission from [56]. Copyright 2020 American Institute of Physics.)

photons and the microwave magnetic modes in structures such as ultrathin films, multilayers, nanowires and 1D magnonic crystals of nanowires [59].

References

[1] Cottam M G and Tilley D R 2005 *Introduction to Surface and Superlattice Excitations* 2nd edn (Bristol: IOP Publishing)

[2] Demokritov S O and Slavin A N (ed) 2013 *Magnonics: From Fundamentals to Applications* (Berlin: Springer)

[3] Gubbiotti G (ed) 2019 *Three-Dimensional Magnonics* (Singapore: Jenny Stanford)

[4] L'vov V S 1994 *Turbulence under Parametric Excitation: Applications to Magnets* (Heidelberg: Springer)

[5] Gurevich A G and Melkov G A 1996 *Magnetization Oscillations and Waves* (Boca Raton, FL: CRC Press)

[6] Stancil D D and Prabhakar A 2009 *Spin Waves: Theory and Application* (Heidelberg: Springer)

[7] Wigen P (ed) 1994 *Nonlinear Phenomena and Chaos in Magnetic Materials* (Singapore: World Scientific)

[8] Cottam M G (ed) 1994 *Linear and Nonlinear Spin Waves in Magnetic Films and Superlattices* (Singapore: World Scientific)

[9] Hillebrands B and Ounadjela K (ed) 2001 *Spin Dynamics in Confined Magnetic Structures* (Berlin: Springer)

[10] Wu M and Hoffmann A (ed) 2013 *Solid State Physics: Recent Advances in Magnetic Insulators–From Spintronics to Microwave Applications* vol 64 (Amsterdam: Academic)

[11] Keffer F 1966 *Handbuch Phys.* **18** 1

[12] Bloembergen N and Damon R 1952 *Phys. Rev.* **85** 699

[13] Bloembergen N and Wang S 1954 *Phys. Rev.* **93** 72

[14] Suhl H 1957 *J. Phys. Chem. Solids* **1** 209

[15] Schlömann E, Green J and Milano V 1960 *J. Appl. Phys.* **31** 3865

[16] Mikeska H J 1978 *J. Phys. C: Solid State Phys.* **11** L29

[17] Kjems J K and Steiner M 1978 *Phys. Rev. Lett.* **41** 1137

[18] Kalinikos B A, Kovshikov N G and Slavin A N 1983 *Sov. Phys.—JETP Lett.* **38** 413

[19] Boone C T, Katine J A, Childress J R, Tiberkevich V, Slavin A, Zhu J, Cheng X and Krivorotov I N 2009 *Phys. Rev. Lett.* **103** 167601

[20] Nguyen H T, Akbari-Sharbaf A and Cottam M G 2011 *Phys. Rev.* B **83** 214423

[21] Rezende S M, Azevedo A and de Aguiar F M 1994 *Linear and Nonlinear Spin Waves in Magnetic Films and Superlattices* ed M G Cottam (Singapore: World Scientific) 335

[22] Chartoryzhskii D N, Kalinikos B A and Vendik O G 1976 *Solid State Commun.* **20** 985

[23] Nguyen H T and Cottam M G 2014 *Phys. Rev.* B **89** 144424

[24] Haghshenasfard Z and Cottam M G 2016 *J. Phys.: Condens. Matter* **28** 186001

[25] Sandweg C W, Kajiwara Y, Chumak A V, Serga A A, Vasyuchka V I, Jungfleisch M B, Saitoh E and Hillebrands B 2011 *Phys. Rev. Lett.* **106** 216601

[26] Demidov V E, Ulrichs H, Urazhdin S, Demokritov S O, Bessonov V, Gieniusz R and Maziewski A 2011 *Appl. Phys. Lett.* **99** 012505

[27] Rodrigues D R, Nothhelfer J, Mohseni J M, Knapman R, Pirro P and Everschor-Sitte K 2021 *Phys. Rev. Appl.* **16** 014020

[28] Fert A, Reyren N and Cros V 2017 *Nat. Rev. Mater.* **2** 17031

[29] Göbel B, Mertig I and Tretiakov O A 2021 *Phys. Rep.* **895** 1
[30] Back C *et al* 2020 *J. Phys. D: Appl. Phys.* **53** 363001
[31] Landau L D and Lifshitz E M 1980 *Statistical Physics* (Oxford: Pergamon)
[32] Pitaevskii L and Stringari S 2003 *Bose–Einstein Condensation* (Oxford: Clarendon)
[33] Anderson M H, Ensher J R, Matthews M R, Wieman C E and Cornell E A 1995 *Science* **269** 198
[34] Davis K B, Mewes M-O, Andrews M R, van Druten N J, Durfee D S, Kurn D M and Ketterle W 1995 *Phys. Rev. Lett.* **75** 3969
[35] Demokritov S O, Demidov V E, Dzyapko O, Melkov G A, Serga A A, Hillebrands B and Slavin A N 2006 *Nature* **443** 430
[36] Demidov V E, Dzyapko O, Buchmeier M, Stockhoff T, Schmitz G, Melkov G A and Demokritov S O 2008 *Phys. Rev. Lett.* **101** 257201
[37] Dzyapko O, Demidov V E, Demokritov S O, Melkov G A and Safonov V L 2008 *Appl. Phys. Lett.* **92** 162510
[38] Chumak A V, Melkov G A, Demidov V E, Dzyapko O, Safonov V L and Demokritov S O 2009 *Phys. Rev. Lett.* **102** 187205
[39] Rezende S M 2010 *J. Phys.: Condens. Matter* **22** 164211
[40] Mohseni M, Qaiumzadeh A, Serga A A, Brataas A, Hillebrands B and Pirro P 2020 *New J. Phys.* **22** 083080
[41] Hirohata A, Yamada K, Nakatani Y, Prejbeanu I-L, Diény B, Pirro P and Hillebrands B 2020 *J. Magn. Magn. Mater.* **509** 166711
[42] Slonczewski J C 1996 *J. Magn. Magn. Mater.* **159** L1
[43] Baltz V, Manchon A, Tsoi M, Moriyama T, Ono T and Tserkovnyak Y 2018 *Rev. Mod. Phys.* **90** 015005
[44] Chumak A V, Serga A A and Hillebrands B 2014 *Nat. Commun.* **5** 4700
[45] Chumak A V, Vasyuchka V I, Serga A A and Hillebrands B 2015 *Nat. Phys.* **11** 453
[46] Haroche S and Raimond J 2006 *Exploring the Quantum: Atoms, Cavities, and Photons* (Oxford: Oxford University Press)
[47] Bittencourt V A S V, Feulner V and Kusminskiy S V 2019 *Phys. Rev.* A **100** 013810
[48] Leo A, Rizzato S, Monteduro A G and Maruccio G 2019 *Three-Dimensional Magnonics* ed G Gubbiotti (Singapore: Jenny Stanford) p 347
[49] Li Y, Zhang W, Tyberkevych V, Kwok W-K, Hoffmann A and Novosad V 2020 *J. Appl. Phys.* **128** 130902
[50] Bai L, Harder M, Chen Y P, Fan X, Xiao J Q and Hu C-M 2015 *Phys. Rev. Lett.* **114** 227201
[51] Cao Y, Yan P, Huebl H, Goennenwein S T B and Bauer G E W 2015 *Phys. Rev.* B **91** 094423
[52] Kaur S, Yao B M, Rao J W, Gui Y S and Hu C-M 2016 *Appl. Phys. Lett.* **109** 032404
[53] Liu Z X, Xiong H and Wu Y 2019 *Phys. Rev.* B **100** 134421
[54] Scully M O and Zubairy M S 2010 *Quantum Optics* (Cambridge: Cambridge University Press)
[55] Wang Y P, Zhang G Q, Zhang D, Luo X Q, Xiong W, Wang S P, Li T F, Hu C-M and You J Q 2016 *Phys. Rev.* B **94** 224410
[56] Haghshenasfard Z and Cottam M G 2020 *J. Appl. Phys.* **128** 033901
[57] Li Y *et al* 2019 *Phys. Rev. Lett.* **123** 107701
[58] Hou J T and Liu L 2019 *Phys. Rev. Lett.* **123** 107702
[59] Haghshenasfard Z and Cottam M G 2020 *Phys. Rev.* B **101** 144438

IOP Publishing

Dynamical Properties in Nanostructured and Low-Dimensional Materials (Second Edition)

Michael G Cottam

Appendix A

Some mathematical topics

Two of the topics that were mentioned briefly in the theoretical methods section 1.4 of chapter 1 are now described here in more detail as a reference tool. The first of these is the tridiagonal matrix (TDM) method as typically used in problems where there are short-range interactions. These cases include, for example, calculations for the discrete-lattice dynamics of phonons (in chapter 2), for spin waves or magnons in exchange-dominated magnetic materials (mainly in chapter 3), for electronic properties involving hopping (in chapter 5), and elsewhere. Next the linear response perturbative method is described, along with demonstrating its formal connection to the quantum-mechanical Green's function (GF) techniques of many-body theory. We have also included an additional section providing an overview of the most useful theoretical techniques.

A.1 The tridiagonal matrix method

Often, in solving dynamical problems involving discrete-lattice systems in which there are short-range interactions between the lattice points or atoms, we may encounter sets of finite-difference equations. In re-expressing these equations in a matrix form, we may obtain tridiagonal matrices (TDMs). In general TDMs are square matrix arrays, which may be either finite or infinite in dimension, in which the only nonzero elements are those along the leading diagonal and the two adjacent diagonals above and below. In other words, these matrices have a diagonal band of elements, but they have only zeros elsewhere. These matrices have several convenient analytic properties, particularly if there is some well-defined pattern to the values of the nonzero matrix elements.

Some early applications of TDMs for the dynamical properties of planar surfaces are to be found in work by Wallis [1] for phonons and by De Wames and Wolfram [2] for magnons or spin waves (SWs), while some formal mathematical properties of these matrices were elucidated by Rutherford [3]. As an initial illustration of the methodology, let us consider the simple set of finite-difference equations

doi:10.1088/978-0-7503-3903-2ch10

corresponding to equation (2.3) for the vibrational modes of a semi-infinite monatomic chain. The equations may easily be recast in a matrix form as $\mathbf{Au} = 0$, where $\mathbf{u}$ is the column matrix with elements given by u_1, u_2, u_3, and so on. Then, after dividing throughout by the appropriate factors, the matrix $\mathbf{A}$ is found to have a tridiagonal form corresponding to

$$\mathbf{A} = \begin{pmatrix} d_1 & -1 & 0 & 0 & \cdots \\ -1 & d & -1 & 0 & \cdots \\ 0 & -1 & d & -1 & \cdots \\ 0 & 0 & -1 & d & \cdots \\ \vdots & \vdots & \vdots & \vdots & \ddots \end{pmatrix}, \tag{A.1}$$

where $d = (2C - m\omega^2)/C$ in the notation of chapter 2, and there is just the one perturbed matrix element $d_1 = d - 1$ (modified due to the 'surface'), which is at the top left-hand corner of the array.

In the above example we have a simple expression for d_1, but for a more general model of a surface, involving possibly several modified parameters, we could write

$$\mathbf{A}\,\mathbf{u} = (\mathbf{A}_0 + \mathbf{D})\,\mathbf{u} = 0 \tag{A.2}$$

with $\mathbf{A}_0$ denoting the unperturbed dynamic matrix

$$\mathbf{A}_0 = \begin{pmatrix} d & -1 & 0 & 0 & \cdots \\ -1 & d & -1 & 0 & \cdots \\ 0 & -1 & d & -1 & \cdots \\ 0 & 0 & -1 & d & \cdots \\ \vdots & \vdots & \vdots & \vdots & \ddots \end{pmatrix}, \tag{A.3}$$

and with $\mathbf{D}$ representing the perturbation matrix due to the presence of the surfaces and/or interfaces. It can be shown that the matrix elements of the inverse matrix $\mathbf{A}_0^{-1}$ are given explicitly by

$$(\mathbf{A}_0^{-1})_{i,j} = \frac{x^{i+j} - x^{|i-j|}}{x - x^{-1}}. \tag{A.4}$$

where x is a complex variable defined by

$$x + x^{-1} = d \qquad (|x| \leqslant 1). \tag{A.5}$$

The above result can be verified, for example, by showing explicitly that it satisfies the necessary property $\mathbf{A}_0^{-1}\mathbf{A}_0 = \mathbf{I}$, where $\mathbf{I}$ is the unit matrix.

Now equation (A.2) can be manipulated into the form

$$\mathbf{A}_0(\mathbf{I} + \mathbf{A}_0^{-1}\mathbf{D})\mathbf{u} = 0. \tag{A.6}$$

If there is no perturbation term ($\mathbf{D} = 0$), the determinantal condition for a nontrivial solution is simply $\det \mathbf{A}_0 = 0$. We see, however, that when $\mathbf{D}$ is nonzero there may also be solutions that arise from

$$\det(\mathbf{I} + \mathbf{A}_0^{-1}\mathbf{D}) = 0. \tag{A.7}$$

In fact, any solutions of equation (A.7) that lies inside the unit circle in the complex plane (i.e. with $|x| < 1$ or equivalently $x = e^{-\lambda}$ when Re $(\lambda) > 0$) will correspond to the *localized* surface excitations and/or interface excitations of the system that decay away from the surface. By contrast, the solutions of det $\mathbf{A}_0 = 0$ will generally be associated with the bulk modes; they correspond to complex values of x lying on the unit circle (with $|x| = 1$, or $x = e^{i\theta}$ with θ being real and representing a phase).

Equation (A.7) provides a very useful result for determining the existence conditions and dispersion relations for the localized modes at surfaces and/or interfaces in a variety of situations. This is because the perturbation matrix **D** has typically only a few nonzero matrix elements and so the dimension of the determinant becomes small. For example, it will sometimes be the case that there is just one matrix element of **D**, such as $D_{1,1}$ as in the example mentioned above, corresponding to the surface of a semi-infinite medium. It is then easily shown that equation (A.7) reduces to $(1 + xD_{1,1}) = 0$. Hence there will be one surface mode provided that the localization condition $|x| < 1$ is satisfied. This condition becomes simply

$$|D_{1,1}| > 1. \tag{A.8}$$

If this is so, then the mode frequency (and hence the dispersion relation) are found by substituting $x = -1/D_{1,1}$ into equation (A.5) and using the definition of d. In the case cited earlier for the longitudinal vibrational modes of a terminated semi-infinite monatomic chain (see equation (A.1) and section 2.1.1) we found that $D_{1,1} = 1$ and so there is no surface mode in this case, as was stated in section 2.1.1.

On the other hand, for the vibrational modes in a semi-infinite diatomic chain (as was discussed in section 2.1.2) we have the two sets of linear coupled equations given in equation (2.14) for the odd- and even-numbered masses, together with equation (2.16) for the 'surface' $n = 0$ mass. The TDM method may be straightforwardly employed in such cases by using the odd equations to eliminate the odd displacement terms in the even set of equations, or vice versa. For example, after some simple algebraic manipulation we find for the semi-infinite diatomic chain that

$$\begin{aligned} &[w_1(w_2 - 1) - 1]u_0 = u_2 && (n = 0), \\ &(w_1 w_2 - 2)u_{2n} = u_{2n-2} + u_{2n+2} && (n \geqslant 1), \end{aligned} \tag{A.9}$$

where we denote $w_1 = 2 - (m_1\omega^2/C)$ and $w_2 = 2 - (m_2\omega^2/C)$. The equations can be re-expressed in terms of a tridiagonal dynamic matrix having the same form as equation (A.3) provided that we make the redefinitions that

$$d = w_1 w_2 - 2 \text{ and } D_{1,1} \equiv d_1 - d = 1 - w_1. \tag{A.10}$$

We may then follow the formal steps described in equations (A.4–A.8) to obtain results for the phonon frequencies in this system. Again, the solutions with $|x| = 1$ yield the 'bulk' modes of the system: specifically, putting $x = \exp(iqa)$ where q is the wave number yields the result $d = w_1 w_2 - 2 = 2\cos(qa)$. This result is found to be equivalent to the expression given in equation (2.15). The surface mode solution comes from using $x = -1/D_{1,1}$, provided equation (A.8) can be satisfied for localization of the mode, and this leads to the result for ω_S quoted in equation (2.17).

Other physical examples where the semi-infinite TDM formalism can usefully be applied arise in the context of the surface SWs in exchange-coupled ferromagnetic and antiferromagnetic systems in chapter 3. Also the method is helpful for evaluating the electronic edge modes in graphene ribbons as indicated in chapter 5.

The calculation for SWs in a Heisenberg ferromagnetic film was given in section 3.2.1. Here the coupled equations, which are expressed in terms of a layer index n, appear in equation (3.9). For a thick film (taking $N \to \infty$) the dynamic equation can be rewritten in the same matrix form as in equations (A.1)–(A.3), but with $\mathbf{u}$ now denoting the column matrix with elements given by the spin amplitudes s_1, s_2, s_3, ... and with the previous definitions of d and $D_{1,1} \equiv (d_1 - d)$ replaced by

$$\begin{aligned} d &= \frac{\left(g\mu_B B_0 - \omega\right)}{SJ} + 2SJ\left[3 - 2\gamma(\mathbf{q}_{\|})\right], \\ D_{1,1} &= -\left(1 + \frac{4(J - J_s)}{J}\left[1 - \gamma(\mathbf{q}_{\|})\right]\right), \end{aligned} \tag{A.11}$$

in the notation of chapter 3. The definition of the complex x in equation (A.5) still applies, and it may easily be verified that the solution for ω for which $|x| = 1$ (found by putting $x = \exp(iq_z a)$) becomes just the same as the previous bulk SW solution in equation (3.12), as expected. Also, by analogy with the discussion for surface phonons earlier in this section, the TDM method indicates that a surface SW exists provided the condition in equation (A.8) is satisfied (ensuring that $|x| < 1$) and then we have a solution corresponding to $x = -1/D_{1,1}$. It is a straightforward algebraic exercise to verify that the results quoted for the surface SW in section 3.2.1 are recovered, noting that the surface parameter Δ in equation (3.14) can be equated to $-1/D_{1,1}$. The case of applying the TDM method to SW modes in bct Heisenberg antiferromagnets (see section 3.2.2) can be treated using a similar approach to that for vibrational modes (phonons) in a diatomic chain. For applications to graphene ribbons with a zigzag edge we have already described the use of the TDM method in section 4.2, where the corresponding expressions for d and $D_{1,1}$ are quoted.

The TDM formalism is straightforwardly extendable to matrices that have a finite dimension N, as needed for applications to finite-length chains and finite-thickness films. In these cases, we again express $\mathbf{A} = (\mathbf{A}_0 + \mathbf{D})$, where $\mathbf{A}_0$ is now the $N \times N$ matrix given by

$$\mathbf{A}_0 = \begin{pmatrix} d & -1 & 0 & \cdots & 0 & 0 \\ -1 & d & -1 & \cdots & 0 & 0 \\ 0 & -1 & d & \cdots & 0 & 0 \\ \vdots & \vdots & \vdots & \ddots & \vdots & \vdots \\ 0 & 0 & 0 & \cdots & d & -1 \\ 0 & 0 & 0 & \cdots & -1 & d \end{pmatrix}, \tag{A.12}$$

and $\mathbf{D}$ is the perturbation matrix, which typically is sparse with only a few nonzero matrix elements. The inverse of $\mathbf{A}_0$ in this case is given by (see [4])

$$(\mathbf{A}_0^{-1})_{i,j} = \frac{x^{i+j} - x^{|i-j|} + x^{2N+2-(i+j)} - x^{2N+2-|i-j|}}{(1 - x^{2N+2})(x - x^{-1})}. \tag{A.13}$$

It follows from the definition of x in equation (A.5) that terms of order x^{2N} vanish as $N \to \infty$, making the above result reduce to equation (A.4) for the semi-infinite case.

A.2 Linear response and Green functions

Here we provide some of the additional steps needed to obtain the formal results quoted in section 1.4.2 regarding the linear response method. There we considered a perturbation involving an operator B coupled linearly to a time-dependent scalar field $f(t)$ as in equation (1.42). Our objective is to calculate the corresponding time-dependent linear response that is produced in a system variable described by any other operator A. Also, we wish to provide the formal connection with the Green's functions of many-body theory. Here we will follow a similar approach and notation to [5].

To introduce the linear response method we use an important result from statistical mechanics that the mean (ensemble-averaged) value of any operator A is worked out from $\mathrm{tr}(\rho A)$ where tr denotes the trace of the following operator product and ρ is the density matrix of the system [5, 6]. The trace of an operator is just the sum of all the diagonal matrix elements if a matrix representation is used.

Before the perturbation is switched on at time $t = -\infty$, the density matrix of the unperturbed system in equilibrium will be given by

$$\rho_0 = \frac{\exp(-\mathcal{H}_0/k_B T)}{\mathrm{tr}[\exp(-\mathcal{H}_0/k_B T)]}, \tag{A.14}$$

for a canonical ensemble [6, 7] with T denoting the temperature, $\mathcal{H}_0$ the unperturbed Hamiltonian and k_B the Boltzmann constant. At any later time t the density matrix of the system will be changed to $\rho = \rho_0 + \rho_1$, where ρ_1 can be deduced using the equation of motion for the density matrix as

$$\rho_1 = \mathrm{i}\int_{-\infty}^{t} \exp\{\mathrm{i}\mathcal{H}_0(t' - t)\}[B, \rho_0]\exp\{-\mathrm{i}\mathcal{H}_0(t' - t)\}\mathrm{d}t'. \tag{A.15}$$

Notice that the right-hand side has been linearized in terms of the perturbation operator B. The required result for the linear response $\bar{A}(t)$ can eventually be manipulated into the form (see [5])

$$\bar{A}(t) = \mathrm{i}\int_{-\infty}^{t} \langle [A(t), B(t')] \rangle f(t')\mathrm{d}t', \tag{A.16}$$

where $\langle \ldots \rangle = \mathrm{tr}(\rho_0 \ldots)$ indicates a thermal average taken with respect to the time-independent (unperturbed) system, while $A(t)$ is just a shorthand for $\exp(\mathrm{i}\mathcal{H}_0 t)A\exp(-\mathrm{i}\mathcal{H}_0 t)$ and similarly for $B(t')$.

With equation (A.16) as the main result needed in linear response theory, we are now in a position to explain the connection with Green's functions (GFs) as follows. For any two operators A and B we may formally define a GF, which we denote by $g(A; B \mid t - t')$ in a time representation, as

$$g(A; B \mid t - t') = -\mathrm{i}\theta(t - t')\langle [A(t), B(t')] \rangle, \tag{A.17}$$

where, specifically, this is equivalent to the two-time, retarded, commutator GF introduced by Zubarev [8]. We are employing the notation that

$$\theta(t - t') = \begin{cases} 1 & \text{for } t > t', \\ 0 & \text{for } t < t', \end{cases} \tag{A.18}$$

for the unit step function. Its appearance as a factor in equation (A.17) is in accordance with causality (in the sense that the stimulus applied through the operator B at time t' produces a response in the operator A at a later time t, so we must have $t > t'$). Equation (A.16) can conveniently be re-expressed as

$$\bar{A}(t) = -\int_{-\infty}^{\infty} g(A;\, B \mid t - t') f(t')\mathrm{d}t', \tag{A.19}$$

where the integration limits for t' now extend over the full range from $-\infty$ to ∞.

The above result provides the response (in A) to the perturbation involving B, but it is in an integral form. It becomes more useful, however, when it is transformed from the time representation to a representation in terms of the angular frequency. It can be shown from the definition in equation (A.17) that the GF depends on the time labels only through the difference $(t - t')$, and so we may define a Fourier transform by

$$g(A;\, B \mid t - t') = \int_{-\infty}^{\infty} G(A;\, B \mid \omega) \exp[-\mathrm{i}\omega(t - t')]\mathrm{d}\omega. \tag{A.20}$$

Here the quantities $G(A;\, B \mid \omega)$ are the frequency Fourier components of the GF defined earlier. Finally, if a similar transformation is made for $\bar{A}(t)$ and $f(t')$ in the time domain to frequency Fourier components denoted by $\bar{A}_\omega$ and $F(\omega)$, respectively, we find that equation (A.19) simplifies to become

$$\bar{A}_\omega = -2\pi\, G(A;\, B \mid \omega)\, F(\omega) \tag{A.21}$$

in the frequency representation. This last result clearly shows the important role of GFs in this perturbative situation. It is customary, for many applications, to define a *response function*, or equivalently a *generalized susceptibility* denoted by $\chi(\omega)$, as the proportionality factor between an applied stimulus to the system, represented by $F(\omega)$, and the response that it produces, represented by $\bar{A}_\omega$. In other words, the definition becomes

$$\chi(\omega) = \frac{\bar{A}_\omega}{F(\omega)}. \tag{A.22}$$

It is then evident that the simple connection established between $\chi(\omega)$ and the corresponding GF is

$$\chi(\omega) = -2\pi\, G(A;\, B \mid \omega) \tag{A.23}$$

in our GF notation.

Two other results for GFs (see e.g. [5]) are worthwhile quoting here for reference. First the GFs satisfy an *equation of motion* of the form

$$\omega\, G(A;\, B \mid \omega) = (1/2\pi)\langle [A,\, B] \rangle + G([A,\, \mathcal{H}];\, B \mid \omega), \tag{A.24}$$

which can be deduced using the operator equation of motion in equation (1.39) and the definitions given in equations (A.17) and (A.20). Second, the correlations in the dynamical system are related to the imaginary part of a corresponding GF by means of an important result known as the *fluctuation-dissipation theorem*. The dynamic correlation functions of relevance here are thermal averages (taken with respect to the Hamiltonian $\mathcal{H}$) of the form $\langle A(t)B(t')\rangle$ and $\langle B(t')A(t)\rangle$ between the two operators. If we introduce the frequency Fourier transform of one of these, say $\langle B(t')A(t)\rangle$, as being $\langle B\, A\rangle_\omega$, then the fluctuation-dissipation theorem can be stated as

$$\langle BA\rangle_\omega = -\frac{2}{[\exp(\omega/k_B T) - 1]}\, \mathrm{Im}\, G(A;\, B \mid \omega). \tag{A.25}$$

The first of the results in the previous paragraph, which gives the equation of motion (A.24) for the GF, can sometimes be used as a method to calculate the GF, as an alternative to deducing it from the linear response result in equation (A.21) or else directly from the GF definition in equation (A.17). The GF equation of motion was employed for the graphene nanoribbon example in section 5.2. The calculation requires the evaluation of the commutators $[A,\, B]$ and $[A,\, \mathcal{H}]$; examples are given in chapters 3 and 5. The fluctuation-dissipation theorem can be employed to calculate the spectral intensities of the excitations in a system and also the scattering cross sections, as in inelastic light scattering.

A.3 Survey of theoretical methods

We conclude this appendix by giving an overview to summarize the range of theoretical methods that have been used (or referred to) for studying the excitations in low-dimensional and nanostructured materials. It is intended that this will put the different methods into context.

Principally, the objective for the theory has meant finding the mode frequencies, including their dispersion properties as a function of the wave-vector components where applicable. In some cases, other properties of the excitations are also of interest, such as the spatial distribution of amplitudes (or intensity) for a particular mode in a nanostructure or the relative total intensity of one mode compared to another mode. In the case of surface or interface modes there are localization conditions to be satisfied, otherwise the mode will not exist. There are also considerations arising in the methodology of employing classical *versus* quantum-mechanical approaches on the one hand. Apart from this, there may be possibilities of using macroscopic (or continuous medium) models *versus* and microscopic (or discrete lattice) models on the other hand. Examples of all of these analytical tools have arisen throughout the chapters of this book, and we summarize them for convenience in the following paragraphs.

Methods using homogeneous equations

Examples abound throughout this book for which the spectrum of excitations and their frequencies are obtained using a homogeneous (typically linear) equation of motion for a system variable. Depending on the context, the variable that characterizes the excitation might be a quantum-mechanical operator, whereupon the equation of motion is just the operator equation (1.39) involving a commutator of that operator with the Hamiltonian $\mathcal{H}$. An obvious example is provided by the SW calculations in chapters 3 and 4, but also in other cases the use of quantum-mechanical creation and annihilation operators has been prevalent (e.g. in the analysis for 2D graphene sheets and ribbons in chapter 5). In other quantum-mechanical approaches, the equation of motion was formulated directly in terms of Schrödinger's wave equation and the system variable was the wave function. This was the case for some situations involving potential wells (as in section 5.2), where the method involved solving for the energy eigenvalues.

In many other circumstances a classical approach might be quite appropriate, as when the wave equation of elasticity is used in sections 2.3 and 7.1, or when the electromagnetic wave equation (based on Maxwell's equations) is employed as for polaritons throughout chapter 6. Both of those given examples use a macroscopic (or continuous medium) approach and differential equations are involved, which may be solved by standard analytical techniques (including the method of separation of variables, where appropriate). In contrast, as mentioned earlier, the analysis of some systems may involve a discrete-lattice model, in which case finite-difference equations are obtained in terms of the spatial variables. These may be more difficult to solve than differential equations, but then the TDM method (as highlighted in subsection A.1) provides a useful tool when the interactions are sufficiently short range.

Methods using inhomogeneous equations

The methods mentioned in the previous paragraphs of solving a homogeneous equation of motion is relevant when calculating just the frequencies and spatial distributions of any excitation, but an *inhomogeneous* equation of motion is more generally involved if we need to derive additional information (such as the spectral intensity or 'strength' of an excitation or its cross section in a scattering experiment). As discussed in section 1.4.2, this is because a driving term has to be added to the equation of motion in order to determine the response to this applied perturbation. The formalism for including a perturbative driving term leads us into linear response theory, which was just described in outline in section A.2. The definition in equation (A.22) is particularly useful, because it shows that the linear response functions are also directly proportional to GFs as seen from equation (A.23). The GF equation of motion is given in equation (A.24); it is an inhomogeneous equation because of the first term on the right-hand side, which represents a correlation function. The methodology involving inhomogeneous equations is more powerful than is the case with homogeneous equations, but it can be more complicated to implement.

Numerical and other methods

The focus in this book has been on the theoretical methods that are mainly analytic in character, in order to emphasize the fundamental physical principles. However, we have also briefly mentioned specialized numerical simulation packages where appropriate, since they can be useful as supplementary tools in specific applications. These tools have included the OOMMF software for simulating micromagnetic low-dimensional systems (see chapters 1 and 4), the DFT and DFPT density functional methods (see chapters 2 and 5) for studying phonons and electronic structure, and other numerical packages for vibrational modes (phonons) as reviewed by Wang *et al* [9].

References

[1] Wallis R F 1957 *Phys. Rev.* **105** 540
[2] De Wames R E and Wolfram T 1969 *Phys. Rev.* **185** 720
[3] Rutherford D E 1947 *Proc. R. Soc. Edinburgh* A **62** 229
[4] Cottam M G and Kontos D E 1980 *J. Phys. C: Solid State Phys.* **13** 2945
[5] Cottam M G and Haghshenasfard Z 2020 *Many-Body Theory of Condensed Matter Systems: An Introductory Course* (Cambridge: Cambridge University Press)
[6] Reichl L E 2016 *A Modern Course in Statistical Physics* 4th edn (New York: Wiley)
[7] Pathria R K and Beale P D 2021 *Statistical Physics* 4th edn (Amsterdam: Academic)
[8] Zubarev D N 1960 *Sov. Phys.—Usp* **3** 320
[9] Wang Y, Shang S-L, Fang H, Liu Z-K and Chen L-Q 2016 *npj Comput. Mater.* **2** 16006

www.ingramcontent.com/pod-product-compliance
Ingram Content Group UK Ltd.
Pitfield, Milton Keynes, MK11 3LW, UK
UKHW051859100926
8556IPUK00005B/692

9 780750 339018